COMPUTER SYSTEMS
FOR PROCESS CONTROL

Earlier Brown Boveri Symposia

COMPUTER SYSTEMS
FOR PROCESS CONTROL

Edited by

Reinhold Güth

BBC Brown, Boveri & Company, Limited
Baden, Switzerland

PLENUM PRESS · NEW YORK AND LONDON

Library of Congress Cataloging in Publication Data

Brown Boveri Symposium on Computer Systems for Process Control (1985: Brown
  Boveri Research Center)
    Computer systems for process control.

  "Proceedings of a Brown Boveri Symposium on Computer Systems for Process Control, held September 2-3, 1985, at the Brown Boveri Research Center, Baden,
Switzerland"—T.p. verso.
    Includes bibliographical references and index.
    1. Process control—Data processing—Congresses. I. Güth, Reinhold. II. BBC Ak-
tiengesellschaft Brown Boveri & Cie. III. Title.
    TS156.8.B76 1985   629.8′95   86-17073

ISBN-13: 978-1-4612-9311-8   e-ISBN13: 978-1-4613-2237-5
DOI:10.1007/978-1-4613-2237-5

Proceedings of a Brown Boveri Symposium,
on Computer Systems for Process Control, held September 2-3, 1985,
at the Brown Boveri Research Center, Baden, Switzerland

© 1986 Plenum Press, New York
Softcover reprint of the hardcover 1st edition 1986
A Division of Plenum Publishing Corporation
233 Spring Street, New York, N. Y. 10013

FOREWORD

The Brown Boveri Symposia are by now part of a firmly established tradition. This is the ninth event in a series which was initiated shortly after Corporate Research was created as a separate entity within our Company; the Symposia are held every other year. The themes to date have been:

1969   Flow Research on Blading
1971   Real-Time Control of Electric Power Systems
1973   High-Temperature Materials in Gas Turbines
1975   Nonemissive Electrooptic Displays
1977   Current Interruption in High-Voltage Networks
1979   Surges in High-Voltage Networks
1981   Semiconductor Devices for Power Conditioning
1983   Corrosion in Power Generating Equipment
1985   Computer Systems for Process Control

Why have we chosen these topics? At the outset we established certain selection criteria; we felt that a subject for a symposium should fulfill the following three requirements:

- It should characterize a part of a thoroughly scientific discipline; in other words, it should describe an area of scholarly study and research.

- It should be of current interest in the sense that important results have recently been obtained and considerable research effort is presently underway in the international scientific community.

- It should bear some relation to the scientific and technological activity of our Company.

Let us look at the requirement "current interest": Some of the topics on the list above have been the subject of research for several decades, some even from the beginning of the century. One might wonder, then, why such fields could be regarded as particularly timely in the 1970s and 1980s. A few remarks on this subject are therefore in order.

Experience shows that scientific progress in most areas does not occur at a constant rate: it comes in waves that are often followed by a standstill. The waves are often sparked by an external event that may come from quite unexpected sources. Our Symposium subjects have always been chosen so as to coincide with such a wave.

Computer systems for process control: Is this a traditional discipline, or is it a modern one? Certainly process control is traditional; the implementation of control systems goes back at least to the beginning of this century. But the utilization of computers is distinctly modern. Today, the growth of research activity in computer science is almost explosive; and so is the flow of results. In our Research Center, Computer Science is the fastest growing of all research departments.

I believe a few words are in order on the identity of this discipline. Is computer science really a science in its own right? Or is it a part of mathematics, or an engineering discipline? This is being debated, sometimes hotly debated. Coming to an agreement on this question is probably not the most urgent need of the day. My personal opinion is that computer science is indeed a science in its own right. While the primordial roots of computer science are to be found in mathematics, computer science has developed its own structure to such a large extent that it should not be regarded as a part of electrical engineering, or a branch of mathematics; it is a science much in the same way that chemistry is a separate science, although it is really based upon the structure of physics. The first computer scientist in today's sense of the word was probably the British mathematician Alan Turing. The most prolific single contributor, however, was undoubtedly John von Neumann. He was the first to formulate clearly that a computer must consist of the following five elements: arithmetic unit, memory, control, input, and output. And he invented the stored-program machine - the concept that instructions and data are stored in the same memory and can be processed in the same arithmetic unit. This feature fundamentally distinguishes the von Neumann machine from its predecessors which had their programs on punched tapes. To us, the stored program seems obvious, even natural. It was not obvious at that time, it was an idea of the greatest significance. Is the stored program a scientific principle, or is it an engineering invention? I feel it must be qualified as an engineering invention; but it is so fundamental that it comes close to being a scientific principle. Von Neumann as a researcher was simultaneously a mathematician, a physicist, and an engineer, and has made major and lasting contributions in all three fields. In all of history there are probably not more than half a dozen people for whom this claim can be made. This, then, is the scientific foundation of what was discussed at the meeting.

As in previous meetings, the number of participants was limited in order to maintain the character of a specialist meeting of restricted size, and it was with much regret that we were forced to disappoint many who wished to participate but who could not be accommodated. We hope that the publication of the present volume is a partial consolation for those whom we did not have the pleasure of welcoming as our guests.

The Symposium was attended by 160 participants from twelve countries. It was both an honor and a pleasure to welcome these scientists and engineers from so many different parts of the world. Their willingness to travel to Baden and spend two full days with us was a challenge as well as an obligation to us as organizers, and we sincerely hope that the expectations which prompted them to attend were fulfilled.

To conclude, we should like to take this opportunity of expressing our sincere gratitude to every Symposium participant. We hope they consider the time spent with us to have been worthwhile. Thanks are due primarily to the authors for having spared no effort in preparing their papers: the contents of this volume reflect the high quality of their work. We thank also the participants in the discussions, both formal and informal, and the editor of these proceedings.

The selection of the theme, the layout of the program and the contact with the speakers were the responsibility of Dr. R. Güth, head of the Computer Science Department at our Research Center. His careful and competent preparation was instrumental in the success of the Symposium. Our thanks also go to Miss R. Niehus and to Mr. H. Wilhelm and his staff for the smooth running of the administrative side of the meeting.

A. P. Speiser
Director of Research

PREFACE

Control systems automate technical processes and make them observable to and controllable by human operators. Process control includes various functions, such as geographically distributed data acquisition and direct control, overall process optimization and plant management, and centralized communication with the human operator.

In recent years, process control has become increasingly sophisticated and its implementation based more and more on computer technology. Modern control systems are complex computer structures configured and programmed according to the process to be controlled. The design of new control systems takes advantage of rapid progress in certain fields of computer science.

This volume provides a survey of recent developments in areas of computer science which are of specific relevance for process control. The volume contains the proceedings of the ninth Brown Boveri Symposium, which took place on September 2 and 3, 1985, at the Brown Boveri Research Center in Baden, Switzerland. The core of the Symposium consisted of thirteen invited papers, written by leading contributors to their respective fields of computer science. The papers are complemented by an edited version of the very informative and lively discussions, based upon tape recordings and written notes.

The body of this volume starts with an Introduction, in which the history of process control and the characteristics of modern control systems are briefly reviewed. Topics of computer science related to process control are identified. The invited papers are structured into three groups corresponding to the Symposium sessions.

The first group is devoted to computer system architecture, covering the structure and organization of computers and distributed computer systems. Architectural principles are discussed in the paper by W.K. Giloi. A taxonomy of computer architecture is outlined, and architectural issues of sequential and parallel computers are addressed. Loosely and strongly coupled multicomputer systems are illustrated by examples. The paper by J. Goldberg provides a survey of the state of the art in fault-tolerant computing. An introduction to the basic notions of fault tolerance is given, and common approaches to the design of fault-tolerant systems are discussed. The scope of concern covers a wide range of unreliability sources, from design issues to operator mistakes. In his paper, G. Le Lann investigates distributed computing under real-time constraints. Approaches to the achievement of desired timing and robustness of computer communication are considered, with special emphasis being put on local area networks. The standardization of local area networks for real-time applications is assessed. In real-time computing, value dimensions and time dimensions of data are of comparable importance. H. Kopetz discusses the characteristics of real-time tasks and outlines principles for the design of real-time computer systems. An overview of a design environment for real-time systems is given. The paper by H. Kirrmann, T. La-live d'Epinay, and H. Stöckler reviews the architecture and implementation of control systems. The state of the art in control-system architecture is illustrated by examples of existing systems. Particular attention is paid to system configuration from pre-fabricated building blocks and also to the underlying concepts of system organization. To choose a configuration that will behave properly in critical situations, the designer must be able to predict the performance regarding response time and throughput. Methods for performance modelling of control systems are the subject of the paper by M. Vitins and K. Signer. An interactive performance-estimation tool is described and its application to the analysis of a large control center is shown.

The second group of papers refers to topics of human-computer interaction, focussing on dialog design and computer graphics. J. Nievergelt, C. Muller, and H. Sugaya identify key issues of dialog design. A model of interaction is presented, and experiments with the construction of interactive computer systems are discussed. Approaches to rapid prototyping of interactive systems are shown. The paper by J. Encarnaçao is devoted to computer graphics for engineering applications. The use of graphical workstations of different levels of performance is examined. Software interfaces and standards for engineering graphics are considered, and trends in the workstation market are estimated.

The third group covers the field of software engineering, comprising methods, languages, and tools for software development and maintenance. B. Liskov and M. Herlihy address the construction of distributed programs, emphasizing program structure and communication. The paper discusses a model of distributed computation, which serves as the conceptual basis for a high-level language for distributed systems. In the course of the discussion, features of the model are compared with alternatives. The paper by Y. Matsumoto concerns requirements engineering and software development. The life cycle of software products is examined, and a flexible method for the development of software is presented. The paper outlines a semantic model, which enables fast prototyping of the product to be developed and gradual refinement in a uniform representation. The paper by K. Frühauf, J. Ludewig, and H. Sandmayr surveys the state of the art in software quality assurance. An introduction to basic notions of software quality is given. Practical experiences are reported, and the impact of quality assurance measures on productivity and software quality is discussed. F. Kaufmann, D. Schillinger and U. Schult present an approach to the graphical construction of control software. Programs are specified in a language that is based on function charts and state-transition diagrams. A programming tool, which supports interactive program construction and performs automatic generation of executable code, is explained. Logic programming is a promising approach to both rapid prototyping of interactive systems and the implementation of knowledge-based systems. In their paper, J. Kriz and H. Sugaya give an introduction to logic programming and the techniques of knowledge representation. The combination of logic programming and procedural programming is discussed, and the implementation of knowledge-based systems is illustrated by an example.

At this point I would like to express my sincere thanks to everyone who contributed to the success of the Symposium. I am especially grateful to the authors, each of whom promptly accepted the invitation to participate and maintained a strong interest up to the preparation of the volume. The expertise of the session chairmen contributed a great deal to the success of the discussion periods and is gratefully acknowledged.

K. Ragaller greatly encouraged my work in the preparation of the Symposium. Special thanks go to J. Kriz, who supported me in the development of the scientific program. Thanks are also due to Miss D. Stadelmann, Mrs. B. Nowatzek, H.R. Aschmann, and U. Schult for their extensive help during the Symposium.

It is a pleasure to acknowledge the support of Dr. A.M. Escudier, who relieved me of the linguistic aspects of the editorial work and also assisted me in the proofreading. The entire camera-ready manuscript for this volume was prepared by Mrs. E. Blatter; the figures were drawn by Miss B. Rüttimann.

Baden, January 1986 R. Güth

CONTENTS

INTRODUCTION

SESSION I
COMPUTER SYSTEM ARCHITECTURE
Chairmen: T. Lalive d'Epinay and R. Karg

PRINCIPLES OF COMPUTER ARCHITECTURE

CHALLENGES AND DIRECTIONS IN FAULT-TOLERANT COMPUTING

DISTRIBUTED REAL-TIME PROCESSING

DESIGN OF REAL-TIME SYSTEMS

SESSION II

HUMAN-COMPUTER INTERACTION

Chairman: H. Sandmayr

SESSION III

SOFTWARE ENGINEERING

Chairmen: H. Sandmayr and O. Klammer

PARTICIPANTS

Dr. Dennis R. Allison
Stanford University
Computer Systems Laboratory
Stanford, CA 94305
U.S.A.

Dr. Urs Ammann
Contraves AG
ESI
Postfach
CH-8052 Zürich
Switzerland

Dr. H.-J. Appelrath
Institut für Informatik
ETH-Zentrum
CH-8092 Zürich
Switzerland

Mr. H.R. Aschmann
Brown Boveri Research Center
Dept. KLR-CS
CH-5405 Baden
Switzerland

Mr. Friedemann Bacher
BBC Brown, Boveri & Co., Ltd.
Dept. BBI-C
Postfach
CH-5401 Baden
Switzerland

Prof. Hj. Baumann
Neu-Technikum Buchs
CH-9470 Buchs / SG
Switzerland

Dr.-Ing. Ulrich Baur
BBC Brown, Boveri & Cie. AG
Dept. IA/TL
Käfertaler Strasse 258
D-6800 Mannheim 1
Germany

Dipl.-Ing. Eberhard Beyer
BBC Brown, Boveri & Cie. AG
Dept. IA/A
Käfertaler Strasse 258
D-6800 Mannheim 1
Germany

Dipl.-Ing. Dieter Brösamle
BBC Brown, Boveri & Co., Ltd.
Dept. ELS
CH-5412 Gebenstorf
Switzerland

M.Sc. Leif Buch
Elkraft
Elkraft A.m.b.A.
Lautruphøj 5
DK-2750 Ballerup
Denmark

Dr. Helmar Burkhart
Institut für Elektronik
der ETH Zürich
Gloriastrasse 35
CH-8092 Zürich
Switzerland

Dipl.-Ing. ETH Ulrich Burri
Software-Schule Schweiz
Morgartenstrasse 2c
CH-3014 Bern
Switzerland

Dipl.-Ing. Manfred Capelle
Rheinisch-Westfälisches
Elektrizitätswerk AG
Dept. KA-AL
Kruppstrasse 5
D-4300 Essen 1
Germany

Mr. Raymond Cettou
c/o IB Grombach & Co. AG
Bodenacherstrasse 58
CH-8121 Benglen
Switzerland

Dr. Paolo Ciompi
I.E.I. - C N R
Consiglio Nazionale delle Ricerche
Istituto di Elaborazione dell'Informazione
Via S. Maria, 46
I-56100 Pisa
Italy

Prof. Dr. Mario Dal Cin
Inst. für Informationsverarbeitung
Universität Tübingen
Köstlinstrasse 6
D-7400 Tübingen
Germany

Dr. D.J. Damsker
Damsker Consulting Services, Inc.
531 Main Street
New York, NY 10044
U.S.A.

Dr. Franz Demmelmeier
Lehrstuhl für Prozessrechner
TU München
Franz-Joseph-Strasse 38
D-8000 Munich 40
Germany

Dr. Dacfey Dzung
Brown Boveri Research Center
Dept. KLR-FN
CH-5405 Baden
Switzerland

Dr.-Ing. Wolfgang Eckelmann
Dept. Techn. Leitsysteme, C 584
Höchst AG
Postfach 80 03 20
D-6230 Frankfurt / Main 80
Germany

Dr. E. Egg
BBC Brown, Boveri & Co., Ltd.
Dept. PA-I
CH-5401 Baden
Switzerland

Mr. Horst Eggers
Mannesmann Demag AG
Wolfgang-Reuter-Platz
D-4100 Duisburg 1
Germany

Dr. F. Eggimann
BBC Brown, Boveri & Co., Ltd.
Dept. E
CH-5401 Baden
Switzerland

Dr. H.P. Eichenberger
General Electric Company
Pelikanstrasse 37
CH-8001 Zürich
Switzerland

Dr.-Ing. Peter F. Elzer
BBC Brown, Boveri & Cie. AG
Dept. ZFL/L3
Eppelheimer Strasse 82
D-6900 Heidelberg
Germany

Prof. Dr. J. Encarnação
Techn. Hochschule Darmstadt
Alexanderstrasse 24
D-6100 Darmstadt
Germany

Mr. Rudolf Fischer
Institut für Elektronik
ETH-Zentrum
Gloriastrasse 35
CH-8092 Zürich
Switzerland

Dipl.-Ing. ETH Rainer Forster
Sektion Automation
Bauabteilung Generaldirektion SBB
Mittelstrasse 43
CH-3030 Bern
Switzerland

Mrs. M. Franc
BBC Brown, Boveri & Co., Ltd.
Dept. ELS-E1
CH-5412 Gebenstorf
Switzerland

Prof. Peter Freeman
University of California
Information & Computer Science
Irvine, CA 92717
U.S.A.

Dr. Heinz H. Frey
BBC Brown, Boveri & Co., Ltd.
Dept. EEQ
CH-5300 Turgi
Switzerland

Mr. Karol Frühauf
BBC Brown, Boveri & Co., Ltd.
Dept. ELS-Q
CH-5412 Gebenstorf
Switzerland

Dr. Kurt Geihs
IBM Scientific Center
Tiergartenstrasse 15
D-6900 Heidelberg
Germany

Prof. Dr. Wolfgang K. Giloi
GMD-TUB First
Hardenbergstrasse 29B
D-1000 Berlin 12
Germany

Dr. Martin Glinz
BBC Brown, Boveri & Co., Ltd.
Dept. CTT-D
CH-5401 Baden
Switzerland

Prof. Dr. Jack Goldberg
SRI International
333 Ravenswood Ave.
Menlo Park, CA 94025
U.S.A.

Dr. Eng. Fabrizio Grandoni
I.E.I. - C N R
Consiglio Nazionale delle Ricerche
Istituto di Elaborazione dell'Informazione
Via S. Maria, 46
I-56100 Pisa
Italy

Dipl. Phys. ETH Peter Grombach
IB Grombach & Co. AG
Postfach
CH-8028 Zürich
Switzerland

Dr. K.-E. Grosspietsch
GMD St. Augustin
GMD-F2G1
Postfach 1240
D-5205 St. Augustin
Germany

Mr. P. Günzburger
Hasler AG / 34
Belpstrasse 23
CH-3000 Bern 14
Switzerland

Mr. H. Gutfeldt
Hasler AG / 34
Belpstrasse 23
CH-3000 Bern 14
Switzerland

Dr. Reinhold Güth
Brown Boveri Research Center
Dept. KLR-C
CH-5405 Baden
Switzerland

Prof. Dr. Volkmar Haase
Inst. für Maschinelle Dokumentation
Technische Hochschule Graz
Steyrergasse 25A
A-8010 Graz
Austria

Dr. H. Halling
Zentrallabor für Elektronik
der Kernforschungsanlage Jülich GmbH
Nukleare Elektronik
Postfach 1913
D-5170 Jülich 1
Germany

Prof. Dr.-Ing. Edmund Handschin
Universität Dortmund
Lehrstuhl für el. Energieversorgung
Postfach 500 500
D-4600 Dortmund 50
Germany

Dr. J. Heger
BBC Brown, Boveri & Cie. AG
Dept. ZFL/L
Postfach 3 51
D-6800 Mannheim 1
Germany

Dr. Ekkehard Heymann
Gelsenwasser AG
Postfach 10 09 44
D-4650 Gelsenkirchen
Germany

Dipl.-Ing. ETH Ulrich Hinden
Holderbank Management & Beratung AG
Projektabteilung
CH-5113 Holderbank
Switzerland

Mr. Gerhard Hock
BBC Brown, Boveri & Cie. AG
Dept. SN
Wallstadter Strasse 53-59
D-6802 Ladenburg
Germany

Dr. Elmar Holler
Kernforschungszentrum
I D T
Postfach 3640
D-7500 Karlsruhe
Germany

Mr. Jørgen Holm
BBC Brown, Boveri & Cie. AG
Dept. AT/E
Neustadter Strasse 62
D-6800 Mannheim 31
Germany

Dr. Peter Hruschka
G E I
Albert-Einstein-Strasse 61
D-5100 Aachen
Germany

Mr. Hansjörg Huser
Brown Boveri Research Center
Dept. KLR-CS
CH-5405 Baden
Switzerland

Mr. Marc Jauslin
BBC Brown, Boveri & Co., Ltd.
Dept. ITE
CH-5401 Baden
Switzerland

Mr. Jørgen Fangel Jensen
Elkraft
Elkraft A.m.b.A.
Lauthruphøj 5
DK-2750 Ballerup
Denmark

Mr. Paul Jervis
Central Electricity
Research Laboratories
Kelvin Avenue
Leatherhead
Surrey KT22 7SE
England

M.Sc. Vagn Jørgensen
Elsam Power Station
Engineering Div.
Elsam,
DK-7000 Fredericia
Denmark

Mr. T. Kälin
Philips AG
Dept. Industrie
Allmendstrasse 140
CH-8027 Zürich
Switzerland

Dr. R. Karg
BBC Brown, Boveri & Cie. AG
Dept. AT/SE
Eppelheimer Strasse 82
Postfach 10 16 80
D-6900 Heidelberg 1
Germany

Mr. Felix Kaufmann
Brown Boveri Research Center
Dept. KLR-CS
CH-5405 Baden
Switzerland

Mr. Rainer Kesenheimer
Brown Boveri Research Center
Dept. KLR-CS
CH-5405 Baden
Switzerland

Mr. Hubert Kirrmann
Brown Boveri Research Center
Dept. KLR-CS
CH-5405 Baden
Switzerland

Mr. Ole Klammer
BBC Brown Boveri Denmark A/S
Dept. ESV-I4
Ved Vesterport 6
DK-1612 Copenhagen V
Denmark

Dr. Günter Koch
Biomatik GmbH
Talstrasse 63
D-7800 Freiburg
Germany

Dipl.-Ing. Karl Koch
Oesterreichische Brown Boveri Werke
Dept. KV
Pernerstorfergasse 94
A-1100 Vienna
Austria

Dr. Raimund T. Kölsch
Software und Management Consulting
Menterschwaigstrasse 9
D-8000 Munich 90
Germany

Prof. Dr. Hermann Kopetz
Inst. für Prakt. Informatik
Gusshausstrasse 30
A-1040 Vienna
Austria

Dr. Jeffrey Kramer
Imperial College
Dept. of Computing
180 Queen's Gate
London SW7 2BZ
England

Dr. Lothar Krings
Brown Boveri Research Center
Dept. KLR-CS
CH-5405 Baden
Switzerland

Dr. Jiri Kriz
Brown Boveri Research Center
Dept. KLR-CI
CH-5405 Baden
Switzerland

Prof. Dr. Albert T. Kündig
Institut für Elektronik
ETH-Zentrum
ETZ J87
CH-8092 Zürich
Switzerland

Dr. Thierry Lalive d'Epinay
BBC Brown, Boveri & Co., Ltd.
Dept. ESL
CH-5300 Turgi
Switzerland

Dr. Norbert Lang
Stadtwerke Bochum GmbH
Hauptabteilung Stromversorgung
Postfach 10 22 50
D-4630 Bochum
Germany

Dipl.-Ing. Jürgen Lantermann
V E W
Hauptverwaltung HEV
Flamingoweg 1
D-4600 Dortmund
Germany

Prof. Gérard Le Lann
INRIA
Domaine de Voluceau
Rocquencourt
B.P. 105
F-78153 Le Chesnay Cedex
France

Mr. G. Leichner
Electric Power Industry Consultants
c/o BBC Brown, Boveri & Co., Ltd.
Dept. ELV-P
CH-5401 Baden
Switzerland

Prof. Barbara Liskov
M I T
Laboratory for Computer Science
545 Technology Square
Cambridge, Mass. 02139
U.S.A.

Mr. Roland Luder
Holderbank Management & Beratung AG
Projektabteilung
CH-5113 Holderbank
Switzerland

Prof. Dr. Jochen Ludewig
Institut für Informatik
ETH-Zentrum
CH-8092 Zürich
Switzerland

Mr. Hero Lüers
Messerschmitt-Bölkow-Blohm GmbH,
LKE 324
Postfach 80 11 60
D-8000 Munich 80
Germany

Dr. Jeffrey N. Magee
Imperial College
Dept. of Computing
180 Queen's Gate
London SW7 2BZ
England

Dr. René A. Marolf
BBC Brown, Boveri & Co., Ltd.
Dept. KLS
CH-5401 Baden
Switzerland

Mr. D.J.L. Martin
BBC Brown, Boveri & Cie. AG
Dept. GK/LE 3
Kallstadter Strasse 1
D-6800 Mannheim 1
Germany

Mr. Hans Matheis
Institut für Informatik
ETH-Zentrum
CH-8092 Zürich
Switzerland

Dr. Eng. Yoshihiro Matsumoto
Toshiba Corporation
Heavy Apparatus Eng. Lab.
1, Toshiba-cho, Fuchu
Tokyo 183
Japan

Dr. R.W. Meier
Brown Boveri Research Center
CH-5405 Baden
Switzerland

Prof. H. Meyr
Inst. für Elektr. Regelungstechnik
RWTH Aachen
Templergraben 55
D-5100 Aachen
Germany

Mr. Peter Mielentz
BBC Brown, Boveri & Cie. AG
Dept. AT/SAP
Postfach 10 16 80
D-6900 Heidelberg
Germany

Dr. Johannes Milde
Brown Boveri Research Center
Dept. KLR-CS
CH-5405 Baden
Switzerland

Mr. Michele Morganti
Italtel
Central Research Laboratory
I-20019 Castelletto di Settimo Milanese
Italy

Dipl.-Ing. Gerhard Motl
Preussische Elekrizitäts-
Aktiengesellschaft
Tresckowstrasse 5
D-3000 Hannover 91
Germany

Dr. Josef A. Muheim
BBC Brown, Boveri & Co., Ltd.
Dept. ESV
CH-5401 Baden
Switzerland

Mr. Carlo Muller
Brown Boveri Research Center
Dept. KLR-CI
CH-5405 Baden
Switzerland

Dr. Klaus-Dieter Müller
KFA Jülich / ZEL/NE
Postfach 1913
D-5170 Jülich
Germany

Dr. Ulrich Nielsen
BBC Brown, Boveri & Cie. AG
Dept. SN/BE
Wallstadter Strasse 53-59
D-6802 Ladenburg
Germany

Prof. J. Nievergelt
Dept. of Computer Science
University of North Carolina
Chapel Hill, NC 27514
U.S.A.

Dr. Kazuhito Ohmaki
Institut für Informatik
ETH-Zentrum
CH-8092 Zürich
Switzerland

Mr. L.M. Panis
BBC Brown, Boveri & Co., Ltd.
Dept. T-V
CH-5401 Baden
Switzerland

Prof. Dr. Rupert Patzelt
Technische Universität Wien
Institut für Elektrische Messtechnik
Gusshausstrasse 25
A-1040 Vienna
Austria

Dipl.-Ing. Georg Pauthner
Techn. Universität Berlin
Institut für Techn. Informatik
Sekr. FR 2-2
Franklinstrasse 28/29
D-1000 Berlin 10
Germany

Mr. Jean-Pierre Pfister
Prospective Engineering Gestion (PEG)
Case Postale 356
2-4 Rue du Lièvre
CH-1211 Geneva 24
Switzerland

Mr. G. Franco Piazza
Dipl. El. Ing. ETH
Autophon AG
Ziegelmattstrasse 1-15
CH-4500 Solothurn
Switzerland

Prof. Dr. Klaus Ragaller
BBC Brown, Boveri & Co., Ltd.
Dept. R
CH-5401 Baden
Switzerland

Dr. Jean-Marie Rayroux
BBC Brown, Boveri & Co., Ltd.
Dept. CH-2
CH-5401 Baden
Switzerland

Dipl.-Ing. H. Reiter
Digital Equipment
Schaffhauserstrasse 144
CH-8302 Kloten
Switzerland

Prof. Dr.-Ing. Ulrich Rembold
Lehrstuhl für Prozessrechentechnik
Informatik III,
Universität Karlsruhe
Postfach 6380
D-7500 Karlsruhe
Germany

Mr. Josef Rödig
BBC Brown, Boveri & Cie. AG
Dept. ZFE
Kallstadter Strasse 1
D-6800 Mannheim 31
Germany

Mr. Willi Roos
BBC Brown, Boveri & Co., Ltd.
Dept. CH-2
CH-5401 Baden
Switzerland

Mr. Willi Roth
BBC Brown, Boveri & Co., Ltd.
Dept. TDC-T
CH-5401 Baden
Switzerland

Dr. Harry Rudin
IBM Research Center
Säumerstrasse 4
CH-8803 Rüschlikon
Switzerland

Dr. Toshiaki Sakaguchi
Mitsubishi Electric Corp.
Central Research Laboratory
8-1-1, Tsukaguchi Honmachi
8-Chome
J-Amagasaki, Hyogo, 661
Japan

Mr. James C. Sanders
Process Control Division
Alcoa Technical Center
Aluminum Company of America
Alcoa Center, PA 15069
U.S.A.

Dr. Helmut Sandmayr
BBC Brown, Boveri & Co., Ltd.
Dept. ELS-E
CH-5412 Gebenstorf
Switzerland

Dipl.-Phys. Rudolf Sartori
Rheinisch-Westfälischer
Techn. Ueberw. Verein
Steubenstrasse 53
D-4300 Essen
Germany

Dr. Rudolf Schild
Landis & Gyr AG
Zentrallabor 3161
CH-6301 Zug
Switzerland

Mr. Daniel Schillinger
Brown Boveri Research Center
Dept. KLR-CS
CH-5405 Baden
Switzerland

Prof. Dr. R. Schnörr
Corporate Executive Vice-President
BBC Brown, Boveri & Co., Ltd.
CH-5401 Baden
Switzerland

Dr. Roger Schönberger
Landis & Gyr AG
Zentrallabor L & G
CH-6301 Zug
Switzerland

Dipl.-Ing. Dieter Schönert
Stahlwerke Bochum AG
Castroper Strasse 228
D-4630 Bochum 1
Germany

Dr. Gerhard Schrott
Institut für Informatik
Techn. Universität München
Arcisstrasse 21
D-8000 Munich 2
Germany

Dr. Herbert Schüller
GRS GmbH
Dept. Prozessrechner
Bereich Systeme
Forschungsgelände
D-8046 Garching
Germany

Mr. Uwe Schult
Brown Boveri Research Center
Dept. KLR-CS
CH-5405 Baden
Switzerland

Mr. H.P. Schulthess
Corporate Executive Vice-President
BBC Brown, Boveri & Co., Ltd.
CH-5401 Baden
Switzerland

Dr. R. Schüpbach
BBC Brown, Boveri & Co., Ltd.
Dept. KLS
Schlossbergweg 10
CH-5400 Baden
Switzerland

Mr. Wilhelm Seifert
BBC Brown, Boveri & Co., Ltd.
Dept. EL-A
CH-5300 Turgi
Switzerland

Mr. Daniel Seunier
Electrobel S.A. - Consulting Engineers
Instrumentation & Control Section
Place du Trône, 1
B-1000 Brussels
Belgium

Mr. K. Signer
BBC Brown, Boveri & Co., Ltd.
Dept. ELS-E2
CH-5412 Gebenstorf
Switzerland

Dr. Luca Simoncini
I.E.I. - C N R
Via S. Maria, 46
I-56100 Pisa
Italy

Mr. John H. Sloane
Aluminum Company of America
1501 Alcoa Building
Pittsburgh, PA 15219
U.S.A.

Ing. J.P. Snyder
P N E M
Project Dept. Power Stations
N.V. Provinciale Noordbrabantse
P.O. Box 7
NL-4930 AA-Geertruidenberg
The Netherlands

Prof. Dr. Ambros P. Speiser
Brown Boveri Research Center
CH-5405 Baden
Switzerland

Dr. H. Stemmler
BBC Brown, Boveri & Co., Ltd.
Dept. IEE
CH-5300 Turgi
Switzerland

Mr. Hans-Peter Stöckler
BBC Brown, Boveri & Cie. AG
Dept. IA/T
Käfertaler Strasse 258
D-6800 Mannheim 1
Germany

Dr. Hirotsugu Sugaya
Brown Boveri Research Center
Dept. KLR-CI
CH-5405 Baden
Switzerland

Mrs. Katharina Tomica
Contraves Zürich
Schaffhauserstrasse 580
CH-8052 Zürich
Switzerland

Mr. G. Tritt
Hasler AG / 34
Belpstrasse 23
CH-3000 Bern 14
Switzerland

Dipl.-Ing. Volker Tschammer
Hahn-Meitner-Institut
für Kernforschung
Glienickerstrasse 100
D-1000 Berlin 39
Germany

Mr. Jef Verheijen
Philips Export B.V.
Sc. & Ind. Equipment Division
Bldg. TQIII-2
NL-5600 MD-Eindhoven
The Netherlands

Dr. Sudhir Virmani
Electric Power Industry Consultants
c/o BBC Brown, Boveri & Co., Ltd.
Dept. ELV-P
CH-5401 Baden
Switzerland

Dr. Michael Vitins
Brown Boveri Research Center
Dept. KLR-CI
CH-5405 Baden
Switzerland

Mr. Ingvar Wetterholm
OKG Aktiebolag
Oskarshamnsverket
S-57093 Figeholm
Sweden

Dr. Hans Widmer
BBC Brown, Boveri & Co., Ltd.
Dept. E-A
CH-5300 Turgi
Switzerland

Mr. I. Wigdorovits
BBC Brown, Boveri & Co., Ltd.
Dept. E-E
CH-5401 Baden
Switzerland

Dr. E. Wildhaber
BBC Brown, Boveri & Co., Ltd.
Dept. KLG-S
CH-5401 Baden
Switzerland

Prof. Dr. Walter Winkler
Höhere Technische Lehranstalt (HTL)
Brugg-Windisch
CH-5200 Windisch
Switzerland

Dipl.-Ing. Horst Wissel
Thyssen Stahl AG
Kaiser-Wilhelm-Strasse 100
D-4100 Duisburg 11
Germany

Dr. Hans-Peter Wollenmann
Elektrowatt
Ingenieurunternehmung AG
Bellerivestrasse 36
CH-8022 Zürich
Switzerland

Mr. Ib Worning
F.L. Smidth & Co. A/S
Vigerslev allé 77
DK-2500 Copenhagen Valby
Denmark

Mr. James E. Yates
Systems Engineering & Integration
Aluminum Company of America
1501 Alcoa Building, Room 4043WPH
Pittsburgh, PA 15219
U.S.A.

Mr. Toshio Yokouchi
Toshiba International Co., Ltd.
Audrey House ELY Place
London EC1
England

Dipl.-Ing. Hubertus Zinke
BBC Brown, Boveri & Cie. AG
Dept. SN
Wallstadter Strasse 53-59
D-6802 Ladenburg
Germany

Mr. Stefan Züger
BBC Brown, Boveri & Co., Ltd.
Dept. ESL-3
CH-5300 Turgi
Switzerland

Mr. Stephan Zürcher
BBC Brown, Boveri & Co., Ltd.
Dept. IT
CH-5401 Baden
Switzerland

INTRODUCTION

R. GUETH

Brown Boveri, Baden, Switzerland

The purpose of a control system is to control a technical process, such that it has desired properties, and to make the process observable to human operators. Control tasks arise in factories, chemical plants, power stations, and in networks for the transport or distribution of water, oil, and electrical energy.

Prior to the introduction of the programmable computer, process control was based on analog controllers and relays. Direct control by computers was introduced in the early 1960s. Around 1970 control systems began taking advantage of the minicomputer, which was then becoming available. In the mid 1970s the microprocessor was introduced and provided the basis for the development of compact computer nodes. Then a new type of distributed control system evolved, utilizing recent data-communication technology. The development of distributed-system architectures significantly enhanced system reliability and adaption to the process.

In the last few years, control systems have become increasingly complex. The control system of a large plant may include several hundred computer nodes. The nodes are geographically distributed and embedded in the plant according to the physical structure of the process. A plant-wide communication system connects the nodes and provides a data link to the central operating room (Figure 1).

As control systems automate costly plants and processes, they must meet stringent reliability specifications. A control system should be available to perform its tasks for long periods without interruption. During run time any fault within the control system must be quickly diagnosed, and faulty parts should be replaceable without affecting the overall operation of the system.

Most functions of a control system are time critical: the results have to be produced

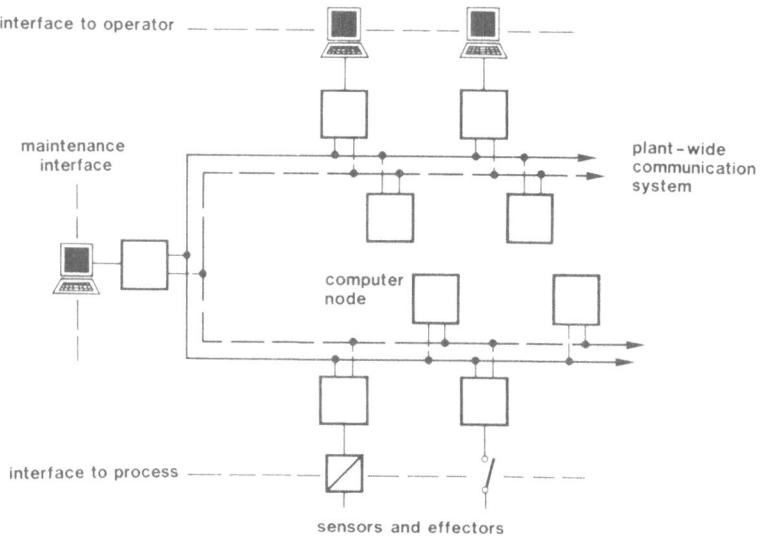

interface to operator

maintenance
interface

plant-wide
communication
system

computer
node

interface to process

sensors and effectors

Fig. 1: Computer System for Process Control

within certain time limits. Furthermore, the functions have to meet their response-time specifications in all critical situations of the process, including the worst case. So computation and communication must be organized such that they always feature predictable timing. A particular consequence of this requirement is that the plant-wide communication system should manage peak traffic with little performance degradation and should ensure that the transmission time of a message does not exceed a specified upper boundary.

The control system of a modern plant performs a large number of different functions, which range from data acquisition and direct control to sophisticated process optimization and plant management. The functions constitute a hierarchy and closely interact with each other (Figure 2). High-speed time-critical functions are at the lower levels of the hierarchy; slower but more complex functions are at the higher levels. High-level functions rely on the lower levels for data acquisition, data reduction, and direct control. Functions at lower levels in turn need the higher levels for coordination and overall optimization.

The interface to the human operator is mainly based on computer graphics, keyboards, and the techniques of human-computer interaction. The display functions include indication of events, presentation of process parts in detail, and bringing to view long-term trends of process variables. As a supplement to computer graphics, voice output is used for the transfer of alarm messages to the operator. Recently, command input

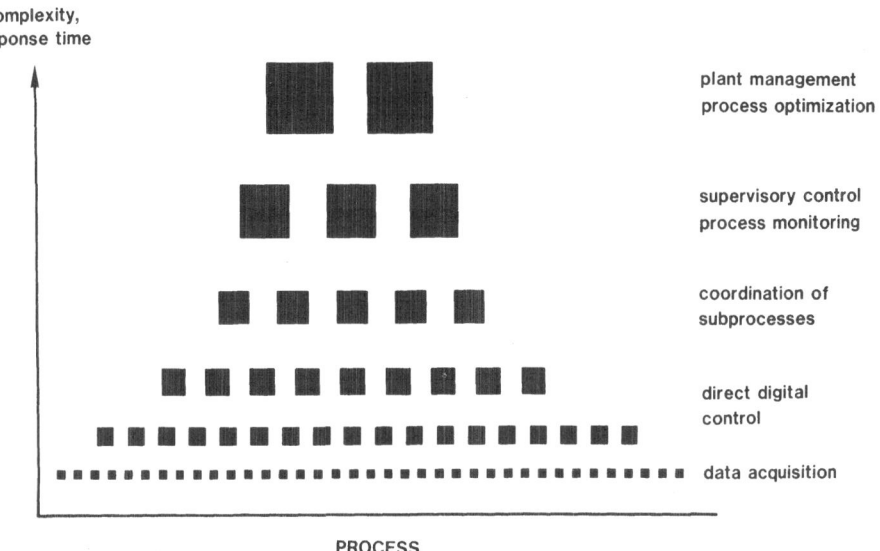

complexity,
response time

plant management
process optimization

supervisory control
process monitoring

coordination of
subprocesses

direct digital
control

data acquisition

PROCESS

Fig. 2: Hierarchy of Control Functions

via speech recognition has been introduced. Since the accident at the Three Mile Island nuclear-power station in 1979, the ergonomical aspects of human-computer interaction have been the subject of increased attention.

A control system has to be tailored to the technical process that is to be controlled. In order to reduce plant-specific engineering costs, effective specification and programming tools are required. Moreover, control-system architecture and engineering tools must facilitate maintenance during a long life time, within which plants and processes may be subject to significant changes and extensions. To facilitate adaptation to the process, installed control systems should be easily modifiable and expandable.

Today, control systems are to a large extent configured from prefabricated hardware and software building blocks. Module families, from which specialized computer nodes and entire computer networks can be configured, have evolved. In order to choose a configuration that will behave properly in critical situations, the design engineer must be able to estimate the response times of different configurations. We recognize a significant need for tools supporting performance analysis. The state of the art in control-software production is characterized by the fact that the construction of complex programs that meet hard real-time requirements is still a critical task. This situation stimulates a tremendous interest in effective software-development tools. The subject of how to reduce software maintenance cost is of particular importance.

We may summarize the characteristics of control systems as follows:

Modern control systems are complex computer structures that interact intensively with a technical process and human operators, and feature

- Geographical distribution
- Reaction in real time
- High reliability and availability
- Easy configuration and programming
- Easy maintenance and extension

The design of computer-based control systems is very challenging as a consequence of the combination of these features.

The design and implementation of control systems benefit from progress in computer science and computer technology, and in recent years, we have witnessed many advances in this area. The enormous accomplishments in microelectronics have resulted in powerful microprocessors, storage chips, and communication devices. We have seen significant advances in the field of computer architecture. The potential of multi-processors is now much better understood. Distributed computer systems and local area computer networks have been realized and studied. Various approaches to fault-tolerant computing have been pursued. Models of computer organization have been developed which deviate from the von-Neumann computer, for example, the data-flow concept.

Powerful personal computers and graphical workstations providing menu-based dialogs and convenient text and picture editing are now available. Devices for computer speech output and speech recognition have been introduced. The area of artificial intelligence has recently attracted most attention, although commercial applications of artificial intelligence methods are still restricted to a few problems. Nevertheless, there are promising indications that knowledge-based expert systems and artificial-intelligence approaches to computer programming deserve careful consideration.

There have been significant advances in the field of software technology. Better design methods, specification techniques, and programming languages have been introduced. The organization of compilers and operating systems is now well under-stood. Advanced models for database systems have been developed. Approaches to program construction departing from the procedural way of programming have been pursued, for instance, functional programming. In the cost-effective development and maintenance of complex software products, however, severe difficulties are still being faced.

The purpose of this Symposium is to provide a forum for the discussion of computer-science topics relevant to process control. The topics have been carefully selected in order to give an overview of recent developments in the field. The core of this volume consists of invited papers, written by experts of international repute. The topics addressed comprise:

- COMPUTER SYSTEM ARCHITECTURE
    - Architectural principles
    - Fault tolerance
    - Distributed computer systems
    - Real-time computing
    - Process-control architectures
    - Performance modelling

- HUMAN-COMPUTER INTERACTION
    - Dialog design
    - Computer graphics

- SOFTWARE ENGINEERING
    - Distributed programs
    - Requirements engineering
    - Software-quality assurance
    - Functional specification
    - Logic programming

SESSION I

COMPUTER SYSTEM ARCHITECTURE

Chairmen: T. Lalive d'Epinay and R. Karg

PRINCIPLES OF COMPUTER ARCHITECTURE

W.K. GILOI

GMD Research Center for Innovative Computer Systems
Berlin (West), Fed. Rep. Germany

ABSTRACT

First a definition of computer architecture is introduced together with the notion of machine data types, to provide a basis for the subsequent discussion of sequential machine architectures and parallel processing architectures. Measures taken in sequential machine architectures to mitigate the von Neumann bottleneck and the semantic gap of conventional machines are discussed, as well as modern trends in processor architecture such as RISC processors and pipeline processors. The discussion of parallel processing architecture centers around the dichotomy of SIMD machines versus MIMD machines and the appropriate control structures in both forms. The notion of data structure architecture is introduced and contrasted with the dataflow architecture. Loosely coupled, distributed multicomputer systems are contrasted with a new generation of highly configurable, strongly coupled, multicomputer systems.

1. WHAT IS COMPUTER ARCHITECTURE?

There has been a school in the past which defined computer architecture as "What the user sees at the machine language level." What the user sees at this level is the class or classes of objects the machine can handle, including the functions the machine can apply to the objects of the class. In other words, what the user sees is the machine data types, using that term in the sense of abstract data types, namely, to denote a class of objects and collection of functions applicable to the objects. "Abstract" in this context means that the nature of the type is defined solely by the function behavior, independently of the internal representation of the objects and functions or

9

other implementational details.

Objects of machine data types, in general, may be defined as a triple*
(<object name>,<object value>,<object attributes>).

In the pure "von Neumann machine" of the past, <object name> used to be a ·physical memory address, <object value> is the content of the addressed memory cell, and <object attributes> is unspecified; that is, there are no access restrictions, and type attributes are inferred from the operation.

In modern computers of the von Neumann variety, object names are virtual addresses. Objects are clustered into domains for which access attributes are specified. Moreover, in tagged architectures (such as the modern LISP machines) we find individual type specifications for all the objects. However, there still exists the feature common to all von Neumann type machines that they can handle only scalar objects. The best-known counter example of a non-von Neumann architecture is the vector machine, a computer architecture that is capable of handling a data structure type consisting of vectors and vector operations.

Confining the notion of computer architecture to the machine data type view only is too restrictive. In the architecture of buildings, to use that analogy, one wants to know more than just what the functionality of the building is. Rather, architecture deals as well with the construction principles, such as post-and-lintel, buttress, space frame, etc., and the resulting structures. Likewise, in computer architecture we have to consider the operational principle of a machine as well as its hardware structure and instruction set.

Consequently, some years ago we introduced a more comprehensive definition of computer architecture which, in the meantime, has proved to be most useful. The definition is given in terms of concepts which first must be defined themselves. This process is iterated until primitives are obtained that need no further explanation. In BNF-like formalism, where ::= denotes "is defined as", (...) denotes n-tuples, and | denotes alternatives, a taxonomy of computer architecture is obtained, whose very first steps are indicated in Figure 1. Further details are given elsewhere.[6] For the purpose of this paper, the definitions given in Figure 1 establish a sufficient frame of reference.

---

* Note that in the world of object-oriented languages (e.g., SMALLTALK), the term "object", in a first approximation, is synonymous with "abstract data type".

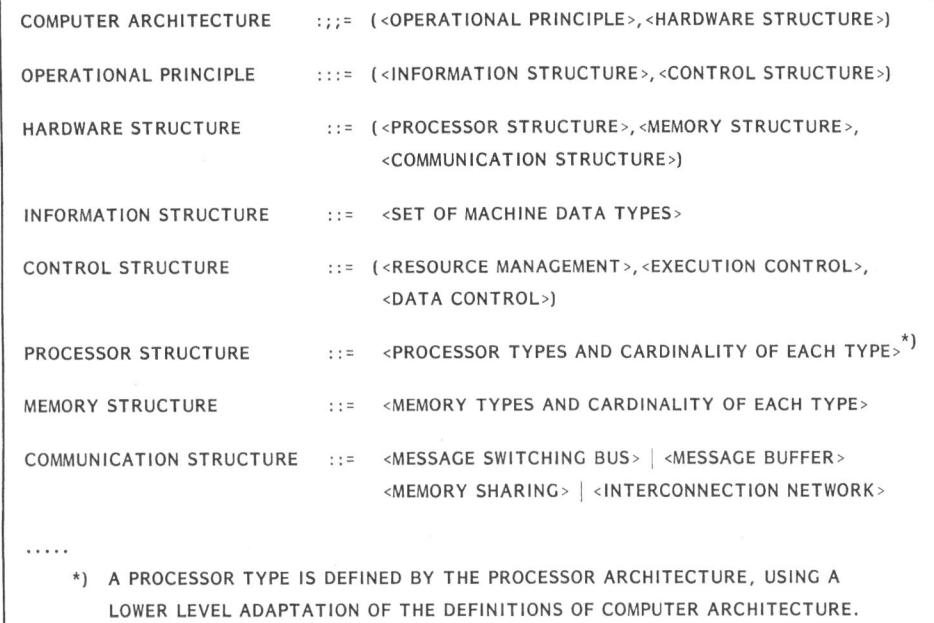

Fig. 1: The First Definitions of a Taxonomy of Computer Architecture[6]

2. SEQUENTIAL MACHINE ARCHITECTURE

2.1 Mitigating the "von Neumann Bottleneck"

The principle of having a processor manipulate the states of single memory words in a word-at-a-time fashion causes an "intellectual bottleneck" for the programmer as well as a physical, performance-limiting bottleneck in the executing machine.[2] Measures mitigating the physical "von Neumann bottleneck" are:

- Instruction pipelining
- Instruction cache
- Hardware-supported vector addressing

Instruction pipelining is illustrated in Figure 2. The idea is to interleave the phases of instruction fetch, instruction interpretation, address calculation, and instruction execution such that several instructions are prefetched, interpreted, and executed as shown in Figure 2: During the execution of the current instruction, the address calculation for its successor is performed, while the next two instructions are inter-

INSTRUCTION PIPELINING:

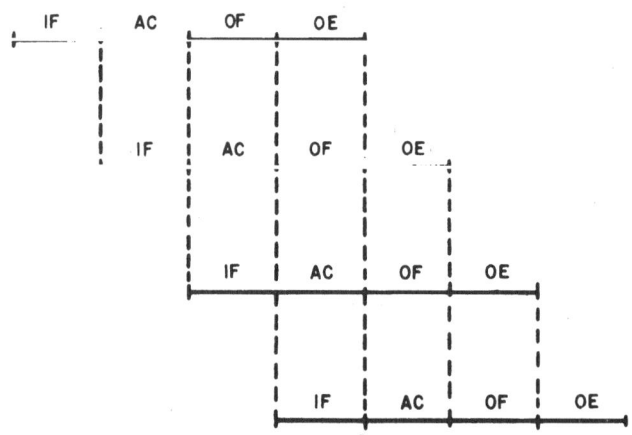

IF : INSTRUCTION FETCH   AC : ADDRESS COMPUTATION
OF: OPERAND FETCH   OE : OPERATION EXECUTION

Fig. 2: Illustration of Instruction Pipelining

preted and fetched, respectively. The effectiveness of that scheme depends on the frequency of branch instructions, since each branch disturbs the look-ahead scheme. Instruction pipelining has become a standard feature in large mainframes, accounting for a performance increase of two or more.

In the case of an instruction cache, a number of instructions of the program under execution are prefetched into the cache and executed from there. A performance gain is obtained if the cache is large enough to accommodate entire program loops, thus resulting in multiple accesses to the same instructions in the cache. In particular in microprocessors such as the Motorola 68020, where the instruction cache is on the chip, small iterative program executions (loops) are performed as fast as if they were programmed in microcode. This property can be extremely valuable in cases of consecutive addressing (e.g., DMA operation, vector processing) or small neighborhood addressing (e.g., image processing).

External caches (which are not on the chip) do not pay in a single processor microcomputer system; the reason being that the "static column memory" that has recently hit the market provides random access to dynamic memory (256K-bit chips) within a "page" of 128 or 256 words at a rate of about 40 nanoseconds per access and, thus, is a match to the fastest available microprocessor.

In processors with hardware-supported vector addressing, a vector element is moved and the counter variable is incremented or decremented and tested for loop exit with one single instruction. As a result, vector processing requires little more than two instructions per operation.

2.2 Memory Management

It has become a standard feature of conventional computers to provide protection of memory domains against unauthorized access. Since the protection is in the access path to each and every memory access, it has to be carried out by dedicated hardware, called memory management, to avoid inefficiencies. There exist two major memory management schemes, namely:

- Memory segmentation into access domains of variable size
- Virtual memory with demand paging and fixed page size

In either case, the instructions of the machine program refer to virtual addresses that must first be translated by the memory management into the corresponding physical address before the memory can be accessed.

In the case of segmentation, the objects referred to must be in memory. The translation of the virtual address into the corresponding physical address is accomplished simply by a look-up or by the segment descriptor table. The segment descriptor, in addition to the physical segment address, contains the segment size specification, so that the hardware can detect and prohibit boundary violations.

In virtual memory with demand paging, not all the pages of the working set of a program under execution need be in main memory. Whenever a reference to a page not in memory is detected, the machine automatically starts a page fault routine to load the missing page from disk to the main memory. Paging, in the first place, is employed to render demand paging more efficient. Moreover, pages may function also as protection domains. In addition, the paged virtual memory can be segmented; thus, demand paging and segmentation with variable segment size can be combined.

Demand paging was originally a scheme to provide the user with a virtual memory space much larger than what was affordable as physical memory. That is, demand paging used to be primarily a means of increasing the system economy. However, with the low-cost, fast 256K-bit memory chip having become reality, with the 1M-bit chip being around the corner, and with the 16M-bit chip being in the planning for the late eighties, memory already has become the cheapest resource of a computer, and the economy argument of demand paging is about to disappear. This leaves the protection issue; and here the memory segmentation solution is simpler and more efficient.

2.3 Closing the Semantic Gap of the von Neumann Computer

As mentioned above, data objects of the classical von Neumann machine carry neither type attributes nor access attributes. Consequently, types, which in typed high-level programming languages are object attributes, become attributes of the operation rather than of the operands. Consequently, scope control and data encapsulation, so essential in modern programming languages, are unknown to the executing hardware. Moreover, type-dependent execution is not possible. This phenomenon is known as the "semantic gap" of the von Neumann machine.

2.3.1 Eliminating the Type Gap

The type attributes associated with the objects of strongly typed high-level programming languages can be handled by the compiler. Therefore, there is no justification for hardware type checking, except for the increased software reliability provided by such a measure (for which the user is not as yet willing to pay the price). The situation is totally different, however, in the case of untyped languages, i.e., languages in which the function executions are type-dependent.

The paradigm of such a language is LISP. The semantic gap renders the execution of LISP programs on conventional von Neumann type computers highly inefficient, since in each function execution the machine first must look up a symbol table for the type of the operands, in order to know what operation to perform.

The semantic gap with respect to types can be eliminated by type tagging. In such tagged architectures, every memory word, in addition to the value of a data object, contains a type tag indicating the type of the object. In microprogrammed machines, the type tags can be used directly to determine the entry into the corresponding microprogram for the type-dependent operation execution.

The problem with type tagging is that it leads to a nonstandard memory word size and processing width; for example, of 40 bits (32 bits for data value representation and 8 bits for tags). Such a nonstandard format prohibits the use of "stock hardware" such as standard 32-bit microprocessors. Rather, one is confined to the use of bit-slice processors -- a technology of the seventies.

One way to avoid the problem with nonstandard formats is called the BEBOP approach. Here, the most significant bits of the 32-bit address of a data object are used for type-tagging the object. In effect this means that the address space (16 gigabytes in the example of a 32-bit address) is subdivided into typed segments of equal size. For example, if the five most significant bits are used as type tags (32 types is about the minimum needed), it means that the size of each segment is 128 megabytes. Needless

to say, the BEBOP approach can be employed only if the memory is virtual memory
with demand paging. But even then this approach is quite wasteful of disk memory,
because of the large amount of unused memory in each segment.

In our opinion, a better approach, that allows the use of stock hardware and yet
avoids the disadvantages of the BEBOP approach, is to use tagged pointers. In LISP,
where access to data objects is usually provided through the CDR-CAR mechanism,
this level of indirection exists anyway. In practice, this may mean using the five most
significant bits, for example, of a 32-bit address as tags. Consequently, the address
space for the user is reduced to 128 megabytes, which is still more than adequate.
The tagged pointer approach can be employed in connection with both virtual as well
as physical memory. The new, powerful, 32-bit wide, microprogrammable "unslice"
processors, which will be available soon, will be most efficient in performing the type
interpretation and, thus, the type-dependent function execution.

2.3.2 Eliminating the Access Gap

A first step towards the enforcement of scope-control rules at the hardware level has
been the introduction of memory management based on memory segmentation. The benefit of such a measure is twofold:

- Unauthorized access of a program to a segment can be prohibited, and
- The integrity of segment boundaries can be guaranteed

For a more sophisticated scope control, however, the granularity of protection provided by memory segmentation is insufficient. A more sophisticated protection model is
given by the access matrix, which specifies for each "subject" (procedure) of a
program the access rights it possesses for each "object" (procedures, data objects) of
the program.[11] The protection matrix model is usually implemented in the form of
capability addressing, where a capability is a pair
        (object_identifier, access_right_specification).
A subject is allowed to access an object only if it has the appropriate capability.

Capability addressing can be realized without additional overhead by a simple extension of the memory management functions.[3] Normally, the memory management hardware for segmented memory consists of a fast segment descriptor table memory, a
comparator for comparing each address against the segment size specification, and an
adder for adding the displacement to the segment base address in order to obtain the
physical memory address. All that must be added in order to turn a segment descriptor into a capability are some additional bits in the descriptor for access right specification plus some logic to interpret those bits. Access right control can then be
performed in parallel to the address translation without any additional time. The only

price to pay for the simple scheme is that in the case of shared objects each sharer must now have its own segment descriptor for the shared segment, i.e., a larger segment descriptor table memory is required. The various descriptors of shared segments differ only in the access right bits.

2.4 Architecture of "von Neumann Processors"

It was mentioned above that the definitions introduced for computer architecture can also be employed for defining processor architectures. That is, a processor is characterized by its operational principle and its corresponding hardware structure, and so on. At present, two major processor types are in use:

- Sequential processors
- Pipeline/systolic array processors

In a macroscopic view, the sequential processor consists of two major portions: the instruction processor (microsequencer) and the data processor (ALUs). Not too much variety is found in the instruction processors or microsequencers. In contrast, the data processors may vary considerably in terms of number of registers, number of processing paths (number of ALUs that can work in parallel), etc.

Until recently, the architecture of sequential processors had reached a certain standard characterized by a relatively large variety of functions of the processor data types. According to the macrostructure of a sequential processor, its functions can be classified into two categories:

- Instruction interpretation functions (including address transformations)
- State-transformation functions applicable to data objects

Based on the observation that only a small subset of the function set of a conventional register machine has a relatively high frequency of occurrence, while the remaining functions have a low frequency of occurrence or may not be used at all, a recent school of thought demands that the function set of the processor data types be minimal. The resulting processor is called a RISC machine (RISC: reduced instruction set computer).*

---

* The frequently encountered view of RISC as a new computer architecture is not quite correct: RISC is a variant of the von Neumann architecture with a minimal instruction set.

RISC proponents claim two advantages of their approach:

- By simplifying instructions and addressing modes, the average number of microinstruction cycles per operation is reduced and, consequently, performance is increased.
- By reducing the functionality of the processor, its hardware structure is simplified; consequently, "real estate" on the chip is saved that can be utilized for further performance-increasing measures such as increasing the number of registers on the chip.

The first argument appeals to those people for whom the MIPS rate of a computer is the ultimate performance criterion (whatever MIPS may mean). First of all, therefore, RISC has become a sales gimmick. In principle, it is not understood that a machine with a simpler and faster instruction set necessarily must have a higher throughput than a machine with a more complex, and thus more powerful but somewhat slower, instruction set. The RISC proponents, in fact, claim a higher throughput for their machine, compared to a conventional register machine of same speed. The higher throughput is attributed to the fact that the minimal instruction set of the RISC processor is fine-tuned to the requirements of the conventional compiled (von Neumann type) high-level programming languages.

In terms of protection, however, RISC machines are a step back to the dark ages of the "pure" von Neumann machine. This issue has recently been discussed by D. Nelson who characterizes the RISC machine by Ralph Nader's famous "unsafe at any speed" slogan.

The real estate issue we deem even less valid. Technology has reached the 1 to 2 micrometer range and is heading towards the "sub-μ range". The limiting factor for the complexity of a VLSI circuit will increasingly be pin limitation and/or heat limitation, but not real-estate limitation. The trend goes to more, not less, functionality in order to utilize the real estate available, e.g., by adding memory management, instruction cache, floating-point arithmetic, and parallel-processing capabilities to the chip.

3. PARALLEL COMPUTER ARCHITECTURE

3.1 Explicit and Implicit Parallelism

As pointed out in Section 1, the different forms of parallel-processing architectures can be discussed in terms of the control structures and information structures (machine data types) suitable for parallel processing, as well as the appropriate hardware

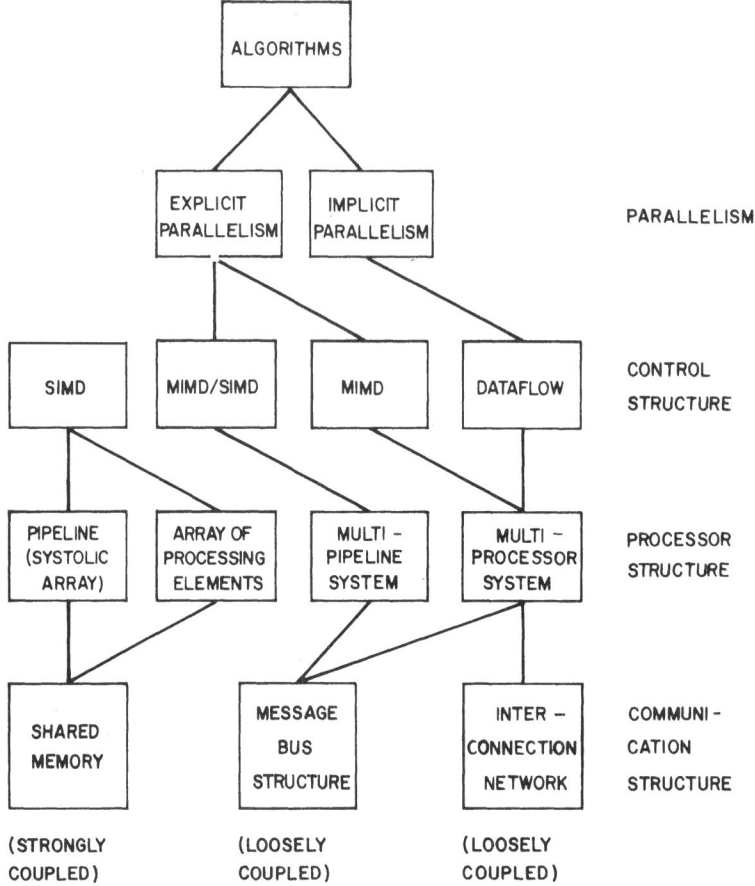

SIMD : SINGLE INSTRUCTION STREAM – MULTIPLE DATA STREAM

MIMD: MULTIPLE INSTRUCTION STREAM – MULTIPLE DATA STREAM

Fig. 3:   Major Forms of Control Structures and Hardware Structures of Parallel
           Computers

structures for their implementation. Some major forms of control structures and hardware structures are listed in Figure 3.

The starting point for a taxonomy of parallel-processing architectures is the parallelism in the algorithms to be executed. In general, an algorithm is a partial ordering of
operations constituted by the data dependencies among the operations. In each state
of algorithm execution, the parallelism is the number of operations that are not data
dependent and, thus, can be executed independently of each other.

We consider a set of data items and a transformation on it defined such that

1) In each step several data items are subjected to the same operation, and
2) By the nature of the transformation there are no data dependencies between the operations performed on the different data items.

The parallelism inherent in such a transformation is called explicit, because it is known a priori that the operations of each step of the transforming algorithm are data independent. Conversely, if the data dependencies between the operations of a transforming algorithm are not known a priori, but must be determined for each step as a result of a data-dependence analysis, we speak of implicit parallelism. In the first case, the explicit knowledge of the parallelism can be utilized to have built-in control structures in the hardware that provide for the parellel execution of the data-independent operations.

The best-known example of explicit parallelism is array processing where the elements of the array, or a subarray thereof, are subjected independently of each other to the same operation. However, many other examples exist, e.g., the unification in logical programs based upon the OR-parallelism.

3.2 SIMD Versus MIMD

The simplest control structure for parallel processing is the SIMD mode of operation, usually realized in the form of a vector machine operating with a pipeline processor. Clearly, pipeline machines are SIMD: Initiated by a single instruction, many data items (e.g., vector elements) are streaming to the pipeline processor, to be transformed there. The addresses needed for the data fetch are not referenced in move instructions but calculated by address generators.

Compared to an equally fast sequential machine (i.e., a scalar processor operating at the same clock rate), the pipeline processor exhibits a decisive performance gain resulting from the following two factors:

- The parallel processing gain in the pipeline, which for long vector asymptotically reaches the number of pipelining stages; and
- The "SIMD gain", which results from the saving of move instructions (the SIMD gain is typically a factor of 2 to 3).

In order to achieve the same performance by a MIMD machine, the number of processors needed must be equal or greater than the performance gain of the pipeline machine (provided, the application allows for a linear performance increase in the

MIMD machine). Consequently, if a SIMD machine and a MIMD machine of equal performance are both realized with the same technology, the SIMD machine is the more cost-effective. MIMD machines, on the other hand, allow for a much higher degree of parallel operation and offer a much higher flexibility of application. The cost disadvantage of the MIMD machine can be reduced or eliminated by using slower and, therefore, "cheaper" technology, the reduced execution speed being compensated for by a higher degreee of parallel operation. However, there are some problems with very large MIMD machines: As yet, nobody has ever built one and, consequently, the appropriate program development environments for very large MIMD machines have yet to be developed.

The performance of a SIMD vector processor is ultimately limited by the access bandwidth of the data memory (a pipeline processor can readily be designed such that it operates at the maximum data rate the memory can maintain). Therefore, measures are taken to obtain a very high effective memory bandwidth, e.g., through long memory words, equally wide buses, and the use of interleaved memories.

The performance bottleneck of large MIMD machines is primarily given by the limited bandwidth of the communication structure interconnecting the "nodes" (processor-memory combinations) of the system. For years, interconnection networks realized in the form of permutation networks have been at the focus of interest. The interconnection properties and complexity of such networks have become well-understood; however, very little work has been done with respect to investigating the technical feasibility of such networks in terms of packaging, connectors, pin limitation, driving power, attainable data rate, and so on.

Figure 4 lists the number of interconnection lines and control lines for the major classes of interconnection networks capable of connecting 1024 sources with 1024 destinations. As the table shows, using decentralized control, which would be a necessity in fault-tolerant systems, leads to a prohibitively high number of control lines. But even in the much simpler case of centralized control, the number of connections needed has an order of magnitude of several thousands to tens of thousands. To obtain a sufficiently high transmission bandwidth, bit serial connections must be ruled out in favor of at least byte-wide interconnection paths, in which case thousands of ribbon cables are required.[5]

Such a structure may be acceptable for the kind of experimental prototype systems that are presently under construction at several research institutes. However, an industrial product must first of all be manufacturable, testable, and maintainable. Therefore, what one wants here is printed circuit backplanes and, maybe, a few cables rather than a "thicket" of thousands of ribbon cables.

COST OF 1024 BIDIRECTIONAL CONNECTIONS

CLASS	NUMBER OF SWITCHING ELEMENTS	NUMBER OF DATA LINES		NUMBER OF CONTROL LINES			
				CENTRAL CONTROL		DECENTRAL CONTROL	
		BIT SERIAL	16 BITS	PER STAGE	PER ELEMENT	ADDRESS	TRANSPUTER
A	1.048.576	2.097.152	33.5 M	–	$2.097.152^1$ $3.145.728^2$	10.5 M	4.2 M
B	9.728	20.480	327.680	38^1 57^2	19.456^1 29.184^2	204.800	40.960
C	5.120	11.264	180.224	20^1 30^2	10.240^1 15.360^2	112.640	22.528
D	11.264	32.768	524.288	30^1 40^2	30.720^1 40.960^2	–	–

^{1} SWITCHING ELEMENT WITH 2 CONNECTIONS
^{2} SWITCHING ELEMENT WITH 4 CONNECTIONS

CLASS A: CROSSBAR SWITCH
CLASS B: BENES NETWORK
CLASS C: N-CUBE; BASELINE; BANYAN; OMEGA; FLIP; DELTA
CLASS D: DATA MANIPULATOR; INVERSE DATA MANIPULATOR

Fig. 4:   Number of Interconnection Lines and Control Lines for Major Classes of Interconnection Networks, Connecting 1024 Sources with 1024 Destinations

The architecture of the SUPRENUM supercomputer, a large MIMD multiprocessor system presently under development at GMD-TUB FIRST, is a result of the considerations above. Rather than using a permutation network, the interconnection structure is given in the form of a matrix of ultra high speed, bit-serial row and column buses as depicted in Figure 5. Each bus operates on the basis of the slotted ring protocol at a gross data transmission rate of 560 megabits per second.[12]

The bus controller handles the first 4 levels of the OSI network model. Its hardware consists of several custom ECL gate array chips, ECL memory, and a microprogrammable flow control unit. The controller hardware is too expensive to be affordable in each and every node of the system. Therefore, clusters are formed, each cluster consisting of up to 16 nodes, a node supervisor, and the controller for the slotted-ring bus matrix. The cluster bus is a fault-tolerant, 32-bit-wide DMA parallel bus. Since the slotted-ring bus matrix allows for alternative routing in the case of a failure, it also exhibits fault-tolerance properties. Each node has 4 megabytes of private memory (there is no global memory in the system) and is capable of performing double-precision floating-point operations at a rate of up to 4 MFLOPS (million float-

W.K. Giloi

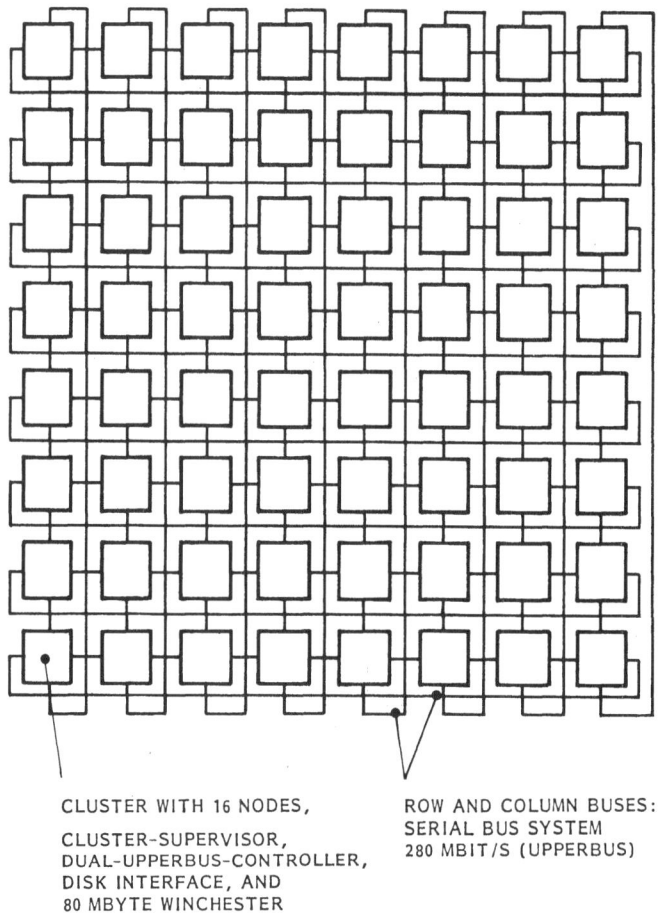

CLUSTER WITH 16 NODES,
CLUSTER-SUPERVISOR,
DUAL-UPPERBUS-CONTROLLER,
DISK INTERFACE, AND
80 MBYTE WINCHESTER

ROW AND COLUMN BUSES:
SERIAL BUS SYSTEM
280 MBIT/S (UPPERBUS)

Fig. 5:   Structure of the SUPRENUM MIMD-Multiprocessor Architecture (Configuration with 1024 Nodes)

ing-point operations per second). The entire system can be configured between 64 and 2000 MFLOPS.[9]

3.2 SIMD Data Structure Architecture

Data structure architectures are computer architectures whose hardware is capable of manipulating complex machine data types with data structure objects (DSO). In a data structure architecture, the DSOs are referred to by a symbolic name, wheras substructures or single data items of a DSO have no names but are accessed by the execution of selection functions. Consequently, data structure architectures are object-oriented: The machine recognizes DSOs as entities that can be accessed only through the application of the functions of the type they belong to. The structure of the DSOs is determined by the behavior of the functions of the type, the internal representation of DSOs being solely a matter of the hardware and not visible at the machine language level. Here we recognize the similarity with the notion of abstract data types.

The machine language of a data structure architecture exhibits a level of abstraction higher than that of the von Neumann languages.[10] Transforming the state of a complex DSO takes only one instruction (SIMD) and is executed in a single computational step, consisting in the initiation (by software) and execution (by hardware) of the appropriate transforming function. The result may be a scalar or a DSO again. No FOR-loops or other "combining forms"[2] are required. The resulting programming style is highly functional, even when the underlying programming language is procedural.

In order to render efficient a data structure architecture, it will be equipped with fast, dedicated type-oriented coprocessors to perform the functions of the data structure types. In general, the DSOs of a machine data type may have a logical structure that differs from the DSO's internal representation in memory, the "storage structure". Consequently, a mapping must be performed between the logical structure of the DSO and the storage structure of its elements. Since this mapping is in the access path to each and every data item, it must be performed at high speed by dedicated hardware in the type-oriented coprocessor, called address generator. The logical structure of a DSO as well as its other attributes are specified in an object descriptor, i.e., the object descriptor provides the information needed by the address generator to calculate the address sequence as required by the logical structure of the DSO and the behavior of the function performed.

Data structure architectures can be designed such that they incorporate the concepts of data abstraction and data virtualization at the hardware level. Since the data items of a DSO are accessible only through the application of functions of the data type they belong to, it is solely the function behavior that determines the structure of the DSO, which can be an appropriately defined, application-oriented virtual structure. In contrast, the internal storage structure of the values of a DSO may be chosen so that it can be most efficiently accessed and processed by the hardware. These features render data structure architectures "object-oriented" or "abstract data type oriented". In order not to overload the machine language with too many primitives, application-specific data structure types may be constructed in an additional software layer, such as a language interpreter, while the set of machine data types is restricted to the types that provide most appropriate and sufficiently general representations of the objects and functions of the application-specific types. Since memory addresses do not exist at the machine language level, the advantages of information hiding, data encapsulation, and access control are already obtained at the hardware level.

Figure 6 gives a general idea of the structure of a data structure architecture.[10] The two salient features of the architecture, which evolved over a decade of research in the area of data structure architectures,[8] are:

- There is a strict separation between the processing sites of scalar data types and data structure types: Scalar types are handled by the (conventional) CPU, while data structure types are handled by dedicated, type-specific coprocessors operating on dedicated, type-oriented memory.
- Run-time checking is performed only on the objects of data structure types, while scalar types are processed in the conventional, low-overhead "von Neumann" manner.

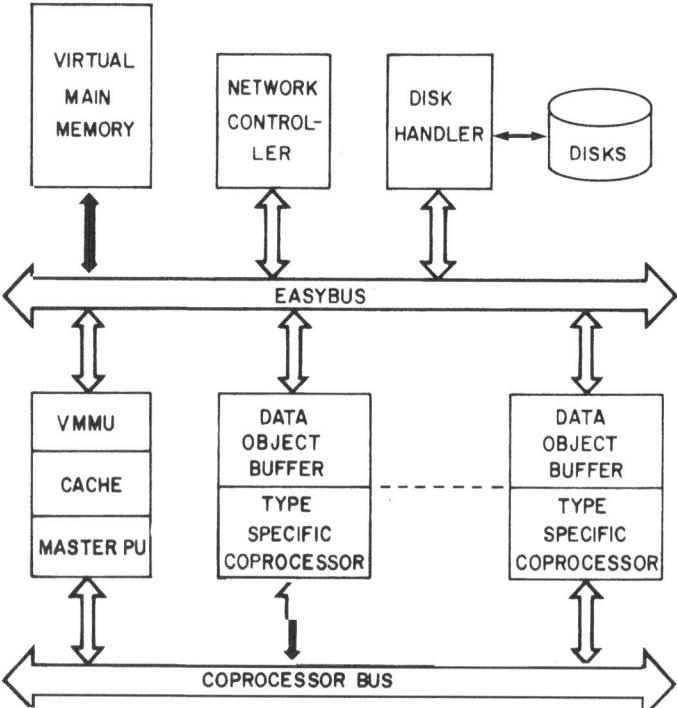

Fig. 6: Data Structure Architecture - General Structure

So far, the concepts of data structure architecture have been put to use with great success in a family of highly cost-efficient machine vision systems of very high performance. Here, the machine data types are "iconic types", that is, picture matrices subjected to a set of standard image analysis operations. Most of these operations are functions performed on a local neighborhood of pixels and, therefore, deviate from the usual sequential array processing algorithms as performed by vector machines. Consequently, the concept of data object virtualization mentioned above comes to fruition.

The proof that the concept of data structure architecture can be applied on a large variety of data structures and not just on vector processing (vector machines are a simple and special case of a data structure architecture) is the Parallel Prolog Machine presently under design at GMD-TUB FIRST. Here, the machine data type is the type LIST, to represent Prolog-specific data types such as:

- List of goal statement headers
- List of terms in a goal statement
- List of definition templates

A specific form of parallelism in Prolog programs, which we call <u>backtracking parallelism</u>, avoids the data dependency problems connected with the AND parallelism and the exponential explosion of unification steps connected with the exploitation of the OR parallelism. Equally importantly, the backtracking parallelism leads to a unification algorithm that can readily be pipelined. As a result, the most frequent operation of a Prolog program, the unification of the goal statement, can be carried out in a SIMD pipeline fashion, much in the same way as array processing.

3.4. Dataflow Architecture

Dataflow can be viewed as an "intelligent" control scheme that dynamically determines the inherent parallelism of the program under execution by analyzing the data dependencies in the program. Guided by the firing rules of the dataflow scheme, the machine dynamically marks the activities of a program as to their executability. The penalty for such a powerful control scheme is the overhead involved, because the machine must not only maintain a list of the activities of a program and their executabilitiy status but must repeatedly search through that list to find executable activities.

In addition to its prohibitively high control overhead, the dataflow solution exhibits another serious problem, which is its inability to transform structured data objects efficiently. According to the firing rule semantics of the dataflow schema, the tokens that indicate the availability of the operands of an operation are consumed with the execution of the operation. Consequently, if multiple use is made of the elements of a data structure object (DSO) in a transformation performed on the DSO, either elaborate copying will take place, or tricks must be used to obtain a memory organization that is accessible in a conventional fashion without violating the single assignment principle.[1]

For all these reasons, the dataflow control scheme cannot compete in efficiency with the data structure architecture principle of exploiting explicit parallelism. This leaves the dataflow principle as a control scheme for exploiting parallelism that is only implicitly given. In all cases where structure data types with the explicitly known parallelism of the operations of the type are given, the data structure architecture approach is far more cost-effective. Therefore, the prediction often voiced in the early days of dataflow research, that the dataflow principle would evolve as the general parallel processing principle, providing the underlying control structure for large MIMD machines used in signal and image processing or array processing in general, will never come true. After we discovered that even the unification process in logical programming, which is as far away from array processing as anything can be, can be performed in a pipelined SIMD fashion, we do not even share the belief, so common among our Japanese colleagues, that dataflow is the natural control structure for parallel inference machines.

4. LOOSELY COUPLED AND STRONGLY COUPLED MULTICOMPUTER SYSTEMS

4.1 Distributed Multicomputer Systems

A distributed multicomputer system consists of a collection of processing nodes; a processing node being a processor-memory configuration including all the other resources (disks, peripherals) needed for a node to run autonomously. The nodes of the distributed system cooperate according to the principle of "cooperative autonomy", that is, the nodes must be cooperative inasmuch as a request received from another node must be honored; they are autonomous inasmuch as they are free to decide when to honor the request.

In a distributed system, the internode communication necessary for the internode cooperation is based upon the exchange of messages. In principle, messages may be exchanged via "mailboxes" in a shared memory (strongly coupled system) or via a message switching bus (loosely coupled system). The loosely coupled system has the advantage that it can be realized without centralized resources. There are two reasons for ruling out centralized resources:

1) Centralized resources are technically not feasible because of the spatial distribution of the nodes (e.g., in a local area network);

2) Centralized resources are undesirable because of fault-tolerance requirements imposed on the system.

Because of unavoidable propagation delays in the transmission of messages, a message concerning the state of the sender node can be obsolete when it is received: By the time the receiver interprets the message, the sender state may have changed again. Consequently, the internode communication (IPC) protocol must be designed such that a node can function correctly without having a precise, up-to-date knowledge of the state of other nodes.

Distributed multicomputer architectures have the potential of

- Being highly modular, readily extensible systems
- Offering high performance through the parallel operation of highly costeffective nodes
- Being fault tolerant

In order to maximize the performance and, thus, the cost-effectiveness of the system, the potential for parallel processing provided by the multiplicity of nodes should be exploited to its fullest. That is, parallel operations should not be unnecessarily restricted, beyond the "natural" restrictions caused by data dependence or resource availability. Since each node is basically an autonomous computer running its own operating system, the units of parallel execution should be the program units managed by the operating system, that is, processes.

In order to meet the requirement of maximum parallelism, interprocess cooperation must be based upon a "no wait synchronization" policy in a "client-server" relationship.[7] In this model, at any instant of time a process can request a service from another process. After having issued the request for service, the "client" (the requestor) can go on with its work until it needs the result of the service requested. Only at this point is a (single) synchronization point needed, which causes the client to wait if the server has not yet rendered the service. Such a scheme actually implements the dataflow schema at the granularity of cooperating processes (which, in our opinion, is the only granularity where dataflow is efficient anyway). It can be shown that all problems related to the no-wait synchronization mechanism (e.g., the orphan problem) can be readily resolved in an object-oriented architecture that provides the system with complete control over the use of data objects.[7]

4.2 Strongly Coupled Multicomputer Systems

Strongly coupled systems are simpler and less expensive to realize and, thus, more cost-effective than their loosely coupled equivalents. The strongly coupled multicomputer system is therefore preferable in all cases where the application allows the multicomputer system to be concentrated in one cabinet and where no other requirements, e.g., fault tolerance, dictate a loosely coupled system.

The advantages of a strongly coupled multiprocessor architecture are fully exploited only if a communication structure is provided that allows each processor to access the shared communication memory at maximum speed, i.e., whithout wait states. It is, therefore, crucial to accomplish the following objectives:

- Develop a correct bus arbitration scheme with minimal arbitration time
- Determine appropriate bus access time and optimize bus control logic
- Minimize bus access cycle by pipelining bus access and arbitration

Figure 7 shows a block diagram of a strongly coupled multiprocessor system designed according to these principles.[4] The core of the system is a very high-speed arbitrated bus called EASYBUS (extensible allocated synchronous bus). The EASYBUS allows random access of the shared memory at a rate of eight million words per second even in the case that each access is performed by a different processor (in the case of block transfer, the access rate is doubled).

The system depicted in Figure 7 features the distribution of system functions over a hierarchy consisting of a UNIX-based, front-end master system, supported by

- Coprocessors for type-specific functions (cf. Section 3.3)
- Peripheral processors for I/O functions, graphics, networking, etc.
- Subsystem for complex application-oriented task

Coprocessors or peripheral processors and subsystems differ inasmuch as the former, after initiation by the master CPU, execute a firmware routine, while the latter perform complex tasks under the supervision of their own operating system. The system in Figure 7 is a flexible workcell controller in which the UNIX-based front-end computer provides the user interface, the program development environment, and the database management for the entire system. Subsystems exist at present in the form of VMEbus control subsystems and high-performance vision systems; a LISP subsystem to accelerate expert systems written in Common LISP is under development at FIRST.

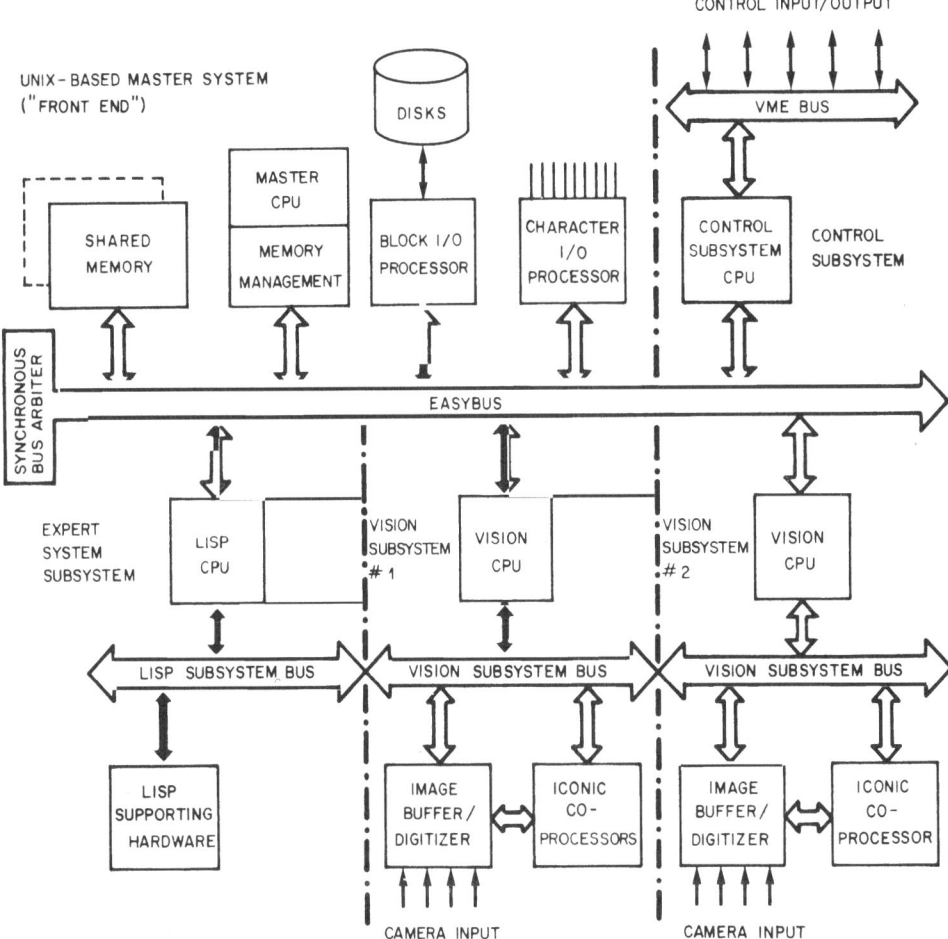

Fig. 7:   Block Diagram of a Configurable, Strongly Coupled Multicomputer Architecture (Flexible Workcell Controller)

REFERENCES

1.   Arvind, and Kathail, V., "A Multiple Processor Dataflow Machine that Supports Generalized Procedures," Proc. 8th Internat. Symposium on Computer Architecture, May 1981, IEEE Catalog No. 81CH1593-2, pp. 291-301.

2.   Backus, J., "Can Programming Be Liberated from the von Neumann Style? A Functional Style and Its Algebra of Programs," Commun. ACM, Vol. 21, No. 8, 1978, pp. 613-641.

3.   Behr, P., and Giloi, W.K., "Object Orientation in High-Performance, Fault-Tolerant Distributed Systems," Proc. 17th Hawaii Internat. Conf. on System Sciences, January 1984, pp. 173-178.

4. Bruening, U., Giloi, W.K., Kallerhoff, R., and Ruesse, M., "UNIMA - A Flexible Framework for Multicomputer Architectures," paper submitted for publication.

5. Ermel, W., "Untersuchungen zur technischen Realisierbarkeit von Verbindungsnetzwerken für Multicomputer Architekturen," Ph.D. Thesis, Technical University of Berlin, 1986.

6. Giloi, W.K., "Towards a Taxonomy of Computer Architecture Based on the Machine Data Type View," Proc. 19th Internat. Symposium on Computer Architecture, IEEE Catalog No. 83CH1889-5, 1983, pp. 6-15.

7. Giloi, W.K., "Obtaining a Secure, Fault-Tolerant, Distributed System with Maximized Performance," in Hardware Supported Implementation of Concurrent Languages in Distributed Systems, G. Reijns, ed., North-Holland, Amsterdam, 1984.

8. Giloi, W.K., "Advanced Object-Oriented Architectures," in Future Generation Computer Systems, Vol. 1, No. 3, February 1985.

9. Giloi, W.K., "Rationale and Concepts for the SUPRENUM Supercomputer Architecture," to be presented at the IFIP Working Conference on Highly Parallel Computers for Numerical and Signal Processing Applications, 1986.

10. Giloi, W.K., and Berg, H.K., "Introducing the Concept of Data Structure Architectures," Proc. 1977 Internat. Conf. on Parallel Processing, IEEE Catalog No. 77CH1253-4C, pp. 44-51.

11. Landwehr, C.E., "Formal Models for Computer Security," ACM Computing Surv., Vol. 13, No. 3, 1981, pp. 247-278.

12. Zuber, G., "UPPERBUS - A Bit-Serial, Fiber Optics-Based Computer Link Bus with 280/560 Mbit/s," Internal Paper, GMD-TUB FIRST, 1985.

DISCUSSION

Chairman: T. Lalive d'Epinay (Brown Boveri, Baden, Switzerland)

P. Freemann (University of California, Irvine, CA, USA)
It was asserted that virtual memories will no longer be needed because of large physical memories. In my view, virtual memories exist for two reasons. First, to make possible larger address spaces than there are physical address spaces. I can certainly see your assertion there. The second reason for virtual memories is to make life easier for the programmer. How will the objectives of virtual memory relating to ease of programming be handled, that is to say, very large address spaces, large databases, dynamic linking and loading, and portability?

W.K. Giloi
First of all, as we all agree, the issue of economy will disappear. The second point is very valid. The second purpose of virtual memory is to have a uniform address space, a very large address space no matter where the data are, in main memory or on the disk. But with appropriate file managers, as you have in UNIX for instance, I would say it is equally convenient for the programmer if he does not have such a large space, let us say 64 Mbytes or 128 Mbytes or whatever you will plug into your computer. You have your file management system and can load your files, load your programs, execute them, etc.

R. Patzelt, (Technical University of Vienna, Vienna, Austria)
I agree with your assertion that we can expand memory to almost any size. However, the data-transfer rate between memory and processor cannot be increased so easily. The data rate is limited by the word length, and the transfer of longer words requires more drivers, connector pins, and so on. According to such restrictions, there must be a compromise. I do not think it will be possible to transfer 128-bit words instead of 16-bit words without significant additional cost.

W.K. Giloi
There are two aspects: the memory capacity and the word length. With more and more bits per chip, increasing the capacity will not really be a problem. It is not even a power problem with currently available static column dynamic RAM, or with fast static CMOS RAM. In terms of the word length of transferred data, of course, you are right. There are connector pin limitations, there is real-estate limitation for the drivers, and there is power limitation. However, 32-bit and 64-bit words are being used.

U. Rembold (University of Karlsruhe, Karlsruhe, Fed. Rep. Germany)
I have a question concerning the flexible workcell controller you presented. Where do you get the software for all those vision and control functions?

W.K. Giloi

A very good point. Frankly, I do not know because the world is not there yet, but efforts are being made. General Motors, for example, has developed a language that combines robot control and vision. I anticipate that this kind of integration of various function classes into one language will go on. But we are far from really having standards in this area. This is something for the process-control people or the manu-facturing people to come up with. We are a little ahead of the curve with an architec-ture like that. Nevertheless, there is a lot of interest because many people realize that they will soon need the power and flexibility of such an architecture. Right now we use it mainly in inspection workstations, and there primarily for computer vision. The demand for computing power in machine vision is almost unlimited. The more computing power you deliver, the more problems can be attacked. A specific example is computer-board bottom-side inspection, where you look whether the pins of the chips come through the holes or not. Some of the recognition tasks that occur in visual-board inspection have not even been attacked yet because it is necessary to carry out hundreds of millions of operations per second in a reasonable period of time.

T. Sakaguchi (Mitsubishi Electric Corporation, Amagasaki, Japan)

Parallel computing is essential to realize high-performance systems. If parallel comput-ing is implemented in VLSI, high regularity of system structure seems to be very advantageous. Systolic arrays feature very high regularity of their structure. Could you comment on the chances of systolic array architecture in the process-control domain?

W.K. Giloi

Systolic arrays are a form of processor architecture, not computer architecture. Systolic-array processors are very useful in SIMD architectures for array processing (in fact, a systolic array can be viewed as a multi-dimensional pipeline) as well as for other operations (e.g., sorting) performed on large, structured data sets. We use them as special purpose processors in our machine-vision systems.

CHALLENGES AND DIRECTIONS IN FAULT-TOLERANT COMPUTING

J. GOLDBERG

SRI International, Menlo Park, California, USA

ABSTRACT

Two decades of theoretical and experimental work and numerous recent successful applications have established fault tolerance as a standard objective in computer system design. As with the objective of correctness, and in contrast to the objective of high speed, satisfaction of fault-tolerance requirements cannot be demonstrated by testing alone, but requires formal analysis. Most of the work in fault tolerance has been concerned with developing effective design techniques. Recent work on reliability modeling and formal proof of fault-tolerant design and implementation is laying a foundation for a more rigorous design discipline. The scope of concerns has also expanded to include any source of computer unreliability, such as design mistakes in software, hardware, or at any system level.

Current art is barely able to keep up with the rapid pace of computer technology, the stresses of new applications and the new expansion in scope of concerns. Particular challenges lie in coping with the imperfections of the ultrasmall, i.e., high-density VLSI, and the ultralarge, i.e., large software systems. It is clear that fault tolerance cannot be "added" to a design and must be integrated with other design objectives. Simultaneous demands in future systems for high performance, high security, high evolvability and high fault tolerance will require new theoretical models of computer systems and a much closer integration of practical design techniques.

1. INTRODUCTION

Fault-tolerant computing has its roots in the earliest theoretical and practical work of the modern computing era,[62] and for the past fifteen years has been the subject of extensive research and development effort. Early efforts were motivated by special, high-stress, high-value applications, such as satellites, passenger aircraft, and telephone switches. Recent objectives have included highly available database services for banks and airlines and safe control of aircraft and chemical reactors. Interest in fault tolerance is spreading to almost all application areas, with motivations that range from protecting life and property to reducing maintenance cost.

The technical scope of fault-tolerant computing has also broadened from an early emphasis on hardware faults to a general concern for sources of system malfunctions of all kinds, including design and programming faults. The field is evolving rapidly to meet numerous challenges, including: (1) new device technology with novel fault modes and special difficulties in testing, (2) rapid growth in system size, (3) novel computer architectures, and (4) increasing interaction with other system objectives, such as security, safety and performance.

Numerous successful experimental systems and commercial products have encouraged users to include fault tolerance as a specific system requirement. One result has been a substantial increase of interest in the field among system designers. The literature of fault-tolerant computing contains a substantial body of theory and practical techniques, but these are not widely disseminated among designers and many of the concepts are evolving in response to new fault phenomena and system requirements.

This paper will serve to introduce the reader to the fundamental ideas in fault-tolerant computing, to current design practice, and to the literature of the field. It presents a set of concepts and terms in their classical usages and discusses current thinking about extensions of the basic ideas of fault tolerance into new areas. References are included to other useful reviews.[6,2,55,53,37,21]

2. FAULT-TOLERANCE CONCEPTS AND APPROACHES

We start with an informal discussion of the basic objectives and ideas of fault tolerance. We then define the concepts more precisely and describe major technical approaches and design issues.

2.1 Fault Tolerance, Reliability and Dependability

We may define <u>fault tolerance</u> as the autonomous capability of a computer system to reduce or eliminate the effect of undesired events on its intended service. The usual motivation of fault tolerance is to increase computer reliability. In ordinary usage, <u>reliability</u> has the general meaning of trustworthiness or dependability. In Reliability Engineering it also has a very particular definition, i.e., the probability that a system will provide a specified service for a stated time. There are other specific terms that support the first, general meaning, but they are strictly distinguished from the formal definition; for example, maintainability, availability, and safety.

In order to reduce the confusion surrounding the multiple meanings of reliability and to allow an unbiased extension of the notion of fault tolerance, Laprie[37] recently suggested the use of the term <u>dependability</u> for the first, general meaning, as follows:

> "Computer system <u>dependability</u> is the quality of the delivered service such that reliance can justifiably be placed on this service."

This suggestion seems to have been accepted by leaders in the fault tolerance field, and will be followed in this survey.

When computers are embedded in physical or information systems, the major objectives in computing dependability are correctness, availability of service, and safety. In current design practice, those concerns are often focused on different parts of a system; thus (1) correctness is usually a concern in the application software, because of its complexity and relative immaturity compared to other design elements; (2) responsibility for availability of service is usually assigned to hardware and the operating system; and (3) responsibility for safety is often assigned to special "safety subsystems", which may include both physical and computer elements.

In this traditional picture, fault tolerance has been seen primarily as a hardware issue, mainly concerned with availability, and to some extent with correctness. Historically, there was sound justification for this partitioning. For example, tolerance of hardware faults was a major technical challenge, and it required strong focusing of research effort. With the recent broadening of dependability objectives to include correctness and safety, the traditional partitioning of concern and design techniques may weaken as designers search for improved effectiveness and efficiency, and as advances are made in the art of designing and verifying complex systems.

2.1.1 Fault Tolerance for Dependability Enhancement

The following properties are traditional objectives for enhancement of dependability through fault tolerance:

- No erroneous output
- Protection for the integrity of internal data
- Continuously available und undegraded service
- Rapid recovery from system failure
- Safe control of critical systems
- Safe shutdown of system operation
- Infrequent system failures
- Low maintenance costs

The time frames for these objectives may be extremely varied, ranging from minutes (in the case of landing an aircraft) to years (for communication systems and unmanned satellites). The objectives themselves may be hierarchically related; thus, continuous availability of a control computer is necessary for safe control but far from sufficient.

Practical fault-tolerant systems are subject to several limitations:

- The intrinsic fault rate of a fault-tolerant computer will be higher than that of the computer of equivalent performance because of the increased number of components, but the potential dependability should be much better, perhaps by several orders of magnitude.

- Fault-tolerant designs generally have a limited capacity in the number of permanent faults or the rate of fault occurrence that may be tolerated. Beyond this capacity, the rate of system failures typically will be greater than for nonredundant systems.

- A central issue in design is development of valid dependability predictions; this requires accurate dependability models, verification that the models are consistent with the design, and valid data about fault occurrences.

Despite these limitations, many fault-tolerant computers have been designed, with a wide range of dependability goals, and several designs are in widespread use. Current research is attempting to overcome current limitations and expand the scope of concerns of fault-tolerant design. An indication of the range and vitality of that research is given in the Appendix, which is the scope of topics for the 1986 IEEE Symposium on Fault-Tolerant Computing.

2.2 Undesired Events and Their Abstractions

There are countless manifestations of incorrect behavior in computing systems. They may have many origins, both physical and human, and may be described at many system levels, e.g., mechanical, electrical, switching, logic and program. They are generically referred to as undesired events. In fault-tolerant computers, an abstract model of an undesired event is developed at the logic or program levels and processed in a way that will restore the system's performance to its intended value.

In the model, a computing system is assumed to be composed of computing elements, each of which has an intended function. If an element is incapable of performing that function, it is said to have a fault or fault condition. The term fault source refers to the phenomenon that gave rise to the fault, and the word error is used to describe incorrect behavior at the logic or program level that may result from the fault condition. For example, we say that the excess dust may be a fault source in a disk unit: it may put the unit into a fault condition, and the data returned may contain errors. Another example is the entry by an operator of an incorrect system tape; in this case, the fault source is human error, and the fault is an incorrect operating system. The notions of "fault" and "error" thus provide convenient abstractions for the complex physical or logical phenomena that constitute undesired events.

In some cases, particularly where human actions are involved, the usages of the terms fault and error are not so clear-cut; for example, it is also common to speak of faulty data or a program error, which we feel are incorrect usages. We find that the notion of function is helpful in achieving consistent usage of these terms.

2.3 Functions, Faults and Errors

We assume that the elements of a computing system are designed to perform a specified function, that is, a well-defined transformation of data from a defined input domain into output data. The function may be combinational or sequential, and the data may be single symbols or strings. We recognize that an undesired event may prevent an element from correctly implementing its function for all input values. We say that the element is faulty or contains a fault, and that, for some input data, the output data will be erroneous or will contain an error. If the underlying fault cause persists, the element will continue to be faulty, but the existence of errors depends on the input data; for example, for some input data, a faulty logic gate or program may produce correct results!

Some erroneous behavior in an element will usually be allowed at the interface with other elements and at the user level, but there will usually be a criterion of acceptability, which, if violated, will lead to the rejection or elimination of the element from an operational system. The element is then said to have <u>failed</u>.

Errors may occur not only in the output of a computing element but also in the new value of its <u>internal state</u>. Examples are: the state of a counter, the content of a file, or the value of an instruction in a program. The consequences of such an error can be viewed in several ways: (1) since the resulting incorrect internal data may cause the element to generate future output errors, the error has enlarged the fault condition of the element (i.e., the size of the class of input data for which erroneous outputs are produced), or (2) it is simply the first of an infinite sequence of errors resulting from the change of the element's sequential state function.

The choice of viewpoints is a matter of convenience. In hierarchical systems, it is convenient to visualize a chain of undesired events as an <u>alternating sequence of faults and errors</u>. Thus a fault in one level may give rise to an error in the data that is passed to a higher level. If this data is stored in a way that changes the function of an element at the next level, i.e., the element becomes faulty; and so on.

The period between the occurrence of a fault and its manifestation in an error is called the <u>latency period</u>. The duration is dependent totally upon the function and the input sequence. Some authors refer to the existence of "latent errors". This is possibly misleading in that it may be understood to refer to actual signal values that are "waiting" to be invoked. It would be better to think of a <u>latent error</u> as the <u>set of incorrect transformations from input to output</u> that are made possible by a particular fault condition. Large latency periods are hazardous in fault-tolerant computers because there is an increased probability of multiple errors and hence increased probability that fault-tolerance mechanisms will be defeated.

We note that our model differs from some others. For example, Kopetz[30] employs (quite consistently) only the notions of failure and fault, with definitions for fault manifestations that consequently differ from those given here.

2.4. Fault Mode: Phenomena and Effects

Faults can have many origins and can take many forms. Significant fault classes, distinguished on the basis of origin, are <u>physical faults</u>, <u>design faults</u> and <u>interaction faults</u>.

A second significant distinction is <u>fault multiplicity</u>. This is clearly a function of the granularity of view; for example, in a memory that contains several faulty cells, the multiplicity of faults may be important to the effectiveness of an error-correcting code, but not to the evaluation of the memory as an element in a system configuration.

A key issue for multiple faults is their distribution in time and space and their mutual correlation.

2.4.1 Physical Faults

Physical faults can result from undesired chemical, mechanical, or electromagnetic activities that originate externally or internally to the computer system.[55](chap. 2). Faults also may originate during manufacture as well as during operation.

The resulting malfunctions can have different temporal characteristics, as follows:

Permanent Faults: A computing element suffers an irreversible malfunction.

Transient Faults: A computing element suffers a malfunction for a limited period of time. Two kinds of transient fault are often significant:
- Random Transient Faults, where successive faults are uncorrelated
- Intermittent Faults, where the originating force appears with some periodicity, e.g., with variations in temperature or physical proximity to a source of noise.

The pattern of errors resulting from these faults depends in part on the pattern of the applied data. Thus a transient error may result from a permanent fault as well as from a transient fault! Confusion of a transient fault for a permanent fault may lead to discarding usable elements unnecessarily.

The time behavior of errors as a function of input data can be very complex indeed; for example, a fault in a computing element (e.g., an unclamped parasitic capacitance) may change its temporal behavior. This might convert a combinational element into a sequential element, or change the time response of a sequential element. Whether or not the element malfunctions in such cases may depend upon the time pattern of input data. These phenomena greatly complicate the problem of fault diagnosis.

Distribution errors are errors in the distribution of information among system elements. A particularly insidious kind are the consistency errors, in which two intended receivers of information from a single source somehow obtain different values. This is a source of confusion in diagnosing a redundant system, because it is usually assumed that redundant computations employ the same input data (hence differences in outputs would seem to indicate processing faults). Another form of such errors is in the synchronization of remote clocks in distributed systems. There has been considerable recent work on this problem under the rubric of "The Byzantine Generals Problem",[35,13] in which the metaphor of dishonest reports from armies in a battlefield is used to describe inconsistent replications of data in a distributed system.

2.4.2 Design Faults

Mistakes in design, e.g., in logic design and programming, may be considered as a source of faults in that the designs physically do not realize the intended function. An obvious quality of design faults is that all elements that employ a faulty design will be faulty, which makes element replacement worthless as a fault-tolerant action in this situation.

There is a common confusion of the notions of fault and error in discussing programs. This confusion results from the fact that a program is an item of data for the process that produces it and a component of a system for the machine that interprets or executes it. Viewed as data, a defective program is erroneous; viewed as a component of a computing element, it is faulty.

To be consistent with previous usage, we can say that a programmer who writes an incorrect program (presumably because he employed faulty reasoning) has made a programming error that results in an erroneous program. When the program is executed it is a faulty program that produces program errors or software errors. This model applies, of course, to any design, whether software or hardware. Approaches to tolerance of design faults will be discussed in the section on software fault tolerance.

2.4.3 Interaction Faults

Incorrect action by operators and maintenance persons is recognized as a very significant source of system errors. For example, in the Bell System ESS No. 1, 30 % of system outages originated in human procedures.[55](p. 462) In such cases, it is the system consisting of the persons and the computer that can be said to be faulty. Most system designs attempt to minimize depending on correct human operation, but there is often less attention paid to the assurance that incorrect operations will be detected and tolerated. Typical unprotected interaction errors include loading the wrong program or hardware component and disconnecting or ignoring alarm indicators. Seemingly innocuous and trivial errors, such as entering the wrong time of day, can have very profound effects. These phenomena are often ignored or assumed away by system designers.

2.5 Fault Avoidance and Fault Tolerance

Responsible designers and builders are generally aware of sources of faults such as physical degradation, noise, incorrect construction, and design mistakes, and they seek to forestall their occurrence by careful, conservative design. Representative design practices are: design review and proof of correctness of design, use of high-quality components, interconnection and packaging techniques, use of conservative circuit margins, and extensive physical testing (for uncovering and removing faults). With an eye to fault-tolerant design, such design practice has been called <u>fault avoid-ance</u>;[6] if it were to accomplish its objectives perfectly, faults would never occur.

In fault-tolerant computers, the possible occurrence of faults is conceded, and some functions are included in the design that serve to suppress or mitigate the conse-quence of faults. Inevitably this results in some redundancy of equipment or time over the amount of equipment or time that would be needed by a design that ignores fault occurrence. Of course, every design contains some redundancy relative to a highly optimized design, but in the context of fault tolerance, the term <u>redundancy</u> always refers to the first kind of deliberately provided redundancy.

2.6 The Basic Paradigm of Fault Tolerance

A general scheme or paradigm of fault-tolerant design has evolved in the course of numerous development efforts. The scheme may be stated with sufficient generality to cover both physical and design faults. For the present we will present a version that is most meaningful for physical faults; design fault tolerance will be discussed in a later section.

The fault-tolerance design paradigm for computing elements that contain physical faults is as follows:

- <u>Requirements Definition</u> establishes specific requirements for dependability and performance, e.g., for reliability, availability, levels of degraded performance, and times for various responses.

- <u>Fault-Tolerant Design</u> establishes a computer organization and a computational strategy for tolerating faults that comprehends both the application and fault-tolerance processes. This is sometimes referred to as design to avoid so-called <u>single-point failures</u>, i.e., failures that defeat the fault-tolerance mechanisms.

- <u>Error-Isolating Design</u> partitions the computer logically into units that are ob-servable, testable and, for some design approaches, reconfigurable. Also, partition the computer physically (not necessarily in the same units as above) so that faults in different partitions are uncorrelated, i.e., so that the faults resulting from a single fault source (e.g., an electrical anomaly or a defective batch of components) will be limited to a single unit.

- <u>Resource Provision</u> provides sufficient computing channels and resources to meet performance and dependability requirements.

- Design Validation evaluates the design with respect to performance and dependability requirements.

The computational strategy of fault tolerance has the following elements:

- Error Detection: During operation, observe selected units to determine the existence of errors. The oberservation can be passive, wherein the input to the observable unit is simply the data provided by the application, or active, wherein special test data is applied in order to expose faults and thus reduce latency time.

- Fault Diagnosis: During operation, after detection of an error, search for and identify the faulty unit responsible for the error. This is sometimes called "fault isolation", but this term should be avoided because it can be confused with the objective of isolating errors by design partitioning. Fault diagnosis is considerably more difficult for multiple faults than for single faults; it is therefore very important to detect the existence of a fault, diagnose its nature and location, and eliminate its impact on the system, as quickly as possible.

- Fault Recovery: During operation, after faulty units have been identified, act to reduce the number and magnitude of the resulting errors, for both the output data and for internal state data. Several actions are possible:

  - Forward Fault Recovery: Assure the earliest possible correct production of future outputs and internal states. Techniques include (1) fault masking, in which the output results of several units are combined so as to suppress individual errors (e.g., by majority voting or error-correcting code logic) and (2) substitution, in which a fault-free unit is substituted for the faulty unit.

  - Backward Fault Recovery: Return the computation to a previous error-free state and repeat the operation using fault-free units.

- Fault Elimination: Concurrently with or following fault recovery, reconfigure the computer so as to eliminate the use of faulty units.

- Performance Optimization: Concurrently with or following fault elimination, organize computations within the new configuration so that it achieves its optimum performance potential; in other words, minimize the degradation of performance resulting from the reconfiguration.

2.7 Application-Specific Emphases

Variations in the specific dependability requirements and in the nature of the computations in a given application lead to different emphases in the use of the general paradigm. Typically a design is developed that is responsive to selected requirements and fault modes. Users should be aware of the assumptions and limitations of each special design. Unfortunately, the label Fault-tolerant is used indiscriminantly and may be understood by nonspecialists to imply greater generality of coverage than actually exists.

Following are some of the more common classes of emphasis.

2.7.1 Signal Processing

In many signal processing applications, there is either little internal state information, or the information is quickly restored by the processing of new input data. In such cases, special steps to restore internal data may not be necessary.

2.7.2 Nonautonomous Operation

In cases where human maintenance service is available and time-criticality is slight, error detection and masking may be adequate to allow continuous, correct operation for the time required by maintenance personnel to repair failed elements. This would obviate the need for automatic reconfiguration.

2.7.3 Database Transaction Processing

In this economically very important application, response-time demands are usually not so severe as to justify forward-error-correction. Most important is to preserve the consistency of databases and to protect data from loss or corruption. In many cases, users will accept some interruption in service (at some economic penalty) and allow operation of the system at reduced throughput during repair operations. The key fault-tolerance mechanism in this environment is backward-error-correction. Backward-error-correction may be complicated where there are several intercommunicating processes. Concern about failures of major memory units leads to replication of databases, and this, in turn, requires mechanisms for safe and consistent updating of multiple copies of data that are in some cases remotely located. These mechanisms may be very slow. Different designs exist for detecting hardware faults at the hardware or program levels. A common additional requirement is that performance capability should be expandable over a wide range with little reduction in efficiency. This requirement is being pursued in various multiprocessor and networking schemes.

2.7.4 Continous Control Processing

In this demanding application, whose economic significance is becoming increasingly important, neither interruption of service nor generation of significant rates of erroneous output are acceptable. These requirements lead to the use of forward-error-correction, and in the case of limited maintenance service, to reconfiguration as well. For simple control applications, there is often very little internal state information. That information may depend only on very recent information about the state of the system under control; thus, following an error or interruption, state information may be quickly regenerated from new sensor information by the natural working of the application process. A special circumstance in such applications is the need for special actuators and for special treatment of sensor data, as follows:

- To avoid dependency on a single error masking output port, which might be a single point of failure, multiple outputs of a fault-tolerant computer will be conveyed to individual actuators that have capability for masking faults, both of incoming data and of internal actuator faults. This arrangement may also be used to tolerate faults in the communication medium connecting the computer and the actuators.

- Special processing is generally required to counter the numerous kinds of errors in sensor data. In most cases, this processing is performed prior to execution of control algorithms, so that if the latter are executed redundantly, all channels will employ the same version of input data.

- Sets of sensors that employ different physical modalities may be used to overcome some sensor errors. Each modality will require its own processing. Algorithms have been developed to extract state information from these redundant types of data. This is a form of redundancy known as analytic redundancy.

Another special problem is the assignment of output values (by the actuators) when system service is interrupted. Typical policies are: continuation of the last output and reversion to an assigned nominal value. In advanced systems, a secondary computation may be called, which is designed to be simpler, and hopefully more dependable.

2.7.5 Integrated Systems

Applications are in sight in which transaction processing and control processing may be integrated, e.g., integrations of engineering, planning, and manufacturing. A natural approach is to use a combination of several specialized systems, but inevitably performance or efficiency considerations will require integration of processing and fault-tolerance techniques.

2.8 Dependability Analysis

Fault tolerance usually is employed for applications in which there is a severe dependability objective and in which the user demands proof, before actual installation, that the dependability objectives will be satisfied. Because system failures in fault-tolerant systems are intended to be extremely rare events, such assurance cannot feasibly be obtained from the classical procedure of whole-system life testing. Instead, it is necessary to test parts of the system and then employ a model to predict how well the system employing those parts will meet its objectives.

Development of such a model-based prediction requires the following actions:

- Define the performance and dependability objectives, e.g., reliability, availability or maintainability.

- Construct a mathematical model of the system that expresses those objectives as aspects of system behavior and fault occurrences for the set of fault types that are contemplated.

- Verify that the model is consistent with the design of the system.

- For the components that are employed in the physical realization of the design, measure the properties and predict the rate of occurrence of the faults that the system has been designed to tolerate.

- For the components and the recovery procedures employed, estimate the completeness (called the <u>coverage</u>) and the speed of the recovery procedures. (A more precise definition of coverage is the conditional probability that, given the existence of a failure, the system is able to recover to the intended state. A second meaning for coverage is discussed in the section on design-for-testability).

- Apply the fault occurrence and recovery data to the model to predict the expected system reliability.

- Test each physical production unit to ascertain that the construction is consistent with the design and that the components are fault-free (some degree of uncertainty may be unavoidable).

Predictions generated by this procedure are subject to many errors of measurement and logic. For example, for very-high-dependability designs, dependability calculations are extremely sensitive to the coverage parameter; unfortunately, the value of coverage is notoriously difficult to estimate accurately. Good coverage analysis requires an understanding of all processes of fault occurrence, detection and recovery. In practice, such understanding requires a combination of logical analysis, simulation and measurement. Model verification, i.e., verification that the design is consistent with the dependability model, is particularly difficult because of the complexity of fault effects in fault-tolerant designs. Recent research in program verification addresses this problem.[44]

Several model types are employed, including Markov state models for representing sequential behavior of systems under fault occurrence events and fault recovery actions, and fault trees, for representing the causal dependencies of system failures. The former is superior for its convenient treatment of sequential behavior; the latter is superior for its convenience in the analysis of complex coverage issues. Petri nets, a dynamic form of Markov model, has seen recent use.[39,8]

There is a substantial body of theoretical dependability analysis based on the Markov model and several practical working tools.[57,43,61,12] The use of fault trees in fault-tolerant design is more limited.[23,38] Much more work is needed to develop a full range of practical techniques. A major limitation to the effectiveness of the models is the paucity of information about actual fault modes and frequencies for modern components and systems. Model interpretation is also a difficult problem because of the numerical instabilities that can result from imprecise failure data and coverage estimates.

3. DESIGN FOR FAULT TOLERANCE

A body of techniques for fault-tolerant system design has evolved and has been applied in research testbeds and in commercial practice. Over the years, the complexity of computer elements has increased dramatically, and the design techniques have evolved accordingly. This section reviews the classical architectural schemes, basic design issues, supporting techniques, and common forms of misdesign.

3.1 Classical Architectures

Over the approximately twenty-five years of fault-tolerant research and development, several schemes of fault tolerance have evolved that have met various criteria for generality and suitability for current technology. The following briefly lists the most significant techniques. They have been used at both the system and subsystem levels.

Multichannel Voting: Processing is replicated in three or more channels, and system output is obtained by a majority vote on the channel outputs. Variations include:

- TMR Triple-Modular Redundancy: Three channels, all permanently operating. Significant commercial examples include the August Systems and Triconex process-control computers.

- NMR N-Modular Redundancy: Like TMR, but with N usually meant to signify more than three channels (usually five).

- Hybrid Redundancy: TMR or NMR but with spare channels that are substituted for failed channels. Examples include the SIFT[64,19] and FTMP[25] research computers.

Dual Redundancy: A pair of channels observes the same inputs and is able to drive the same outputs. Errors are detected by self-checks or by parallel operation and comparison, and a decision is made as to which channel should drive system outputs. Variations include:

- Both channels process the same data; one is dedicated to driving the system output, and the other serves only for error detection.

- Both channels process the same data, each checking its own results; control is initially assigned to one channel, and is switched to the second after a significant failure. An important commercial example is the Tandem transaction processing computer.[55(chap. 11)]

- Both channels process the same data, each checking the other; system output requires agreement; fault tolerance is achieved mainly by roll-back and restart, or by internal reconfiguration.

- Channels operate concurrently, but on different data, until a channel error indicates a need for redundant processing. A variant on this version is n-channel redundancy, using any number of self-checking channels.[28,31]

Dual-Dual Redundancy: Four channels, all reading the same inputs, are connected in parallel-operating pairs and one pair is selected to drive system outputs. When the channels of a pair disagree, indicating a fault, the second pair is selected. A significant commercial example is the Stratus transaction processing computer.

The several techniques can be combined and applied throughout a system; for example, processors, buses and memory channels all can be dualized with independent and local control of element selection, or the processors may be triplicated with only the memories duplicated, as in the C.vmp computer.[55] An ingenious combination of error-correcting coding and multichannel processing is found in the (N,K) architecture,[33] in which each computer word is redundantly encoded in several symbols, and each symbol is stored in a different channel.

3.2 Design Issues

The foregoing schemes should be understood as broad sketches of fault-tolerant strategies. In some cases they have been implemented directly in hardware at a single system level. In other cases, they have been implemented as virtual systems, within monolithic or distributed systems. Their implementation requires consideration of complex design tradeoffs with other behavioral objectives. This section reviews some of the most significant design issues.

3.2.1 Hierarchical Fault Tolerance

Well-designed systems employ hierarchical organization to enhance design verification, design reusability, and other desirable qualities. Fault-tolerance functions also benefit from hierarchical structure, and can be applied in various forms at all the familiar levels, e.g., logic network, register, process, and network. Applying these functions at multiple levels enables a designer to find efficient tradeoffs of conflicting objectives, such as breadth of fault coverage, which is served best by mechanisms at high levels, and precision of fault location, which is served best at low levels.

The philosophy of hierarchy has been applied to fault-tolerant design in the principle of recursive fault-tolerant design,[52] in which each level of a system has all the fault-tolerant functions of the system as a whole, i.e., detection and correction for its own and its supporting levels. This principle has been employed in limited form by interposing a "fault tolerance layer" between layers of an existing, hierarchically structured system.[41]

Closely related is the issue of transparency of fault-tolerance mechanisms. In most cases, programmers will prefer to let fault-tolerant processes operate without visibility at the program level. In some cases, programmers may wish to set the level of protection provided for particular computations or control the particular way in which a system responds to errors, e.g., in making choices about how to meet a hard deadline.

3.2.2 Self-Testability

Self-testability, i.e., the thoroughness with which a computer can test and diagnose its own faults, places a fundamental limit on its reliability. This is a consequence of the fact that fault-tolerance actions are based on determinations of (1) which components are faulty and fault-free, in reconfiguration, or (2) the maximum number of faulty elements present, in error masking.

The accuracy of reliability predictions also depends upon that thoroughness, because life predictions are invariably based on an assumption about how many faults exist when the system is placed into (or restored to) operation. Usually, this number is assumed to be zero, but if the diagnostic program is incapable of determining with absolute confidence that there are no more than N faults, then it must be assumed that at least N faults are present at the start of operations.

High levels of testability are difficult to achieve and are becoming increasingly difficult to achieve as gate density and number of processors increase with advances in technology and architecture. Great differences can be achieved by careful design and the use of highly iterative structures, which are usually much easier to test than irregular structures.

3.2.3 Distributed Fault Tolerance

Distributed processing is increasingly employed because of its capability for smooth growth in capacity and because it utilizes limited bandwidth well in geographically distributed systems. It also offers opportunities for improvements in fault tolerance because the physical isolation of individual nodes helps limit the propagation of errors from a given fault and because it provides a convenient architectural framework for pooling and configuring spare resources. Some significant problems that must be solved to achieve these potential benefits are (1) achieving a correct and current view of the fault status of all elememts, despite time delays and errors in communication among the processors; (2) achieving consistent replicas of data in multiple data stores; (3) robust encapsulation of internode communications; (4) correct and economic recovery of consistent distributed data states, for both simple errors and crashes; and (5) optimum distribution of loads under fault conditions.

Solutions to these problems exist, but they tend to be very serial in processing and coarse in the size of replicated objects. Considerable research attention is being applied to these problems.[34,36,54,40,9,59,27,11]

Most research in distributed processing is concerned with the consequences of faults, and not with the determination of which processors in a system are faulty. Occasionally it is assumed that processors detect their own faults, but this surely does not cover the case where the processors's fault prevents it from correctly diagnosing itself or reporting its diagnosis to other processors! The problem of mutual diagnosis of a set of fault-prone processors was studied in an early, now classical, work by Preparata, Metze and Chien.[49] Resolution of ambiguous reports was found to be very complex computationally. Study of this problem has continued.[24]

J. Goldberg

3.2.4 Data Consistency

Schemes for fault detection that are based on comparing the results of two or more
computations usually assume that the computations are derived from the same initial
data. It is often assumed that multiple versions of data in a fault-tolerant computer
may be created and distributed reliably by the usual methods, e.g., fan-out or
busing, using parallel or serial transmission. These methods are fault-prone, and it is
not a trivial matter to avoid or tolerate the very subtle errors that may result. Various algorithms have been developed to solve this problem, including interactive
consistency and unforgeable signatures.[47,13]

3.2.5 Data Integrity

In many applications, e.g., commercial transaction processing, protection of a database has a much higher value than uninterruptable or error-free service. Two major
techniques for this objective are (1) maintenance of multiple copies in separate data
stores and (2) mechanisms to assure that all data modifications are atomic (in the
sense of indivisibility), i.e., are revoked unless allowed to run to completion.[54]
Other techniques employ internal redundancy in the data structures to allow error detection and correction.[60]

3.2.6 Performance-Reliability Trades

Incorporation of fault-tolerance functions in a computer inevitably affects its performance. Functions such as synchronizing redundant processes, testing, error recovery
and reconfiguration require substantial resources, and often cannot be performed concurrently with application functions. If there are physical constraints on the amount
of available computing resources, then allocation of resources to fault tolerance will
reduce the amount of resources available for application processing. Those functions
that are not performed concurrently with the applications will inevitably increase
processing time, which can lead to instability of plant control in process control
applications.

A rough measure of the cost of a system is the product of the amount of equipment
used and the time required to perform a reference service. The amount of overhead
or resource competition may then be expressed as the redundancy ratio, i.e., the
ratio of the equipment-time costs for the fault-tolerant and the functionally equivalent
non-fault-tolerant computer. This ratio can vary greatly with the severity of the
reliability objectives and the fault rate, e.g., from roughly 1.5, for an idealized,
mainly memory type system with infrequent, simple faults and short mission time, to

at least an order of magnitude higher, for complex systems with frequent, complex faults and long mission lifetimes. There is no accepted rule for estimating this ratio as a function of performance and dependability requirements, although there are numerous particular tradeoff analyses. Absence of a good general rule is a handicap in current system planning.

Another aspect of the interaction of performance and fault tolerance is the increasing appearance of system requirements for specified levels of performance as a function of the severity of fault condition. For example, the system may be required to provide specified levels of performance after the occurrence of the first, second, and third subsystem failures. Similarly, the amount of time required to recover from a fault may be specified for different levels of fault occurrence. The function of performance versus fault level is called performability.[45,16]

3.2.7 Granularity of Application

As in other software and system design disciplines, the designer of a fault-tolerant system has a choice of granularity, i.e., the size of the system element or the duration of the basic time interval to which his techniques will be applied. The negative effects of fine-scale granularity are (1) a higher burden on performance in size or time delay, (2) higher intrinsic fault rate due to the increased number of system components or control decisions (but this is not necessarily translated into higher system failure rate--hopefully, adding these elements will result in a lower system failure rate!).

The positive effects of fine-scale granularity include (1) longer lifetime, because only a small fraction of a system will be discarded for each component failure; (2) faster recovery from errors, because less time will have elapsed from the time of error discovery; (3) and reduced likelihood that the fault-tolerance mechanisms will be defeated by multiple faults. Multiple faults increase the difficulty of diagnosis when there are multiple faults within a module and may cause confusion when a second fault occurs during recovery from a previous fault.

These conflicting tendencies must be resolved within the context of specific requirements for dependability and constraints of size and performance. Advances in VLSI technology make possible dramatic reductions in the granularity scale of fault-tolerance techniques in future systems. As observed in the foregoing discussions of hierarchical design, some faults are best tolerated at high levels of granularity; thus the advent of VLSI makes possible a widening of the granularity range that is available to the designer.

3.2.8 Hardware or Software Implementation

The designer has considerable choice of hardware and software techniques for implementing the various fault-tolerance functions. Compared with hardware implementation,
software implementation results in slower performance, greater flexibility during the
lifecycle and greater use of standard hardware components. The performance disadvantage of software techniques makes them more appropriate for use at high system
levels and for applications that allow a low duty cycle of fault detection and recovery.
Both techniques have their contributions in a well-tuned, multilevel system.

3.2.9 Partial and Supporting Techniques

The following briefly lists the many individual techniques employed to support system-
level schemes or to provide local or specialized fault tolerance:

Fault Diagnosis and Design for Testability: There is an extensive art in fault diagnosis, mostly aimed at discovering and locating permanent faults in logic networks. Recent advances for arbitrarily structured logic include the use of serial techniques
(the so-called level-shift scan-design technique) for obtaining access to interior
elements of high-density logic modules and self-testing schemes based on efficient test
generators and recognizers. The latter are comprehended under the general notion of
design for testability.[65] The use of iterative structures is a more fundamental approach to testability.[15,5]

Despite these advances, the art has not kept pace with advances in complexity of
devices and systems. At higher system levels, testing is usually limited to the exer-
cising of high-level functions with hopefully representative data, but this is a very
weak form of testing. Low-level testing is more thorough, but the size of modern
logic networks makes thorough testing prohibitively expensive in time. This is an
important economic problem during manufacture, and a serious performance problem
during operation. The use of highly regular structures in modern VLSI design offers
some encouragement.

Some attention has been given to the problem of correctly diagnosing transient and
intermittent faults,[43,58] but far from the amount given to permanent faults, which
have been more significant economically and more tractable technically.

The problem of distinguishing transient faults from permanent faults, discussed previously, is an instance of the general problem of determining what fault mode is responsible for observed errors, which is very difficult in stressful environments,
where new component types or packaging methods are employed, and where design
errors may be significant. Recent work in artificial intelligence has studied techniques

for reasoning about fault modes.[14,17,18]

The term coverage, used earlier to describe success in recovery, is also used to describe the completeness of a test procedure in detecting faults. Recovery-type coverage is related to detection-type coverage in that proper recovery from a fault requires good fault detection and recovery actions; thus good detection coverage is necessary but not sufficient, for good recovery coverage. These two uses of the term have caused much confusion.

Error-Correcting Code: Coding techniques are very effective for memory and interconnection systems and are available for multiple errors and for errors that are peculiar to memory and transmission faults, such as burst errors and asymmetrical errors, etc.[63]

Self-Checking Logic: Self-checking logic was an early, ingenious combination of error-detecting coding and multichannel processing at the logic level.[63] Each binary logic variable is represented by two of the four possible states of a pair of bits, e.g., (01) and (10). The two bits are realized by independent logic elements, whose functions are dual; for example, the self-checking AND function is realized by the combination of an AND and an OR gate, each driven by the same self-checking data signals. The logic of error detection is very simple (merely check for invalid symbols), and may be applied at any point in a network. This early scheme is still attractive for supporting the error-detection-and-retry technique for tolerating transient faults, and VLSI technology has made its cost more tolerable.[46]

Run-Time Assertion Checking: Assertions about the proper values and relationship of program variables provide a good, high-level check for errors due to both equipment and design faults.[4]

Watchdog Timers and Processors: Faults that result in errors in the control of program flow may be detected with fair coverage by simple timers that monitor distinguished processes, and with high coverage by simple processors that check for allowed branching sequences.[42]

Checkpoint Recovery: Data sets are recorded periodically to allow a computation to return to a valid state following detection of an uncorrectable error. This procedure has been generalized in the Recovery-Block technique, and techniques have been developed for minimizing the amount of data that must be maintained to allow recovery.[51]

Background Error Detection: Test data is interleaved with application data in order to minimize the latency time of errors.

Element Reconfiguration: Various techniques are available for removing faulty elements from a working configuration. These include hardware-based signal switching, such as in crossbars or other multiport networks, or software techniques, such as table-driven process reassignment, or address modification. Currently, reconfiguration is used at the level of major elements, such as processors, memories or channels, but current research is examining the technique for application to VLSI devices.[32,48]

Special techniques are being explored for use in large-scale multiprocessors. These include (1) interconnection schemes for redundant processor sets[50] and (2) redundant interconnection networks.[1]

3.3 Design Mispractice

The recent growth in requirements for fault tolerance has led some engineers to try to add fault-tolerance mechanisms to existing design, or to enhance designs-in-progress with nominal fault-tolerance design features. Some of the common manifestations are

- Minimal Capabilities for Error Detection and Fault Location: It is assumed that errors will be detected by program tests (to be written by the application programmer), parity checks, standard machine-error flags or time-outs. Often omitted are provisions for detecting errors in actual computations and exposure of latent faults.

- Incomplete Provisions for Reconfiguration: Various multiprocessor schemes, possibly enhanced by special interconnection structures, may be offered as coverage for fault-tolerance requirements. Often omitted are issues of protection for the reconfiguration controller, possible common fault modes, and fault diagnosis.

- Inadequate Protection from Distribution Faults: Buses are very convenient as a means of interelement communication, and designers may be tempted to ignore their many fault modes, some very subtle, when multiple buses or direct-link connections would be a costly alternative.

- Nonrobust Synchronization: Design errors in intersystem synchronization are very common even ignoring the possibility of faults. Fault-tolerant synchronization is a difficult design problem, and some solutions may be very expensive in time.

We may hope that such design shortcuts will become ancient history as the state of awareness and practice of fault-tolerant design become more widespread.

4. EXTENSION OF FAULT-TOLERANCE CONCEPTS

The notions and definitions of fault tolerance presented in the foregoing section evolved over a number of years from an original concern with hardware faults. As concern turned from logic design issues to system issues, device-based definitions of fault and error were extended to cover computer elements at all system levels. There were further suggestions,[20] that the notion of faults and fault tolerance be broadened to cover other sources of computing error, such as programming mistakes and operator blunders.

Significant effort has been applied to the extension of the fault-tolerance paradigm to programming, under the notion of software fault tolerance. Recently, the notion of fault tolerance has been extended to both the computer and the system under control. These two extensions will be discussed in the following paragraphs.

4.1. Software Fault Tolerance

It will be recalled that the fault-tolerance paradigm, presented in the foregoing section, stressed the importance of ensuring that faults are isolated and that faults that occur in different component parts are not correlated. The effectiveness of the various fault-tolerance mechanisms discussed subsequently strongly depends on these assumptions. It is obvious that the assumption of uncorrelated faults among a set of spares further assumes that all their designs are perfect, i.e., they are fully consistent with the specified function (usually, the spares will employ the same design, but that is not essential). Clearly, if the erroneous computation of a module is due to an incorrect design, all equivalent designs will produce the same error.

The problem of erroneous design is a possibility throughout a computer, both in software and hardware. The most prevalent cases are in software, because it usually is the most immature part of a system; that is, on the average, design decisions in software are exposed to less scrutiny and are exercised less frequently than design decisions in hardware. This immaturity may tend to increase in future VLSI designs. Presently, concern for hardware design errors is reflected in the use of sets of computers that are made by different manufacturers in multicomputer systems (as in the space shuttle).

The terms fault-tolerant software and software fault tolerance have been coined to describe the use of redundancy to tolerate faulty software designs.[51] They should not be confused with the name software-implemented fault tolerance (the name of both a machine and a technique), which refers to the use of software to implement fault tolerance algorithms, whether of hardware or software origin. The general hope of fault-tolerant software design is that the faults that do occur in the several versions will be independent and only rarely will produce correlated errors for the same input data. If so, the improvement in software reliability would be very substantial.

Basically, the paradigm of software fault tolerance is a direct instantiation of the general paradigm, as follows:

- Specify the desired function of a software element.

- Generate several designs that implement that function, and take care to avoid the use of the same design approach so that the effects of design mistakes will be minimally correlated. This is referred to as <u>design diversity</u>.

- In execution, combine the results of elements using the several designs so as to recover from the effects of errors in the elements.

The most popular and well-studied forms of software fault tolerance are:

- <u>N-Version Software</u>: The outputs of all versions are combined so as to generate the best result, e.g., by voting, selection of the median, or some other appropriate rule. The scheme usually has the versions executed in parallel, but this is not essential.[10,26]

- <u>Recovery Blocks</u>: Outputs of versions are evaluated by a programmed <u>acceptance test</u>. Unacceptable results are rejected. The scheme usually has the versions executed serially, but this is not essential.[51,2]

Current schemes have been criticized for the following potential weaknesses:

- There may be a high correlation of faults in the several versions, because of natural errors of thought in solving particular problems or because of ambiguities in specifications.

- The impact on software dependability will be small, because the technique does not address faults in requirements or specifications.

- The costs of producing, maintaining and executing multiple versions are excessive.

These criticisms are not fundamental and may be alleviated by developing good engineering methods.[22] Recent experiments have shown that correlations between faults in independently produced programs are not negligible,[29] but that real improvements are obtainable in practice.[3] Further experimental research is in progress.[7]

Some work has been done on applying the technique to operating system software, and to providing architectural support to minimize performance degradation.[56] In some systems, design diversity is applied to both hardware and software. For example, in the space shuttle, four processors of one design execute identical software for critical functions, while a fifth processor of different design and manufacture executes different software for the same functions. Examples of the use of hardware design fault tolerance within computer systems at a lower level are not known.

4.2 Safety

4.2.1 Basic Issues

Some users of computer-based systems prefer that the computer stop operating rather than produce erroneous results. In other cases, the users may prefer erroneous data (within limits) to interruption of service. In applications in which a computer is used to control a system that has critical operating states and is affected by forces other than the computer neither of these actions may be acceptable; this case is considered to be a problem in safety, where the concern may be for danger to life or an intolerable economic loss.

These concerns have been expressed (sometimes in a cryptic manner) as policies for so-called fail-safe or fail-soft computer control. For example, the policy "fail-operational, fail-safe (or fail-soft)" signifies that the first two failures of a computer subsystem should leave the system output unaffected, while a third failure may cause the computer system to stop operating or to have reduced performance, but in such a way that the system under control is left in a safe state (this is more easily defined than accomplished!).

Clearly, in cases where the system being controlled (called, generically, the plant) contains considerable kinetic energy, e.g., in chemical or mechanical form or in the form of a very large flow of messages, simple stoppage of computing service is unacceptable. Where the energy is in "potential" form, i.e., where a large amount of energy may be released by a computer control action , an inappropriate or premature output, e.g., because of a "false alarm", is also unacceptable. In these cases, not only must correct observations be made of the plant's environment and internal state and correct values be computed, but the computed values must be delivered at proper times. That is, the output values must be correct and adequate computing power must be available in order to produce the output at the required time.

4.2.2 Design Issues

These concerns go beyond the problems of building a reliable computer, although they are clearly related. Safe design of a computer-controlled system must consider not only interactions between the computer and the plant, but also interactions of these elements with the environment, with other plants, and with human users. These interactions include both the flow of information and the models that are used to understand that information.

Proper control must consider (1) errors in observing and predicting the state of the system being controlled and the external processes that influence it, (2) the consequences of erroneous control, including the cost of false alarms, and (3) the need for global rather than merely local optimization in the avoidance of hazards. Also, the plant itself should incorporate some protection from erroneous computer action. Each of these objectives presents significant technical challenges.

Some research is underway to clarify these issues, and to determine if there are general principles for computer design that go beyond current practice in fault-tolerant design.[38] One outstanding question is whether there are general principles (analogous to the principle of recursive fault-tolerant system design) by which safety requirements may be propagated into the specifications of the computer's subsystem.

4.2.3 Separation of Safety Functions

There are some familiar approaches to the design of safe systems such as the use of a special, physically separate subplant (the "safety system") that acts to prevent the primary, or "production", subplant from entering a hazardous state. For such plant designs, a special computer may be assigned to monitor the state of the primary subplant and to activate the protection subplant. In many cases, the decision to activate may be complex, but the safety action is simple, e.g., turn off power to the primary subplant. In other cases, the shutdown action may be complex; for example, given a hazardous situation, the safety computer may have to guide the plant through a complex control space, in which the proper path may involve balancing one hazard source against another. The dependability requirement for the control computer is thus not merely uninterrupted service, but appropriate and correct computation. This case illustrates that the conventional policy of chosing control paths that have a strictly monotonic decrease in instantaneous hazard may not be the best policy in all circumstances.

The separation of functions between production and safety computations simplifies the analysis of safety issues and reduces the probability that safety will be compromised by the hardware and design faults of the production subsystem. Nevertheless, correct design of control under all reasonable plant hazards and computer fault conditions requires careful verification.

4.2.4 Integrated Control

The disadvantage of partitioned design is that actions tend to be very costly, e.g., a total shutdown of the plant, or the initiation of costly alarm procedures. The cost of safety actions might be reduced greatly if they could be introduced into the control process at early stages, in which hazard avoidance actions are inexpensive and the effects are easy to evaluate. Such actions could be introduced at discrete points in the normal control process, perhaps in the form of sets of concentric "fences" surrounding hazardous control states. They would allow a certain form of experimentation, in which the response of the system would be observed so as to validate the model of the plant and the environment that is currently used for control. A concomitant disadvantage of this approach is that the safety processes are more closely intertwined with the production processes, and thus are more subject to design errors and common modes of failure. Improved verification techniques will allow a higher degree of such integration.

4.2.5 Human Factors

The principle of mutually suspicious design is clearly appropriate for the elements of a critical control system. Thus, the plant should not follow arbitrary commands by the computer, and the computer should not assume that its model of the plant is perfect. This is especially true for the interaction between operators and the control computer. Operators contribute the ability to deal with unexpected events and sensory powers that may be unavailable to the computer. They are also subject to unique forms of error (bias, mind-sets, and the like). Redundancy, in the form of multiple operators has been employed, and in some critical applications, communication between the operators is constrained, in order to limit coherent errors. Systematic design for safety should include fault-tolerant protocols for operator-system interaction.

4.3 Fault Tolerance and Security

The goal of secure computing is to protect a computer's information and information processing service from unauthorized access and interference. Security in computer systems is similar to fault tolerance, in that it is a desired property of service that may be compromised by undesired events, with possibly serious consequences. Other similarities are: (1) it requires a policy that must be defined for each application (2) it may be enhanced by proper design, and (3) because of its critical nature, users may demand a high degree of design verification.

In addition to the formal similarities, which has already led to a sharing of specification and verification techniques, there are certain operational interactions. For example, any equipment or design faults in a system that is subject to security attacks (called <u>penetrations</u>) may increase the vulnerability of a system to penetration, in fact, a penetrator may attempt to diagnose a system fault in order to be able to take advantage of it. A further connection between the subjects is that a penetration may be seen as an attempt to induce a particular kind of fault in a system.

These interactions and similarities suggest that security, fault tolerance and safety should be viewed as very closely related design objectives. Recent advances in formal design verification has encouraged some researchers, e.g., P.G. Neumann (in an unpublished note) to suggest that the properties of safety, fault tolerance and security may be integrated in the design of multilevel machine hierarchies. His vision is that each level of the hierarchy will have functionality that serves all these objectives. Such an approach may lead to more economical, secure, and dependable safety control.

4.4 New Paradigms for Very Large Systems

The scale of computer systems that have been the concern of fault-tolerant computing technology has gradually grown from the single computer, whether a microcomputer or a mainframe computer, to large-scale computer systems, including highly parallel multiprocessors and highly distributed, heterogeneous systems. It is appropriate to examine the adequacy of the standard paradigm of fault-tolerant design to see if it covers the needs of such very large systems.

The present paradigm for fault tolerance assumes that the system under concern has been made intrinsically very reliable through the use of fault-avoidance techniques and that it has a clearly defined behavioral specification. Both assumptions will be considered:

<u>Assumptions on Intrinsic Reliability</u>: Present techniques are effective when faults are rare occurrences and the state of a system is knowable to a high degree. In the very large systems of the future, it may have to be assumed that all modules are faulty to some extent, both physically and in design, and that fault-tolerance functions will not have all the information needed to achieve perfect recovery. Further, certain critical events may occur that will require major accommodation in system configuration and control, such as major breakdowns and significant system modifications. Such major challenges may require new strategies to prepare a system for change, to protect critical functions and to recover from damage.

Fault tolerance in future systems may, therefore, be expected to be an on-going activity, pursuing goals with imperfect information and continuously preparing to respond rapidly to a major crisis. The objective of fault tolerance will be integrated with other system requirements such as security and evolvability. New techniques will be needed for such characteristics.

Assumptions on Well-Specified Behavior: Precise specifications of behavior are very difficult to construct for even the smallest systems. For future large systems, it may be expected that high-level goals will be stated imprecisely, and in a way that provides considerable flexibility for their realization. A requirement for future fault-tolerant systems may, therefore, be that the system be able to find ways to achieve high-level goals in the face of major changes in external conditions and available internal functionality.

5. CONCLUSION

We have reviewed basic concepts in fault-tolerant computing and some common approaches and techniques for the design of fault-tolerant systems. We have described how the initial objective of reliable computing has evolved into the notion of dependable computing, with a widening of fault sources, from hardware to design, programming and operation, and a broadening of concerns to include other system objectives, such as performance, safety and security.

Integration of fault tolerance with other system properties, extension to cover the fault modes of advanced technologies and the features of new architectures, and meeting demands for very long life in very complex systems constitute major challenges to researchers in fault tolerance. Meeting these challenges will require infusion of concepts and techniques from the entire field of computer science and engineering.

6. ACKNOWLEDGEMENTS

This survey has benefited greatly from numerous conversations with the author's friends and colleagues at the Computer Science Laboratory of SRI International and in the field of fault-tolerant computing.

REFERENCES

1. Adams, G.B. III, and Siegel, H.J., "The Extra Stage Cube: A Fault-Tolerant Interconnection Network for Supersystems," IEEE Trans. Comput., Vol. C-31, No. 5, May 1982, pp. 443-454.

2. Anderson, T., and Lee, P.A., Fault-Tolerance Principles and Practice, Prentice-Hall, Englewood Cliffs, N.J., 1981.

3. Anderson, T., Barrett, P.A., Halliwell, D.N., and Moulding, M.R., "An Evaluation of Software Fault Tolerance in a Practical System," Proc. 15th Symposium on Fault-Tolerant Computing, 1985, pp. 140-145.

4. Andrews, D.M., "Using Executable Assertions for Testing and Fault Tolerance," Proc. 9th Symposium on Fault-Tolerant Computing, June 1979, pp. 102-105.

5. Armstrong, J.R., and Gray, F.G., "Fault Diagnosis in a Boolean n-Cube," IEEE Trans. Comput., Vol. C-30, August 1981, pp. 587-590.

6. Avizienis, A., "Fault Tolerance: The Survival Attribute of Digital Systems," Proc. IEEE, Vol. 66, No. 10, October 1978, pp. 1109-1125.

7. Avizienis, A., Gunningberg, P., Kelly, J.P.J., Strigini, L., Traverse, P.J., Tso, K.S., and Voges, U., "The UCLA DEDIX System: A Distributed Testbed for Multiple-Version Software," Proc. 15th Symposium on Fault-Tolerant Computing, 1985, pp. 126-134.

8. Beounes, C., and Laprie, J-C., "Dependability Evaluation of Complex Computer Systems: Stochastic Petri Net Modeling," Proc. 15th Symposium on Fault-Tolerant Computing, 1985, pp. 364-369.

9. Bernstein, P.A., and Goodman, N., "Concurrency Control in Distributed Database Systems," ACM Comp. Surv., Vol. 13, June 1981, pp. 185-221.

10. Chen, L., and Avizienis, A., "N-Version Programming: A Fault-Tolerant Approach to the Reliability of Software Operation," Proc. 8th Symposium on Fault-Tolerant Computing, June 1978, pp. 3-9.

11. Chou, T.C.K., and Abraham, J.A., "Load Distribution under Failure in Distributed Systems," IEEE Trans. Comput., Vol. C-32, No. 9, September 1983, pp. 799-808.

12. Coste, A., Doucet, J.E., Landrault, C., and Laprie, J-C., "SURF: A Program for Dependability Evaluation of Complex Fault-Tolerant Computing Systems," Proc. 11th Symposium on Fault-Tolerant Computing, 1981, pp. 72-78.

13. Cristian, F., Aghili, H., Strong, R., and Dolev, D., "Atomic Broadcast: From Simple Message Diffusion to Byzantine Agreement," Proc. 15th Symposium on Fault-Tolerant Computing, 1985, pp. 200-206.

14. Davis, R., Shrobe, H., Hamscher, W., Wieckert, K., Shirley, M., and Polit, S., "Diagnosis Based on Description of Structure and Function, in Proc. AAAI-82, 1982, pp. 137-142.

15. Boelens, O.C., Friedman, A.D., and Menon, P.R., "Fault Location in Iterative Logic Arrays," Proc. 4th Symposium on Fault-Tolerant Computing, 1974.

16. Furchtgott, D.G., and Meyer, J.F., "A Performability Solution Method for Degradable Nonrepairable Systems," IEEE Trans. Comput., Vol. C-33, No. 6, June 1984, pp. 550-553.

17. Genesereth, M., "The Use of Hierarchical Models in the Automated Diagnosis of Computer Systems," Technical Report, Stanford Univ. Computer Science Dept., December 1981.

18. Georgeff, M.P., "An Expert System for Representing Procedural Knowledge," in Joint Services Workshop on Artificial Intelligence in Maintenance, Vol. 1, 1983.

19. Goldberg, J., "SIFT: A Provable Fault Tolerant Computer for Aircraft Control," in Proc. Info. Proc. 80, Tokyo, Japan, 1980.

20. Goldberg, J., "New Problems in Fault-Tolerant Computing," _Proc. 5th Symposium on Fault-Tolerant Computing_, 1975, pp. 29-34.

21. Goldberg, J., "The Problem of Confidence in Fault-Tolerant Computer Design," in _Informatik-Fachberichte 78, Proc. GI/NTG Conference: Architektur und Betrieb von Rechensystemen_, Springer-Verlag, Berlin, 1984, pp. 347-361.

22. Goldberg, J., "Perspectives in Fault-Tolerant Software," in _IEEE COMPCON 85_, 1985, pp. 264-269.

23. Hecht, H., and Hecht, M., "Use of Fault Trees for the Design of Recovery Blocks," _Proc. 12th Symposium on Fault-Tolerant Computing_, 1982, pp. 134-139.

24. Holt, C., and Smith, J.E., "Self-Diagnosis in Distributed Systems," _IEEE Trans. Comput._, Vol. C-34, No. 1, January 1985, pp. 19-31.

25. Hopkins, A.L. Jr., Smith, T.B. III, and Lala, J.H., "FTMP--A Highly Reliable Fault-Tolerant Multiprocessor for Aircraft," _Proc. IEEE_, Vol. 66, No. 10, October 1978, pp. 1221-1239.

26. Kelly, J.P.J., and Avizienis, A., "A Specification-Oriented Multi-Version Software Experiment," _Proc. 13th Symposium on Fault-Tolerant Computing_, June 1983, pp. 120-126.

27. Kim, K.H., "Approaches to Mechanization of the Conversation Scheme Based on Monitors," _IEEE Trans. Software Eng._, Vol. SE-8, No. 3, May 1982, pp. 189-197.

28. Kirrman, H.D., and Kaufmann F., "Poolpro-A Pool of Processors for Process Control Applications," _IEEE Trans. Comput._, Vol. C-33, No. 10, October 1984, pp. 869-878.

29. Knight, J.C., Leveson, N.G., and St.Jean, L.D., "A Large Scale Experiment in N-Version Programming," _Proc. 15th Symposium on Fault-Tolerant Computing_, 1985, pp. 135-139.

30. Kopetz, H., "The Failure Fault (FF) Model," _Proc. 12th Symposium on Fault-Tolerant Computing_, 1982, pp. 14-17.

31. Kopetz, H., and Merker, W., "The Architecture of MARS," _Proc. 15th Symposium on Fault-Tolerant Computing_, 1985, pp. 274-279.

32. Koren, I., and Breuer, M.A., "On Area and Yield Considerations for Fault-Tolerant VLSI Processor Arrays," _IEEE Trans. Comput._, Vol. C-33, No. 1, January 1984, pp. 21-27.

33. Kroll, T., "The (4,2) Concept Fault-Tolerant Computer," _Proc. 12th Symposium on Fault-Tolerant Computing_, 1982, pp. 49-54.

34. Lampson, B.W., "Atomic Transactions," in _Lecture Notes in Computer Science_, Vol. 105: _Distributed Systems-Architecture and Implementation_, Springer-Verlag, Berlin, 1981.

35. Lamport, L., Shostak, R., and Pease, M.C., "The Byzantine Generals Problem," _ACM Trans. Program. Lang. Syst._, Vol. 4, No. 3, July 1982, pp. 382-401.

36. Lamport, L., "Using Time Instead of Time-Outs in Fault-Tolerant Systems," _ACM Trans. Program. Lang. Syst._, Vol. 6, No. 2, April 1984, pp. 256-280.

37. Laprie, J-C., "Dependable Computing and Fault Tolerance: Concepts and Terminology," _Proc. 15th Symposium on Fault-Tolerant Computing_, 1985, pp. 2-11.

38. Leveson, N.G., "Software Safety in Computer-Controlled Systems," IEEE Computer, Vol. 17, No.2, February 1984, pp. 48-55.

39. Leveson, N.G., and Stolzy, J.L., "Safety Analysis Using Petri Nets," Proc. 15th Symposium on Fault-Tolerant Computing, 1985, pp. 358-363.

40. Liskov, B., "On Linguistic Support for Distributed Programs," IEEE Trans. Software Eng., Vol. SE-8, No. 3, May 1982, pp. 203-210.

41. Lu, L.Y., "A Virtual TMR Node," Proc. 15th Symposium on Fault-Tolerant Computing, 1985, pp. 286-292.

42. Lu, D.J., "Watchdog Processors and Structural Integrity Checking," IEEE Trans. Comput., Vol. C-31, No. 7, July 1982, pp. 681-685.

43. Makam, S.V., and Avizienis, A., "ARIES81: A Reliability and Life-Cycle Evaluation Tool for Fault-Tolerant Systems," Proc. 12th Symposium on Fault-Tolerant Computing, 1982, pp. 267-274.

44. Melliar-Smith, P.M., and Schwartz, R., "Formal Specification and Mechanical Verification of SIFT: A Fault-Tolerant Flight Control System," IEEE Trans. Comput., Vol. C-31, No. 7, July 1982, pp. 616-630.

45. Meyer, J., "Closed-Form Solutions of Performability," IEEE Trans. Comput., Vol. C-31, No. 7, July 1982, pp. 648-657.

46. Nicolaidis, M., "Evaluation of a Self-Checking Version of the MC 68000 Microprocessor," Proc. 15th Symposium on Fault-Tolerant Computing, 1985, pp. 350-356.

47. Pease, M., Shostak, R., and Lamport, L., "Reaching Agreements in the Presence of Faults," J. ACM, Vol. 27, No. 2, April 1980, pp. 228-234.

48. Oruc, A.Y., and Prakash, D., "Routing Algorithms for Cellular Interconnection Arrays," IEEE Trans. Comput., October 1984, pp. 939-942.

49. Preparata, F., Metze G., and Chien, R., "On the Connection Assignment Problem of Diagnosable Systems," IEEE Trans. Electronic Comput., Vol. EC-16, December 1967, pp. 848-854.

50. Raghavendra, C.S., Avizienis, A., and Ercegovac, M.D., "Fault Tolerance in Binary Tree Architectures," IEEE Trans. Comput., Vol. C-33, No. 6, June 1984, pp. 568-572.

51. Randell, B., "System Structure for Software Fault Tolerance," IEEE Trans. Software Eng., Vol. 2, June 1975, pp. 220-232.

52. Randell, B., "Recursively Structured Distributed Systems," Technical Report, Computing Laboratory, Univ. Newcastle upon Tyne, May 1983.

53. Rennels, D., "Fault-Tolerant Computing--Concepts and Examples," IEEE Trans. Comput., Vol. C-33, No. 12, December 1984, pp. 1116-1129.

54. Shrivastava, S.K., and Panzieri, F., "The Design of a Reliable Remote Procedure Call Mechanism," IEEE Trans. Comput., Vol. C-31, No. 7, July 1982, pp. 692-697.

55. Siewiorek, D.P., and Swarz, R.S., The Theory and Practice of Reliable System Design, Digital Press, Bedford, Mass., 1982.

56. Slivinski, T., "Study of Fault-Tolerant Software Technology," Technical Report, Mandex, Inc., Report to NASA Langley Research Center, 1984.

57. Stiffler, J., Bryant, L.A., and Guccione, L., "Care III Final Report, Phase I," Technical Report, NASA Langley Research Center, CR159122, November 1979.

58. Stiffler, J., "Robust Detection of Intermittent Faults," Proc. 10th Symposium on Fault-Tolerant Computing, 1980, pp. 216-218.

59. Svobodova, L., "Resilient Distributed Computing," IEEE Trans. Software Eng., Vol. SE-10, No. 3, May 1984, pp. 257-267.

60. Taylor, D.J., and Black, J.P., "Principles of Data Structure Error Correction," IEEE Trans. Comput., Vol. C-31, No. 7, July 1982, pp. 602-608.

61. Trivedi, K.S., Probability and Statistics with Reliability Queueing and Computer Science Applications, Prentice-Hall, Englewood Cliffs, N.J., 1982.

62. von Neumann, J., "Probabilistic Logics and the Synthesis of Reliable Organisms from Unreliable Components," in C.E. Shannon and J. McCarthy, eds., Automata Studies, Princeton Univ. Press, 1956, pp. 43-98.

63. Wakerley, J.F., Error Detecting Codes, Self-Checking Circuits and Applications, Elsevier, New York, 1978.

64. Wensley, J.H., Lamport, L., Goldberg, J., Green, M.W., Levitt, K.N., Melliar-Smith, P.M., Shostak, R.E., and Weinstock, C.B., "SIFT: The Design and Analysis of a Fault-Tolerant Computer for Aircraft Control," Proc. IEEE, Vol. 66, No. 10, October 1978, pp. 1255-1268.

65. Williams, T.W., and Parker, K.P., "Design for Testability," IEEE Trans. Comput., Vol. C-31, No. 1, January 1982, pp. 2-15.

DISCUSSION

Chairman: T. Lalive d'Epinay (Brown Boveri, Baden, Switzerland)

F. Demmelmeier (Technical University of Munich, Munich, Fed. Rep. Germany)
Could you comment on the research in software fault tolerance?

J. Goldberg

There is increasing intensity of work in techniques for software fault tolerance, for example, at the University of Newcastle, UCLA, University of Virginia, and I believe there are many workers in Europe, particularly in Germany. This is a controversial subject because we do not have a science for it. It remains an art. The reasons for faults are mainly human in this case because people make mistakes in their reasoning, and these become embodied in bad specifications or bad programs. Despite the fact that this is not amenable to scientific methods right now, there seems to be an art developing on how to make sure that, for example, you eliminate coherent mistakes when programming efforts proceed. You look for the mistakes that both designers make unconsciously. I think if enough people work hard to develop a certain branch of software engineering, we can make respectable progress. Unfortunately, promises have been made that are not realistic on the reliability improvement of such systems. It is a very controversial issue. One has to be aware of promises and at the same time be a bit hopeful for good practical techniques.

H. Kirrmann (Brown Boveri, Baden, Switzerland)
You seem to be disappointed about the prospect of correctness of programs. What are the reasons?

J. Goldberg

Let me temper my pessimism. I think, it is still useful for certain roles in the system. For example, if we can design our systems from very simple modules, perhaps we can use proof for the correctness of the interaction between those modules. I guess the reason for my pessimism is that user appetite grows faster than our tools for proof can increase their power. At SRI we are working on user tools for proof of correctness of systems. So we are working hard on that technology, but to try to deal with programs of very large size - hundreds of thousands of lines - such as are now coming out, is simply beyond the current technology. Growth in high-level languages will help, as languages mature. As we use more abstract representations of programs, then the opportunities for mental mistakes in our programs will be diminished.

R. Patzelt (Technical University of Vienna, Vienna, Austria)
Is it possible to define the notion of "dependability" clearly?

J. Goldberg

There is some good work in that direction by J.C. Laprie in Toulouse. He defined it as follows: "Computer system dependability is the quality of delivered service such that reliance can justifiably be placed on this service." Also the IFIP Working Group 10.4 has been meeting to discuss his definition of dependability. There is a mathematical theory for reliability in which the word "reliability" has a very special meaning, namely the probability that a system will perform as specified for a given amount of time. However, when people speak about reliability, they speak about it in an informal manner. To avoid such confusion, Laprie has suggested the word dependability. And I like that. I think it is a good suggestion.

DISTRIBUTED REAL-TIME PROCESSING

G. LE LANN

INRIA, Le Chesnay, France

ABSTRACT

Design issues raised by the achievement of safeness, liveness and timeliness properties for distributed real-time computing systems are first investigated. This is accomplished through an overview of problems and known solutions related to faults/intru-
sions, variable delays, concurrency, global physical time, process scheduling and
through an examination of interdependencies which exist between problems and of
possibly conflicting or complementary approaches embedded in known solutions. Examples of such interdependencies and conflicting or complementary solutions are given.

These design issues are then investigated in greater detail for a particular class of
processing functions, that is, interprocess communications. Merits and limitations of
various algorithms utilized to achieve atomic and reliable message transfers, end-to-
end flow control and time-constrained message scheduling are identified. Special
attention is given to the multiaccess control problem in local area networks where
finite bounded message transfer delays must be guaranteed.

With reference to the work conducted within the ISO/OSI and the IEEE 802 Committees, emerging standards for interprocess communication are assessed as well as their
applicability to real-time local area networks. Finally, observable trends of the real-
time local area network market are discussed.

1. INTRODUCTION

The design of correctly functioning distributed real-time computing systems requires complete knowledge of the causes of the problems to be addressed, namely, faults/intrusions, variable delays and concurrency. This is examined in Section 2. Specific algorithms can be devised to circumvent these difficulties and to construct systems which exhibit safeness, liveness and timeliness properties. Examples of complementary or conflicting solutions are given for the problems of time-constrained resource sharing and reliable clock synchronization.

Section 3 focuses on a specific class of processing functions of special interest to designers of distributed real-time systems, that is, interprocess communications. In particular, the problems of atomic broadcast and guaranteed bounded access delays in decentralized multiaccess local area networks are investigated in some detail.

Finally, Section 4 explores the applicability of emerging standards to real-time local area networks as well as observable market trends.

2. GENERAL DESIGN ISSUES

2.1 Definitions

A computing system is distributed if the following conditions are fulfilled:

- There exists a multiplicity of hardware/data/software entities
- Entities are asynchronous
- A multiplicity of entities contributes to the performing of every system function
- For each function, contributing entities are strictly equal and none of them has knowledge of the global system state for that function

Detailed analysis of the implications of this definition has been given elsewhere.[19,21]

Let us refer to a software entity as a process. A system is said to be real-time if the following conditions are fulfilled:

- All system processes are bound to meet finite physical deadlines with respect to their starting dates and termination dates
- A process not meeting a deadline constitutes a system failure

For process dates, one can specify latest deadlines only, thus expressing the promptness properties required from the system. In addition to this, one can also specify earliest deadlines, thus expressing the expected timeliness properties.

2.2 Objectives

The difficulty of designing correctly functioning distributed real-time computing systems has several causes. Some of them can be attributed to one particular aspect of
these systems (either distributed or real-time). Others ensue from the union of both
aspects (distributed and real-time). Of course, the level of difficulty to be attained
directly depends on what is meant by "correct functioning". In the most general
sense, a correct system is a system which delivers expected outputs in time (as
prescribed in the system specifications), this being guaranteed under sufficiently
general assumptions regarding the computing system itself and its environment.

More rigorously, a correct system is a system which exhibits safeness, liveness and
timeliness properties. Safeness and liveness are logical properties. Timeliness refers
to physical properties. Safeness properties express the fact that nothing "wrong" can
happen (e.g., a system is deadlock-free, an invariant is always satisfied). Liveness
properties express the fact that something "good" eventually happens (e.g., process
termination). Timeliness properties express the fact that something "good" does happen "in time", i.e., within some physical time interval (e.g., resource allocation
within a time window).

In order to possess such properties, systems must be provided with specific algorithms. An algorithm acts as a dynamic filter on any set of events applied onto a
system. For the sake of simplicity, let us assume that an event is a state change. Let
$\underline{M}$ be the set of all possible merges between state change sequences as produced by a
given set of events. Some of these merges correspond to global system states that are
incorrect. The precise purpose of an algorithm is to "filter out" unacceptable merges.
In other words, only those merges exhibiting some desired property $\underline{p}$ will be instantiated through the utilization of $\underline{Ap}$ (algorithm for property $\underline{p}$), this restricting $\underline{M}$ to
some subset $\underline{Mp}$. As future possible sets of external events are not all predictable in
general, $\underline{Ap}$ must be conceived as a dynamic algorithm, this being in contrast with
designs based on a priori analysis and static algorithms.

Although it is relatively easy to conceive algorithms which enforce each of the three
properties taken separately - safeness (s), liveness (l), timeliness (t) - it is much
more of a challenge to conceive such algorithms so as to guarantee that $\underline{Ms}$, $\underline{Ml}$ and $\underline{Mt}$
have a nonempty intersection. It can be speculated that, for given assumptions re-
garding a system and its environment, there must be a "best choice" of algorithms
that results into the largest intersection of $\underline{Ms}$, $\underline{Ml}$ and $\underline{Mt}$. Unfortunately, such a
"best choice" methodology does not exist yet. Solutions are known, which may not be
optimal.

Let us now turn our attention to the various causes of problems and to solutions related to the design of distributed real-time systems.

2.3 Faults and Intrusions

Faults and intrusions have at least one thing in common: they can lead to a system failure, i.e., a deviation from the system specifications.[18] For example, an erroneous value read from a faulty sensor can lead to an inappropriate halting of a system (which is a failure for a continuously available system). Similarly, confidential data obtained by an intrusive process can be used to defeat a system, immediately or at some later time. Faults and intrusions influence the way state information is produced, propagated and consumed by processes.

State information may not be propagated at all to some processes, or propagated too slowly, which is a failure for a real-time system. State information may be propagated in time but it can be erroneous. In this case, there are two possibilities. Either the error is not detected, thus leading to a potential failure, or the error is detected. If error recovery cannot be conducted in time, a timing failure occurs.

It is widely admitted that fault/intrusion avoidance is not sufficient as a method to construct dependable systems, this being especially true for distributed systems. The complementary approach, fault/intrusion tolerance, makes use of redundancy. However, it should be noticed that redundancy is used to help to reconstruct an original state, should a fault occur, whereas redundancy is used to prohibit the reconstruction of an original state, should an intrusion occur. Most recent work in this area deals with fault/intrusion assumptions which are far more complex and realistic than the simple failstop or Trojan horse models. A number of authors[6,8,17,27,29] have given examples of elaborate voting or encryption algorithms.

One major difficulty with distributed real-time systems is that fault/intrusion tolerance algorithms must be distributed, i.e., they should not rely on the existence of a particular fault-free process, and they must cope with physical time. Issues of global state observation and global physical time are then relevant to designers of such algorithms.

2.4 Global State Observation

Leaving aside faults and intrusions, one could expect that it should be possible to construct the global state of a distributed system in some particular location so as to allow processes to observe it and to update it consistently. Doing this would be equivalent to designing a centralized system. Conducting such a design by applying principles devised for conventional centralized systems will definitely lead to a catastrophe. This would be just like designing the Navstar Global Positioning System by applying Newtonian physics principles only.[5] Conducting such a design by applying principles established for distributed systems will definitely lead to very poor performance, for concurrent asynchronous processing would be almost totally impossible.

In distributed systems, processes must be individually provided with some system state information. Unfortunately, individual copies of an ideal global system state cannot be assumed to be neither identical nor communicated at the same time to all processes, because of variable delays and concurrency. The only alternative consists in devising algorithms which provide processes with partial but correct state information (in the sense of correct value and correct timing).

2.4.1 Variable Delays

In every multiplexed or shared asynchronous system, delays must be considered as variable. The actual value of a delay for a given operation depends on many variable factors, among which one finds the current system load and the system multiplexing ratio, plus external physical conditions (e.g., temperature) for communication delays.

Even when queuing delays are not accounted for, actual delay values cannot be predicted in a distributed system simply because each operation executes across processor boundaries. Depending on the outcome of such decisions as routing of intra-operation messages or process/processor binding, the duration of a message passing operation or of some distributed operation might vary greatly. Recall that pipelined or parallel fully synchronous systems are not distributed systems according to our definition.

Safeness and liveness properties can be enforced in spite of variable delays via the utilization of logical timestamping algorithms, whose purpose is to generate partial/total orderings of events to be observed by distributed processes. Examples of such algorithms are the sequential numbering/windowing algorithm found in the HDLC and in the ISO/NBS transport protocols. Another example would be the timestamping algorithm used in the three-way handshake protocol.

However, logical timestamping algorithms work correctly under the condition that an upper bound is known for variable delays. This raises some problems with concurrently executing asynchronous processes.

2.4.2 Concurrency

In a distributed system, an operation is viewed as an indivisible and atomic set of suboperations which executes across processor boundaries. An operation is atomic if all of its suboperations complete successfully or none of them completes and if internal states are not visible to concurrently executing operations.[9] Concurrent executions of operations yield arbitrary suboperation interleavings, which destroy the necessary atomicity property. Again, safeness and liveness properties can still be obtained out of logical timestamping algorithms. Most of the relevant work in this area has been conducted in the framework of distributed databases[1] where the notions of internal consistency for single databases and of mutual consistency for multiple-copy databases are equivalent to the notion of atomic operation. In particular, issues of decentralized synchronization and "reliable" broadcast have generated many reflections and proposals. More recently, similar work has been developed in the specific context of distributed operating systems.

With respect to timeliness properties, the only rigorous way to deal with concurrency is by resorting to some physical timestamp based scheduling algorithm. Indeed, with static or fixed priorities, process starvation can occur, except for the processes of highest priority. This is against the goal of enforcing liveness properties. Pure deadline-oriented scheduling or hybrid (static and time-dependent priorities) scheduling algorithms are the only acceptable solutions. Under specific assumptions with respect to external loads, it should be possible to demonstrate whether all deadlines can be met and therefore whether upper bounds for operation executions exist or not. When not the case, a good scheduling algorithm minimizes a cost function (e.g., the number of processes which miss their deadlines or the average priority value of late processes).

Late processes can be harmful. It is then mandatory in a real-time system to guarantee that late processes turn dead after they have reached their deadlines. This can be achieved through a function, called promptness control, which is usually associated with a scheduling algorithm and whose purpose is to eliminate obsolete clients from system waiting queues. Additional beneficial effects of promptness control are savings with respect to resource utilization.

2.5 Examples of Interdependencies

As indicated previously, solutions to the problems raised by the achievement of safeness, liveness and timeliness properties are not independent, as illustrated below.

2.5.1 Safeness/Liveness Versus Timeliness

In order to enforce atomicity for distributed operations (a safeness property), one can make use of some logical timestamping algorithm.[28,31] In order to enforce liveness properties, one can supplement such algorithms with either some logical clock synchronization scheme[15] or with a deadlock-free lock/unlock protocol.[25] When a timestamp is obsolete or when a particular lock is already set, access to a resource is denied to a requesting process, which has to wait, which might be rolled back or even aborted.

Assume now that processes are bound to complete in predetermined finite time. It is easy to see that safe and live execution orderings might conflict with timely orderings. Atomicity is guaranteed but possibly at the expense of missing deadlines or, conversely, all deadlines are met but possibly at the expense of destroying system data consistency.

This is a common situation with real-time processes which must produce some timely nonrecoverable output signals during their course of execution. Two-phase locking, roll-back and delayed commitment protocols may well be of no use in that case. It seems more appropriate to resort to resource preclaiming along with a scheduling algorithm based on either pure or hybrid time-dependent timestamps. For example, a ticketing algorithm[20] can be used satisfactorily in a distributed real-time system by rendering ticket values time-dependent.

2.5.2 Faults/Intrusions and Global Physical Time

In order to handle faults and intrusions, well-behaving processes must share at least one global variable, that is, physical time. Indeed, a process can only tell that a fault has occurred when some expected correct event has not occurred within a prescribed time interval. A process can tolerate faults provided that a sufficient number of correct events are observed within a prescribed time interval. Finally, intrusions are of no consequence provided that protection rules (e.g., encryption protocols, digital signatures) are changed often enough at unpredictable times for intrusive processes but in synchrony for legal processes.

Various algorithms have been proposed which maintain a number of physical clocks in synchrony. More precisely, given some assumptions concerning the physical properties of the system considered (e.g., intrinsic clock accuracy, lower/upper bounds for message propagation delays and message processing times), an upper bound for relative clock drifting can be shown to exist.[7,15]

A major threat to such algorithms is fault occurrence. Different types of faults of various complexity must be dealt with, from failstop to Byzantine faults. Most complex fault-tolerant algorithms make use of encryption and/or digital-signature schemes, which render them tolerant to intrusions as well.

However, we are definitely faced with the conditions for a recursion. Global time is necessary for handling faults and intrusions. Conversely, the establishment of global time can be defeated by fault/intrusion occurrence. The only way out of this recursion is rigor with respect to assumptions on the one hand and solutions on the other hand. For example, it has been demonstrated that 3c+1 clocks are needed to handle up to c arbitrarily faulty clocks; assuming unforgeable digital signatures, this number reduces to 2c+1.[16] Another algorithm, which also assumes unforgeable digital signatures, imposes no minimum number of good clocks but requires that the network of clocks remains connected.[13] More recent work builds on these results.[4,23] However, more progress is to be expected in this area for most of these algorithms still exhibit some inherent limitations (e.g., intolerance to transient faults or impractical implementation).

Let us now concentrate on those issues raised by the design of real-time communication systems.

3. DESIGN ISSUES IN REAL-TIME COMMUNICATION SYSTEMS

Issues which are felt as being most important with respect to the achievement of desired timing and robustness properties in real-time communication systems are presented, with special emphasis on local area networks (LANs).

3.1 Deterministic Multiaccess Protocols

It is somewhat puzzling to observe that false statements, or at least overly simplified statements, keep being propagated and trusted. This is the case with the controversy concerning token-passing LANs (802.4 and 802.5), which are labelled as "deterministic" LANs, and contention LANs (802.3), which are labelled as "probabilistic" LANs. To begin with, it might be useful to remember that determinism refers to systems such that there exists a finite number of system state transitions for switching from one state (e.g., message submission) to another state (e.g., successful message transmission). Determinism is a logical concept. In other words, the utilization of a deterministic algorithm is not sufficient to meet given timing constraints. Many parameters have to be taken into account in order to compute the exact physical values of expected upper bounds. If these values are too high, determinism does not help at all.

3.1.1 Token Passing LANs Are "Deterministic"

Why is it that 802.4 and 802.5 LANs are considered as being "deterministic"? Let us concentrate on those messages which are first in the waiting queues of the various access units. Let us consider 802.5 token rings first. Can a token ring guarantee that each of these messages will be transmitted in bounded finite time? The answer is yes if static priorities are not used (but what about those units which handle very critical and urgent messages?). The answer is definitely no if static priorities are used. It is well known (from queuing theory) that when static priorities are used, only the clients with the highest priority enjoy guaranteed service. For all other clients, starvation can occur. In other words, some messages might be denied access to a ring forever. Is this a deterministic service?

Let us consider 802.4 token buses now. Access units make use of four timers, one of them (Class 6) corresponding to a guaranteed time interval used to transmit most urgent messages. An important question is how to compute the "good" values of these timers so as to keep token rotation time to a "reasonable" value. These computations must integrate not only some straightforward variables (e.g., maximum number of access units, maximum physical length of the bus, etc.) but also more subtle variables. For example, one must know how often every unit will decide to "leave" the logical ring, how often a unit must "solicit" missing units which are not on the logical ring but which would like to join in, how long is the "insertion" procedure execution when many units collide in response to a "solicit" frame, etc. These variables depend on assumptions made on the nature of the input traffic. Except in very specific cases (which do not represent the vast majority of potential 802.4 bus users), input traffic assumptions are of a probabilistic nature and so is the percentage of time spent in executing the leave/insert logical ring protocols. Therefore, one has to admit that 802.4 busses upper bounds are probabilistic.

Let us look now at the fault handling issue. With token passing LANs, it is more difficult to predict what impact faults might have on access delays than with contention LANs. Such faults as unit crashes or electromagnetic noise do not impact 802.3 LANs very much because no global variable must be protected against these faults, with the exception of physical signals. Conversely, not only such physical signals must be guarded against faults with 802.4 or 802.5 LANs, but also the token variable (MAC level) which, in the case of token rings, carries also the vital priority indicators. The recognition of the need to recover from token losses led to the conclusion that a single unit should be designated as the control (central) unit. Should this unit fail, another is elected as the new control unit. Now the questions: "How does one know how often a token is lost?", "How often does a control unit go down?". Will the answers contain deterministic information or will we be playing with probabilities?

There is another more subtle point which is that 802.4 and 802.5 protocols, which have been designed from the start to eliminate collisions, do not eliminate collisions at all when election of a new control unit is necessary (or when 802.4 "solicit" frames are transmitted). Unfortunately, it is not indicated how such collisions can be fully resolved in some predetermined time in all circumstances. We leave to the reader the task of drawing conclusions from the discussion above.

3.1.2 Contention LANs are "Probabilistic"

3.1.2.1 CSMA-CD Compatible LANs

It might be the case that because CSMA-CD protocols belong to the family of random access protocols, CSMA-CD protocols would be regarded as behaving probabilistically! One might also be tempted to believe that there is only one way to resolve collisions, that is, the Ethernet way. Although it is irrefutable that the Binary Exponential Backoff algorithm is of a probabilistic nature, it is wrong to state that contention LANs must be probabilistic in general. The space of choices is given in Figure 1. As can be seen, although initial accesses can lead to collisions, it is possible to resolve such collisions either probabilistically or deterministically. A large number of algorithms which provide contention LANs with deterministic behavior have been published in the literature.[2/3/12/14/24/26/30/32/33/34]

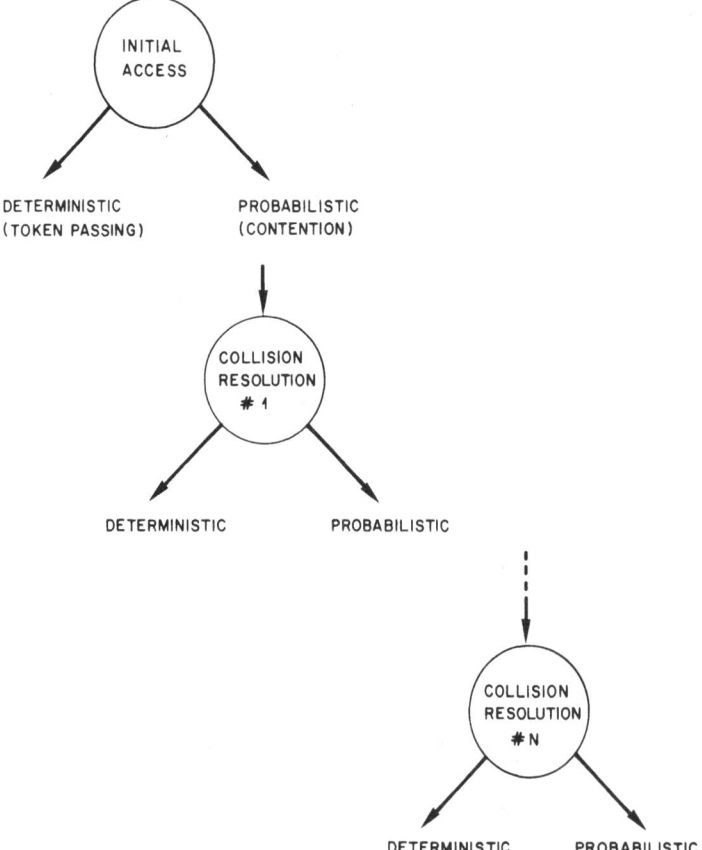

Fig. 1: Decision Tree for Collision Handling

Among these numerous proposals, it suffices to choose those which are 802.3 compatible to obtain a "standard" CSMA-CD LAN which is at least as "deterministic" as token passing LANs. The potential commercial success of this approach lies in the low prices reached by CSMA-CD access units. If the physical "intervention" needed to implement a deterministic collision resolution scheme is limited, in complexity and cost, then deterministic 802.3-like LANs would be very appealing. Being deterministic, such LANs could guarantee that all messages involved in a collision are transmitted in some bounded finite time. Therefore, such LANs could be used to carry all kinds of traffic mixes such as aperiodic data packets and periodic voice packets.

3.1.2.2 High-Speed LANs

Deterministic CSMA-CD can only be used when CSMA-CD achieves efficient channel utilization, i.e., when the ratio of the propagation delay over message duration is small compared to 1 (less than 0.2 is good practice). When not the case, i.e., for large LANs (metropolitan area networks) or for high-speed LANs, neither CSMA-CD nor explicit token-passing protocols are appropriate. The overhead incurred for every token passing operation (token handling protocol execution), for token transmission on the medium and for lost token recovery, is fixed and largely independent of the bandwidth available. Largely independent of the bandwidth as well is the time wasted in reconfigurating physical/logical rings. Acceptable at "low" bandwidth (lower than a few dozens of Mbits/s as an indication), this overhead becomes unbearable at higher speeds, such as those attained with optical transmissions for example.

If we are forced to abandon CSMA-CD and explicit token passing protocols for high-speed LANs, it may appear that the only choice left is the well-known family of synchronous static time division multiplexing protocols. In fact, this is not quite the case. A fair amount of work has been invested in the identification of contention protocols that would achieve a very good utilization ratio of high bandwidth channels. In a recent survey paper, most of these protocols are presented and evaluated comparatively.[10]

3.2 Multilevel Message Scheduling

Establishing finite upper bounds for access delays via the use of some deterministic multiaccess protocol solves only one part of the general problem of how to meet timing constraints attached to message passing tasks. Referring to the ISO/OSI Reference Model, messages handled at all layers of protocols should be scheduled according to specific algorithms. Optimal scheduling is achieved if:

(i)   Finite upper bounds are guaranteed for service times, at all layers, for
      given conditions of utilization, and

(ii) A cost function is minimized when these conditions are exceeded. Examples
      of cost functions which make sense in a real-time system are the number of
      messages which miss their deadlines or the average/maximum message late-
      ness over a given time interval.

Deterministic message scheduling algorithms are needed to enforce property (i) as well
as to guarantee that, in case different costs are associated with competing messages,
least costly messages are discarded on overload or abnormal situations (property
(ii)). Priorities are used to associate costs with messages so that a scheduler can
determine in which order to serve messages present in a waiting queue. As mentioned
in Section 3.1.1, static priorities are generally not appropriate for real-time applications. In fact, static priorities are just a mere approximation of time-dependent priorities, based on the implicit assumptions that it is possible to predict future system
states and that the chosen mapping of priorities over tasks is optimal with respect to
a cost function. In general, such assumptions are only valid for periodic tasks/mes-
sages. Consequently, periodic traffic can be scheduled satisfactorily with such simple
algorithms as rate monotonic methods.[22]

For aperiodic traffic, time-dependent priorities are needed. This is the only way to
achieve optimal scheduling, as defined above, without having to conduct a more or
less rigorous a priori analysis of future system history, which would conflict anyway
with the objective of meeting robustness and flexibility requirements. Time-dependent
priorities can be deadlines assigned to every message when generated by a task.
Deadline driven algorithms can be used to schedule mixes of periodic and aperiodic
messages. In order to break ties in a predictable and optimal manner when competing
messages have identical or nearly identical deadlines, one can devise a hybrid scheduling algorithm F in the form of $F(m) = s(m) + a.dt(m)$, where $s(m)$ is the static
priority attached to message m, a is some weighting factor and $dt(m)$ is the time left
before the deadline of m is reached.

By allowing tasks to assign messages deadlines that are arbitrarily close to current
time, one can naturally obtain bypassing effects in waiting queues. By typing "emergency" messages as such, one can prohibit their destruction (which is triggered for
obsolete messages) even if a prescribed deadline is missed.

General message scheduling models encompass a chain of schedulers, one at each protocol layer for descending message flows as well as for ascending message flows. Although new formal results keep being published in the area of hierarchical hybrid
scheduling, much work remains to be done.

3.3 End-to-End Resource Allocation

In conventional wide area networks or LANs, end-to-end resource allocation translates into end-to-end flow control, whose purpose is to allocate resource units, called credits, at both ends of a connection in such a way that speed matching between a sender and a receiver is reliably achieved. Most often found at layer 2 and layer 4 (using the ISO/OSI terminology), the window mechanism is widely used to perform end-to-end flow control. Credits are allocated on a per connection basis by the receiving end of each connection. In considering the use of window-based protocols for achieving resource allocation in real-time LANs, one encounters some difficulties, as illustrated below with two examples.

3.3.1 Credits

In most cases, credits correspond to one specific type of resource, that is buffer space. But preallocation of buffer space does not guarantee response times (or even throughput). There are at least two other types of resources to be allocated, which are CPU cycles and I/O capacity.

Unfortunately, it is difficult to see how conventional crediting schemes can be used to manage these types of resources appropriately and dynamically. Buffer biased concern in this area is in contradiction with existing LANs performance problems and technological trends. Most existing LANs do not have memory starvation problems. Rather, LAN access units are CPU bounded, rendering the allocation of scarce CPU cycles an even more critical issue. Furthermore, technological know-how and costs have evolved in such a way that it is now cheaper to buy memory space in excess (and waste part of it) than to overdimension CPU capacity.

3.3.2 Connections

Resources (buffers) are allocated on connections only. For connectionless modes of communication, successful message delivery is not guaranteed. Many messages transmitted over real-time LANs being aperiodic, it would be totally unrealistic (because too costly) to maintain permanent connections for these messages.

Opening and closing temporary connections might not be more appropriate either. Not only would it be costly to do so but too slow in many instances. What is needed is guaranteed and timely delivery of datagram-like messages.

Another issue which must be tackled explicitly with real-time LANs is resource allocation among multiple connections maintained by a receiver. In conventional communication systems, the choice of the strategy used is left to the receiver. Clearly, with real-time LANs, the emergency or the priority of a given message can only be expressed by the message source. Therefore, resource allocation at the receiving end must also be under the control of the sources.

These problems can be overcome if, at all layers (in theory) and for all communicating entities, a single system-wide resource allocation and scheduling algorithm is used. What becomes apparent is the need to investigate the relationships existing between distributed operating systems and communication-oriented system software/ firmware more rigorously than has been done in the past.

3.4 Reliable Message Transfer

3.4.1 End-to-End Transfer

Most conventional communication protocols are based upon backward error recovery techniques to ensure reliable message transfer. Typically, positive acknowledgement and retransmission schemes are used. Appropriate for transmission of computer generated data over wide area networks which experience specific types of noise, these schemes could be inadequate for handling real-time traffic over LANs. For example, some messages carry periodically refreshed data. Certain conditions render retransmission of unacknowledged messages totally useless (or even harmful).

Similarly, it may not always make sense to rely on a source to retransmit an urgent message which has been lost or damaged. First, retransmission could take place too late (something really unpleasant has happened in the meantime). Second, the source itself may be destroyed right after issuing the initial urgent message. These are cases where only forward-error-recovery techniques are acceptable.

For real-time LANs, it might be profitable to reexamine the basic design options of conventional communication protocols and to investigate the possibility of using error-correcting codes as well as more sophisticated schemes (e.g., authentication) so as to achieve timely error-free delivery of critical messages and a sufficiently high degree of protection against accidental or malicious intrusions.

3.4.2 Broadcast Transfer

With most LANs, whether passive (e.g., buses) or active (e.g., rings), broadcast communications can be implemented efficiently at the physical level, through the natural propagation of signals generated by every message source. Consequently, the achievement of broadcast communications, at any protocol layer, should be relatively easy with a LAN.

It is somewhat surprising to observe that much confusion has been generated by this issue. The most common mistake consists in assuming that fully reliable broadcast can be obtained at some low level in the protocol hierarchy, thus freeing designers of upper levels from dealing with the problem. For example, one can find proposals "explaining" how the mutual consistency problem, as it arises in multiple-copy databases, can be simply solved by using a CSMA-CD broadcast bus. For proponents of this solution, it seems that supposedly perfect mutual exclusion achieved at the bus level is largely sufficient to guarantee identical orderings of read/update actions on all database copies.

Another mistaken approach consists in utilizing some on-the-fly acknowledgement scheme, as available at the MAC level with token rings, in order to implement "reliable" broadcast. Such schemes can be cleverly utilized to speed up message error recovery, assuming error detection and negative/positive acknowledging can be conducted successfully with some prescribed probability. Of course, fully "reliable" broadcast is not obtained out of such schemes.

In fact, the major weakness of such naive proposals lies in their lack of rigor with respect to fault assumptions and objectives. What is the meaning of "reliable broadcast"? Should a broadcast simply be equivalent to an atomic action? Should message delivery be guaranteed in some finite time? Should concurrent broadcast transfers be ordered identically by all receivers? Are faults considered being of the type failstop or are they Byzantine faults?[4]

Taking a broader view, it becomes obvious that the "reliable broadcast" problem is in fact a particular instance of a much more general problem that has plagued computer scientists for years, that is, the problem of how to achieve consensus in an unreliable distributed system. A formal proof has been given[11] of the intuitive belief that there exists no asynchronous consensus algorithm that can tolerate even a single failstop process. Again, we are asked to investigate synchronization issues very carefully.

4. CURRENT TRENDS OF THE REAL-TIME LAN MARKET

The current real-time LAN market can be divided into two segments, one corresponding to special-purpose LANs and the other to general-purpose LANs.

4.1 Special-Purpose LANs

Real-time LANs which fall into this category are developed for meeting specific needs and/or for connecting specific devices. Three major examples of such LANs are those built for "automation cells" as found in modern factories/plants, those designed for defense applications (e.g., seaborne/spaceborne systems), and finally those used in such large telecommunications networks as ISDNs.

Almost every manufacturer of programmable controllers has developed one or several real-time LANs that are ideally suited to interconnecting specific controllers. Examples are Allen Bradley's Data Highway and Gould Modicon's Modbus. Until recently, standardization has not been a real concern for these manufacturers. The situation is likely to change, however, as it is becoming obvious that programmable controllers, robots, etc. are in fact very similar to ordinary computers with respect to processing and communication activities. Also, interfaces for sensors and actuators are becoming more standardized.

Consequently, the trend in favor of "open" LANs, i.e., LANs which conform to standards, is likely to develop in the future. Standards are set by national and international bodies and also by a few powerful manufacturers, still leading to possibly incompatible options as exemplified by the differences existing between HDLC and SDLC, the IBM link protocol used in Intel's Bitbus.

The case of real-time LANs aimed at defense applications or used in ISDNs is slightly different for it is almost inconceivable that such LANs could be developed without explicit reference to a standard. In defense applications, the need for interoperability among devices produced by different manufacturers makes it mandatory to agree upon standards with respect to LANs. A classical example is MIL-1553-B. Such standards are either those established by national and/or NATO committees or those established by nondefense organizations when the various performance criteria met are appropriate for defense environments.

For LANs used in ISDNs, it is required to conform to standards established by the CCITT. In this area, LANs manufacturers are competing against manufacturers of private branch exchanges (PBXs), assuming the distinction between LANs and distributed PBXs still makes sense. Rolm's CBX and Thomson's LCT 6500 are examples of such products.

In any case, the specific nature of the traffic currently carried by telecommunication networks (analog and periodically sampled traffic) may not be a sufficient incentive to develop specific real-time PBXs or LANs in the future. Again, it is becoming obvious that such devices as digitized telephone sets or telematics terminals are very similar to computerized workstations. Therefore, standards established for the computing industry (ISO) and those promoted for the telecommunications industry (CCITT) are bound to converge as is the case for computing industry standards on the one hand and automation industry standards on the other. Such a convergence is the goal set by the Manufacturing Automation Protocol initiative undertaken by General Motors in 1980.

4.2 General-Purpose LANs

General-purpose LANs are intended to be used in many different environments, satis-fying possibly largely varying performance requirements. Real-time LANs which fall into this category are identical to or are derived from those being considered by the IEEE 802 Committee. One major incentive to choose a general-purpose standard LAN for a real-time application is generalized VLSI implementation of low-level protocols, which has several interesting consequences. Cost reduction is one, as demonstrated by the prices of recent silicon versions of 802.3 protocols. Large-scale testing/relia-bility assessment is another, which cannot be obtained out of limited special-purpose production. Finally, de facto compatibility is another pleasant consequence because of the small number of VLSI manufacturers able to produce 802-compatible circuits.

Costs not being considered, LANs which provide "deterministic" services (802.4 and 802.5) are usually favored over those providing only "probabilistic" services (802.3). However, token rings have some drawbacks in real-time environments. A token ring relies on an active topology (active ring access units). This is inherently less robust than a passive bus. A large number of real-time LANs span short distances (e.g., less than one kilometer). The complexity of a token ring does not seem necessary for such LANs. A passive bus is simpler to manage. Stars, rooted/unrooted trees and meshed topologies are most familiar in a real-time environment. These topologies are bus, not ring, oriented. Also, control of time intervals spent in transmitting messages is more accurate with token buses than with token rings. In particular, starvation is less likely to occur when using the four timers available with 802.4 buses than when using the unique timer and the static priorities available with 802.5 rings. Finally, buses have been put into operation in thousands of locations. This is not yet the case with token rings, whose IEEE 802 approval (1985) comes after official approval of 802.3 proposal (1983) and 802.4 proposal (1984).

Taking costs into consideration, which is usually the case in the real world, conten-tion buses could win against token buses in several instances. Most large computer manufacturers (with one major exception) support 802.3 - like LANs. Such LANs are simple to manage and their VLSI implementation is cheap compared to other 802 LANs. Many users prefer to sacrifice costly "determinism" in favor of cheap "probabilism". Of course, as indicated in Section 3.1, cheap "determinism" is ideal.

It can be predicted that various types of real-time LANs will co-exist on the market and in real-time installations. For example, 802.4 buses, which are currently viewed as the best vehicle for MAP-oriented applications can, in their broadband version, interconnect several 802.3 buses. However, the MAP initiative should not be equated with the exclusive utilization of CATV 802.4 buses. For example, fiber-optics-based LANs are expected to seize a significant part of the market by 1990. Investments made in communication software/firmware/hardware above layer 3 should not be lost by the rapid emergence of new types of LANs.

5. CONCLUSION

Distributed real-time computing systems have a bright future. The major difficulties raised by the design of such systems have been presented and guidelines given. Developers of distributed real-time systems must not only have excellent knowledge of relevant problems and appropriate solutions but also be able to select and integrate complementary solutions into one single design. For example, an efficient communication-oriented architecture could be designed by applying recursively the same unique set of parameterized protocols to the various constitutive elements (VLSI circuits, boards, processors, computers).

Furthermore, developers of distributed real-time systems must be able to reconcile their need for efficient and sometimes specific architectures with the technology available. For example, the integration of protocol layers 1 to 5 and silicon operating system kernels on a single board is a move in the right direction. However, the current lack of integration between communication protocols and basic kernel mechanisms, which are centralized kernels in most cases, needs to be corrected. Undoubtedly, many of these design and architectural challenges will be met in the near future.

REFERENCES

1. Bernstein, P.A., and Goodman, N., "Concurrency Control in Distributed Database Systems," ACM Comput. Surv., Vol. 13, No. 2, June 1981, pp. 185-221.

2. Chlamtac, I., and Franta, W.R., "Message Based Priority Access to Local Networks," Computer Communication, Vol. 3, No. 2, April, 1980, pp. 77-84.

3. Chlamtac, I., Franta, W.R., and Levin, K.D., "BRAM: The Broadcast Recognizing Access Method," IEEE Trans. Commun., Vol. Com-27, No. 8, August 1979, pp. 1183-1190.

4. Cristian, F. et al., "Atomic Broadcast: From Simple Message Diffusion to Byzantine Agreement," Proc. 15th Symposium on Fault-Tolerant Computing, June 1985, pp. 200-206.

5. Denaro, R.P., "Navstar Global Positioning System Offers Unprecedented Navigational Accuracy," Microwave Systems News and Communications Technology, Vol. 14, November 1984, pp. 54-83.

6. Denning, D.E., "Protecting Public Keys and Signature Keys," IEEE Comput., Vol. 16, No. 2, February 1983, pp. 27-35.

7. Dolev, D. et al., "On the Possibility and Impossibility of Achieving Clock Synchronization," Proc. 16th Annual ACM Symposium on Theory of Computing, ACM, April 1985, pp. 504-511.

8. Ehrsam, W.F. et al., "A Cryptographic Key Management Scheme for Implementing the Data Encryption Standard," IBM Systems Journal, Vol. 17, No. 2, 1978, pp. 106-125.

9. Eswaran, K.P. et al., "The Notions of Consistency and Predicate Locks in a Database System," Commun. ACM, Vol. 19, November 1976, pp. 624-633.

10. Fine, M., and Tobagi, F.A., "Demand Assignment Multiple Access Schemes in Broadcast Bus Local Area Networks," IEEE Trans. Comput., Vol. C-33, No. 12, December 1984, pp. 1130-1159.

11. Fischer, M.J., Lynch, N.A., and Paterson, M.S., "Impossibility of Distributed Consensus with One Faulty Process," J. ACM, Vol. 32, No. 2, April 1985, pp. 374-382.

12. Franta, W.R., and Bilodeau, M., "Analysis of a Prioritized CSMA Protocol Based on Staggered Delays," Acta Informatica, Vol. 13, No. 4, 1980, pp. 299-324.

13. Halpern, J. et al., "An Efficient Fault-Tolerant Algorithm for Clock Synchronization," IBM Technical Report RJ-4094, IBM T.J. Watson Research Center, New York, 1983.

14. Kurose, J.F. et al., "Controlling Window Protocols for Time-Constrained Communication in a Multiple Access Environment," ACM Sigcomm, Vol. 13, No. 4, October 1983, pp. 75-84.

15. Lamport, L., "Time, Clocks and the Ordering of Events in a Distributed System," Commun. ACM, Vol. 21, No. 7, July 1978, pp. 558-565.

16. Lamport, L., and Melliar-Smith, P.M., "Synchronizing Clocks in the Presence of Faults," J. ACM, Vol. 32, No. 1, January 1985, pp. 52-78.

17. Lamport, L., Shostak, R., and Pease, M., "The Byzantine Generals Problem," ACM Trans. Program. Lang. Syst., Vol. 4, No. 3, July 1982, pp. 382-401.

18. Laprie, J.C., "Dependable Computing and Fault-Tolerance: Concepts and Terminology," Proc. 15th Symposium on Fault-Tolerant Computing, June 1985, pp. 2-11.

19. Le Lann, G., "Distributed Systems - Towards a Formal Approach," IFIP Congress, August 1977, North-Holland, Amsterdam, pp. 155-160.

20. Le Lann, G., "A Distributed System for Real-Time Transaction Processing," IEEE Comput., Vol. 15, No. 3, February 1981, pp. 43-48.

21. Le Lann, G., "On Real-Time Distributed Computing," invited paper, IFIP Congress, September 1983, North-Holland, Amsterdam, pp. 741-753.

22. Liu, C.L., and Layland, J.W., "Scheduling Algorithms for Multiprogramming in a Hard Real-Time Environment," J. ACM, Vol. 20, No. 1, January 1973, pp. 46-61.

23. Mahaney, S.R., and Schneider, F.B., "Inexact Agreement: Accuracy, Precision and Graceful Degradation," Proc. 4th ACM Sigact-Sigops Symposium on Principles of Distributed Computing, August 1985.

24. Massey, J.L., "Collision-Resolution Algorithms and Random-Access Communications," in Multi-User Communication Systems, G., Longo, ed., Springer-Verlag, Berlin, CISM No. 265, 1981, pp. 73-137.

25. Menasce, D., and Muntz, R., "Locking and Deadlock Detection in Distributed Data Bases," IEEE Trans. Software Eng., Vol. SE-5, No. 3, May 1979, pp. 195-202.

26. Molle, M.L., "Unifications and Extensions of the Multiple Access Communications Problem," UCLA Report No. CSD-810730, July 1981, p. 131.

27. Pease, M., Lamport, L., and Shostak, R., "Reaching Agreement in the Presence of Faults," J. ACM, Vol. 27, No. 2, April 1980, pp. 228-234.

28. Reed, D.P., "Implementing Atomic Actions on Decentralized Data," Proc. 8th ACM Sigops Symposium, November 1979, pp. 66-74.

29. Rivest, R.L., Shamir, A., and Adleman, L., "A Method for Obtaining Digital Signatures and Public-Key Cryptosystems," Commun. ACM, Vol. 21, No. 2, February 1978, pp. 120-126.

30. Rom, R., and Tobagi, F.A., "Message-Based Priority Functions in Local Multi-Access Communication Systems," Computer Networks, Vol. 5, July 1981, pp. 273-286.

31. Thomas, R.H., "A Majority Consensus Approach to Concurrency Control for Multiple Copy Databases," ACM Trans. Database Syst., Vol. 4, No. 2, June 1979, pp. 180-209.

32. Tobagi, F.A., "Carrier-Sense Multiple Access with Message-Based Priority Functions," IEEE Trans. Commun., Vol. COM-30, January 1982, pp. 185-200.

33. Tobagi, F.A., and Rom, R., "Efficient Round-Robin and Priority Schemes for Unidirectional Broadcast Systems," in Local Networks for Computer Communications, A. West and P. Janson, eds., North-Holland, Amsterdam, 1980, pp. 125-138.

34. Towsley, D., and Venkatesh, G., "Window Random Access Protocols for Local Area Networks," IEEE Trans. Comput., Vol. C-31, No. 8, August 1982, pp. 715-722.

DISCUSSION

Chairman: T. Lalive d'Epinay (Brown Boveri, Baden, Switzerland)

D. Damsker (Damsker Consulting Services, New York, NY, USA)
Let me make a comment on the Manufacturing Automation Protocol (MAP). It deals with communication on the higher levels of control and plant management, where time constraints are not very hard. MAP does not yet cover time-critical communication on the data-acquisition and direct-controls. You did not mention PROWAY, which is an industrial standard.

G. Le Lann
I am aware of the work of the PROWAY community. The fact that I quoted MAP does not mean that MAP is the unique solution. MAP is not going to solve all communication problems in process control. There is definitely room for special purpose local area networks that meet very stringent timing constraints, for example, bus systems with a centralized master. But we should be aware of the fact that standards exist and VLSI-implementations of those are being made available.

T. Lalive d'Epinay
I would like to make a comment on data consistency in distributed real-time systems. In some applications, timeliness of data is more important than strong consistency between distributed data. For example, if an analog variable is displayed at distributed nodes, slight inconsistencies do not matter.

H. Kopetz (Technical University of Vienna, Vienna, Austria)
I would like to give two different definitions of consistency, which ought to be distinguished. One is the consistency between the state of the process and the data inside the control system. I call that "external consistency". External consistency is usually very important in process control: if the internal data base is not Δt-consistent with the outside world, the control system runs into critical situations. Secondly, there is the consistency of the data inside the distributed control system. Two nodes may contain values that are Δt-consistent with the outside world, but different from each other. I call that "internal inconsistency". In many process-control applications, internal consistency is far less important than external consistency. However, there are real-time applications where internal consistency is just as important as external consistency. Maybe Gérard Le Lann can say something about the internal consistency problem.

G. Le Lann
I guess you are right. However, we can define consistency as accurately or inaccurately as we want. For example, variables to be displayed at different nodes may differ by Δv and still be considered consistent. We all know that phenomenon from

geographically distributed clocks. We cannot insist on having exactly the same clock value being displayed by all clocks at the same time. So we have to allow a certain discrepancy. Consistency is a matter of specification.

DESIGN OF REAL-TIME SYSTEMS

H. KOPETZ

Vienna Technical University, Vienna, Austria

ABSTRACT

In a real-time application, the value dimension and the time dimension of information are of comparable importance. This paper presents a methodology for the design of distributed real-time systems which requires an early binding of software functions to hardware components such that the timing and reliability properties of a still un-refined design can be analyzed. After a discussion of real-time system requirements some design prinicples for distributed fault-tolerant real-time systems are presented. In the final section an overview of a design environment and a set of software tools which support this methodology are given.

1. INTRODUCTION

The significant decrease in the cost of computer components makes the application of computer technology cost effective in many new areas. The scope of the next genera-tion of real-time control systems will therefore expand both downwards to instru-mentation (sensors and transducers) and upwards to commercial data processing systems. Substantial benefits can be realized by this integration of instrumentation, process control, production control and commercial data processing into a single, coherent, information-system architecture.

In order to achieve this integration, the uniform information-system architecture must meet the requirements both of the real-time environment and the commercial data-proc-essing environment. In this paper we concentrate on the real-time systems require-ments:

1) Time Validity of Information: In a real-time system, an image of the actual state of the environment must be kept in the computer. The information in this image is invalidated by the passage of real time. The validity of this information in a real-time system must be monitored in order to detect outdated information (e.g., the state of a traffic light is only valid for a short period of time).

2) Response Time: A real-time system must respond to a stimulus from the environment within a given application-dependent interval of real time. This response time must be guaranteed even under anticipated fault and overload conditions. Many real-time control systems, e.g., air-traffic control systems or power monitoring systems, are needed most urgently when the environment is in a critical situation.

3) Dynamic Maintainability: An integrated information system cannot be stopped for system enhancement or fault removal. A successful integrated information system is dynamic, i.e., it will be continuously adapted and modified during its lifetime. Therefore the architecture must support incremental development and dynamic maintainability. A flexible schema for the introduction of redundancy (fault tolerance) should be provided.

4) Standardized Components: In order to take full advantage of mass produced microelectronic components and software, complex information processing systems should be built around a large number of identical hardware and software units.

Many of the available methodologies for the specification and design of computer systems are concerned with the functional properties of the system, i.e., the value dimension, and do not address the time aspect.[12] The well-known Structured Analysis and Design Technique, SADT,[9] provides a graphic language for modeling and describing systems from defined viewpoints. The aspect of time is only introduced as an addendum. SREM[1] is based on the stimulus-response method and is thus more suitable to the inclusion of timing properties. Finite State Machine Models[3] can be used in all phases of the development lifecycle, from system definition to detailed design. The MASCOT[7] design method is also aimed at all stages of development, from design to acceptance. These latter methods take an operational approach which is suitable for the inclusion of timing information and are thus related to our design methodology.

Given a set of requirements, both functional and timing, it is proposed to realize a stepwise design of a distributed real-time control system by an operational approach. This makes it possible to analyze the performance properties of the design representation early in the development cycle, even before all details of the design are known, in order to determine if this still unrefined design will meet the specified timing and reliability requirements. Since the timing properties of a design are related to the capability of the underlying hardware, an early binding of software functions to hardware units is required for this timing and reliability analysis.

In this paper, after a discussion of the characteristics of the real-time environment, we present the principles of a methodology for the design of the architecture of a distributed fault-tolerant real-time system which considers the time and value aspects in parallel. The basic elements for the system representation are self-checking components, with or without internal state, which communicate via messages. These elements are closely related to the structural elements of the implementation of the distributed real-time system. This close relationship between the representation of the design and the implementation of the system reduces the semantic gap and thus the implementation effort. The final section of the paper gives an overview of a design environment for the support of this design methodology.

2. THE REAL-TIME ENVIRONMENT

In a typical real-time system, a control object (the control environment) and the control system (the computer) are connected via sensor- and actuator-based interfaces. The control system accepts data from the sensors either at regular intervals or is event driven. It processes the data and outputs the results to the control object via the actuators. The output data influence the control object such that the effects can be observed via the sensors, thus closing the loop as shown in Figure 1.

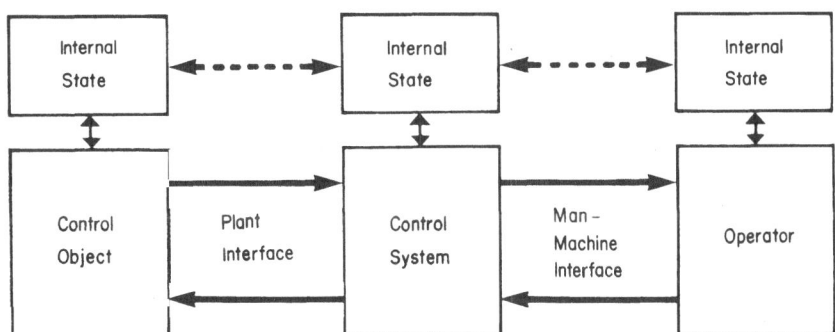

Fig. 1: A Typical Real-Time System

The control system must respond to a stimulus from the control object within an interval dictated by the environment, called the "response time". This response time must be guaranteed even under extreme load conditions and anticipated fault conditions. Typical response times are in the range of one ms to greater than one second. If a serious failure occurs, either in the control system or the control object, such that continued operation of the control object (plant) is not possible, then the system must shut down in a controlled, predetermined manner (failsafe operation).

In a real-time environment, any information must be assessed in both the value domain and the time domain. In the rest of this paper we use the following definitions:

Correct Information
Information is correct if it corresponds with the intentions of the user.

Timely Information
Information is timely if it is available within the intended interval of real time.

Valid Information
Information is valid if it is correct and timely.

In the "non-real-time world", we are concerned only with the value domain, i.e., with correctness. The inclusion of the time domain adds a new dimension to the "validity problem" in real-time systems, as the speed of processing in a given interval of real time, i.e., the system performance, becomes an essential property.

2.1 Real-Time Data Versus Archival Data

A real-time control system has to perform real-time control functions and to collect data for archival purposes. We therefore distinguish between two databases which are needed for these two different pupuses, defined as follows:

Real-Time Database
The set of all data elements needed for the instantaneous real-time control, operator display, alarm monitoring and other real-time functions is called the "real-time database".

Archival Database
The set of all data elements required for archival purposes is called the "archival database".

There are important differences between these two databases from the point of view of time and fault tolerance. The real-time database changes as time proceeds, i.e., the information in the real-time database is invalidated by the passage of real time. If we lose the real-time database, control of the environment is interrupted. After an element has been stored in the archival database, it may not be further modified, i.e., it does not change as time proceeds (it is not allowed to modify "history"). If we lose the archival database, there is no immediate effect on the real-time control of the environment. Experience has shown that in most applications the real-time database is much smaller than the archival database.

2.2 Consistency

In order to respond correctly to an external or internal stimulus, the real-time database within the control system must contain a valid (i.e., correct and timely) image of the external state of the environment.

We introduce the concepts of "strong consistency" and "weak consistency" between the real-time database and the state of the environment at a given point in time.

Strong Consistency

The real-time database of a control system is strongly consistent with the state of the environment if at time t

    (i)   every element of the real time database is defined.

    (ii) the input values of the real-time database reflect the state of the environment by or after t - δ.

    (ii) the state of the environment reflects the output values of the real-time database by or after t - δt.

The interval δt is called the validity time of the information.

Note that the relationship between the state of the environment and the real-time database is symmetric. The same definition can be applied if the environment consists of another information processing subsystem, thus defining strong consistency between two cooperating information processing subsystems which are connected by a two-way communication channel with a bounded delay in each direction. In constrast to this, the atomic broadcast protocols[2] concentrate only on the unidirectional flow of information from one sender to a set of receivers.

Weak Consistency

The real-time database of a control system is weakly consistent with the state of the environment if at time t every value of the real-time database is either

(i)   strongly consistent, or

(ii) undefined

Inconsistency

The real-time database is inconsistent with the state of the environment at time t if there is a value in the real-time database which is defined and not strongly consistent.

The implementation of a real-time system must guarantee that there is no instance during the lifetime of a real-time system where the real-time database is inconsistent. An inconsistency in the real-time database can lead to a catastrophe. Weak consistency can be tolerated by most applications as long as an upper bound on the duration of the weak consistency can be given. Whenever a value is observed in the environment, it will take some time, t_c, before this value is available in the real-time database (and vice versa). If the validity of this information

$$\delta t \leq t_c$$

then it will not be possible to realize strong consistency in this particular case, although it is still possible to realize weak consistency if the information is discarded before it becomes invalid.

This latter technique is used in the MARS system to enforce weak consistency[5] at all times. Every information item is augmented with a validity time attribute. Whenever the real time exceeds this validity time, the information is discarded by the underlying operating system. In a loosely coupled distributed system the local clocks can, at best, be synchronized up to one tick deviation. This inaccuracy of synchronization has to be taken into account[4] when comparing time stamps which originate from different nodes of a distributed system.

As an example consider an "automated car" in front of a traffic light. Waiting longer than δt in presence of "green" shows that the strong consistency between the real-time database of the car and the state of the traffic light can be lost. If we replace the value of the real-time database by the value "undefined" immediately before the validity of this information expires, then we can preserve weak consistency. If we do not discard these values before the validity of this information expires, then we have inconsistency between the state of the environment and the real-time database in the car. Assuming that a traffic light is "green" while in reality it is "red" can lead to catastrophic situation. The duration of the interval δt is determined by the real-time characteristics of the given application, in our case by the duration of the yellow phase of the traffic light. If the transmission delay between the traffic light and the

car is always longer than the duration of the yellow phase (δt), then it is not possible to implement such a system.

2.3 Event-Driven Versus Periodic Systems

Any action of a real-time system is initiated by either a state change (event) or a
time signal. We introduce the following definitions:

Event-Driven System

A real-time system in which the initiation of system activities is primarily realized by
the occurrence of a state change, i.e., event (or interrupt).

Periodic System

A real-time system in which the initiation of system activities is primarily realized by
signals equally spaced in time.

Duty Cycle

The time period between related sequences of actions which are initiated by signals
equally spaced in time.

It can be very difficult to protect an event-driven system from faults in the environment, e.g., spurious events cause the initiation of more system activities than the
system was designed to handle. Any good design must contain mechanisms to protect
the system from such a condition. One common hardware technique for the suppres-
sion of unwanted signals is the provision of a low pass filter.

If the duty cycle is considerably shorter than the required event-recognition time,
even an event-driven action can be delayed until the beginning of the next regular
duty cycle and processed within this duty cycle. In such a case it can be guaranteed
by the design that only a single event of a given type will be active within a given
duty cycle. This technique is used in many programmable industrial controllers. It
has the advantage of an intrinsic flowcontrol and protection of the system from (erroneous) overload conditions caused by a fault in the environment.

In a real-time system the duty cycle (sample rate) is determined by the response-time
requirements of the particular application. The validity time of an observation can be
set equal to the duty cycle plus the transmission delay t_c. It is thus possible to
derive the time parameters from the timing characteristics of the application.

3. DESIGN PRINCIPLES

The primary goal of any design methodology must be the reduction of complexity. Many of the problems of present-day information systems are directly related to the complexity of the design. Although part of this complexity is inherent in the problem to be solved, part is also related to the chosen solution and the representation of the design. It is this latter part which can be reduced by an appropriate design methodology.

In the following section we present the important principles of our methodology for the design of distributed real-time systems:

1) <u>Network Structure with Clustering</u>: The operational structure of a real-time control application can be described by a set of components which produce, consume and process information and by the information flow between these components. We feel that this "operational representation", which makes it possible to express functional as well as timing properties, is the most important representation of a distributed real-time system. This operational structure can be modeled by an information flow graph in which the components form the nodes and the channels the edges. In the majority of applications this information flow graph will not correspond to a tree. It follows that the appropriate structure of a real-time application is the network, not the hierarchy. In order to manage the complexity of a large network the concept of clustering will be introduced. A cluster is a subset of the network with high inner connectivity. Clusters which are composed of components are the basic elements of our design methodology. The partitioning of a given system into disjunct clusters is the first step during system design. The second step is concerned with the decomposition of a cluster into its components.

2) <u>Autonomous Components with Minimal State</u>: A cluster consists of a set of communicating components. A component is a hardware-plus-software unit of given functionality and performance. This early binding of software functions to hardware units is necessary for the early analysis of the timing and reliability properties of a design. The component is also a unit of information hiding, i.e., the inner details of component implementation are hidden from the outside.[8] Components are autonomous, i.e., they are responsible for plausibility checks on the input and output and intelligent behavior under fault conditions. At this level of representation all passive resources (e.g., a database or a process interface) are assigned to (active) components i.e., active-resource controllers.[6]

We distinguish between stateless components and state components. A stateless component stores no (variable) state information between two activations of the component, whilst a state component stores state information between activations. During the design every effort is made to reduce the size of the internal state and the number of state components within a given cluster. This facilitates state restoration after the occurrence of a failure.

Since the timing properties of the components are specified at an early stage of the design, it is possible to check the consistency between the performance requirements and the timing behavior of a given design before the components have been implemented. Later on it has to be verified that the timing properties of the component implementation are consistent with the specified performance properties of the component.

We assume further that components have the self-checking or failstop property,[11] i.e., they operate either as intended or produce no results. The diverse inner failure modes of a component, which are not known at the time of system design, are thus reduced to a single external failure mode of the component, i.e., nonoperation of the component. This makes it possible to perform a reliability analysis of the design before the components have been implemented.

3) Predictable Time Behavior: The minimum time between the occurrence of two events of the same type must be included in the specification of a real-time application. These minimum times form the basis for the performance analysis and error detection. If these minimum times are missing, it is not possible to design a system with predictable performance under heavy load or overload conditions. Every real-time system must contain mechanisms for the detection of input errors in the time domain (i.e., the interval between two events is either too large or too small) in order to protect the system from a faulty environment.

4) Message Orientation: The information exchange between components is expressed only by the exchange of messages. A message is a named unit of information which is formed for the purpose of communication. We consider a message a natural concept for describing the information exchange between two entities. Furthermore, a message can be an implementation unit, thus reducing the gap between specification and implementation.

In our methodology we distinguish between event messages and state messages. Event messages are used for the exchange of information about events, i.e., occurrences at a given point in time. Their semantics corresponds to the "classical" message semantics, i.e., they are queued when received and consumed when read. State messages are used to exchange information about the state of the environment which has been observed at a given point in time and is assumed to hold for a certain interval of time. A new version of a state message overwrites the previous version, i.e., state messages are not queued and they are not consumed when read. Thus a state message can be read an arbitrary number of times in the same way as a distributed global variable. A set of periodic state messages, introduced to express the actual state of the cluster environment, we call such the data field of a cluster.

Since every change of a state is an event, in principle any information exchange can be realized either by a set of (periodic) state messages or by a set of event messages or a combination thereof. However, there are distinct differences between these two methods of communication from the point of view of flow control, error detection and coupling. In an information exchange by event messages, flow control is realized by explicit acknowledgement messages. In the case of periodic state messages, the transmission period of the state message determines the maximum information flow, making the communication traffic predictable even under high load conditions.

In the case of event messages error detection is realized by the sender, and in the case of state messages by the receiver. We feel that error detection by the receiver is more important in real-time applications. For example, if the communication link to a control valve is interrupted, the receiver can detect the loss of communication and switch to a safe state autonomously. Event messages require bidirectional information flow, while state messages require only unidirectional flow. Bidirectional information flow requires synchronization between sender and receiver. This is difficult to implement if group addressing is used and generates a siginificant traffic overhead.

Considering all these aspects, the coupling between sender and receiver is weaker for state-message transmissions than for event-message transmissions. In order to improve the structure of a cluster (decrease intercomponent coupling) and the predictability of the time behavior under heavy load conditions, we prefer state-message transmission over event-message transmission in a distributed real-time system.

5) <u>End-to-End Protocols</u>: Reliable communication can be implemented only with the knowledge and help of the application at the end points of the communication system.[10] The application of time redundancy at the lower levels of the communication protocols increases the communication reliability but destabilizes the timing properties of a system. In real-time systems, this tradeoff between time and reliability is not clear-cut. Time redundancy at the lower levels of the

communication protocols leads to an uncontrolled increase in communication traffic under error and heavy-load conditions. We feel that strict control of communication traffic by the application software is necessary in real-time systems. An unacknowledged datagram with a "no wait send" semantics should be the only communication service available to the application software. The timing properties of the application can then be derived from an analysis of the end-to-end protocols in the application software.

6) Limited Persistence of Real-Time Data: Any observation of the state of the environment is valid only for an application-dependent interval of real time. After an observation has lost its "real-time relevance" it may not be used for any action of the control system. This validity time of information must be derived from the application environment and expressed as an atomic attribute of any real-time information. We therefore introduce this validity time as part of any message. Any real time information has to be processed or updated before it has lost its validity. If such an update does not occur in time, the outdated information will be discarded and replaced by the value "undefined".

One reasonable way to fulfill this requirement is by the provision of a global physical timebase in the whole distributed architecture. This global time can also be used to measure the duration between events observed at different nodes and for the synchronization of distributed system actions.[5]

7) Representation of the Real-Time Database by a Set of Periodic State Messages: In the design the real-time database of a cluster can be represented by a set of periodic state messages, i.e., a data field. Thus every component has immediate (unsynchronized) access to the valid state information of the environment without knowing which component is the sender of the information. At an early stage of the design activity, it is sufficient to specify only the names of these state messages. The specification of the detailed data structure of the state messages can be postponed until a later phase of the design.

In order to make possible a straightforward implementation of this concept, it is necessary to provide a powerful communication system, a group-addressing facility and a validity time for information. It is then possible to guarantee weak consistency at all times, even if a message is lost. Experience has shown that in most real-time applications the size of the real-time database and the updating requirements are such that a communication system with a bandwidth of 10 Mbit/s is sufficient.

8) Controlled Name Spaces: The stepwise refinement of a problem domain must be supported by a corresponding name-space architecture. In our methodology

every cluster has a unique name. At the next level i.e., inside a cluster, we have message names and component names. The message name refers to the semantic content of the message and is thus unique within each cluster. The internal state of a component is considered a special state message with a unique message name.

Component names within a cluster refer to a single component or a group of components. This naming convention makes it easy to implement group addressing since a given component can be referenced by more than one name. Since every passive resource is controlled by a component, i.e., a resource controller, passive resources have to be accessed via their active controller names. The names of objects which are internal to a component are not visible from the outside.

9) Observable and Simple Interfaces: In order to validate the operation of a system, the information flow between the different structural entities must be observable from "outside" without changing the time behavior of the system. The message interface and the naming rules support a passive "outside" observer. Inside a cluster a standard representation of information can be enforced since the translation of a representation to a nonstandard form, which may be required by the environment, can be performed within the interface component. Thus representational problems do not appear inside a cluster, but are resolved in the interface component.

10) Consideration of the Single-Fault Case: On average every physical unit will fail after its characteristic mean time to failure (MTTF). The effects of such failures on the operation of the system must be analyzed during the design phase. This failure-effect analysis is facilitated by the early binding of software to hardware, the self-checking property of components and the message orientation of the communication system. The error-detection latency, i.e., the time between the occurrence of a failure and detection of the associated error must be assessed. This error-detection latency is an important parameter for the reliability analysis and should be as low as possible.

4. DESIGN ENVIRONMENT

The effectiveness of any design methodology can be significantly enhanced if it is supported by an appropriate set of software tools, i.e., a design environment. We distinguish between design tools, analysis tools and management tools. Design tools support the system analyst in the design of the distributed real-time application. Analysis tools can be used for analysis of a given design and comparison of different designs. Management tools support the project management and documentation. In the following section we present a brief outline of the major tools for our design methodology.

4.1 Design Tools

The following design tools will be provided:

4.1.1 Requirements

In addition to the description of the desired system functions, this design methodology for real-time systems requires the specification of application-specific time parameters, e.g., system response time, maximum and minimum times between events, validity time of an observation, and error-detection latency. These times are determined by the application environment and have to be included in the requirements document. The functional requirements and the timing requirements of a real-time application are described in a semiformal manner using the requirements subsystem.

4.1.2 Cluster Definition

Any real-time application must be partitioned into a set of disjunct clusters. From the point of view of information processing, the plant, i.e., the control environment, is also a cluster. Thus there are at least two clusters in any control application, the control environment (the control object) and the control cluster (the control system). The cluster-definition tool supports the definition of the different clusters and the specification of the intercluster interfaces. This tool can be used to describe the operation of the plant as well as the operation of the control system. This is advantageous for the development of simulation models of the plant (control environment).

All questions relating to the representation of information in different clusters are handled in the intercluster interfaces. Within any cluster, the information flow is expressed by state and event messages of standard form.

4.1.3 Cluster Design

The stepwise refinement of a cluster into autonomous components which communicate by the exchange of state and event messages is supported by the cluster-design tool. This tool covers the specification of the cluster data field with its time parameters (cycle time, validity time, timeouts), the development of the transactions within a cluster and the definition of the functions, the timing and the internal state of the components.

The result of the cluster design is the external specification of the components and the message specification of the cluster. After the cluster design has been completed it is possible to analyze this design by the analysis tools specified below.

4.1.4 Component Design

This tool supports the detailed design of the software in an autonomous component, i.e., given the external specification of the functions and the timing of a component, as provided by the cluster design, the development of the individual tasks and protocols of a component is supported such that the external specifications are met.

4.2 Analysis Tools

After a cluster design has been completed it can be evaluated with the aid of the analysis tools. In addition to the consistency check between the requirements and the cluster functions, the timing and reliability analyses of the design can be performed.

4.2.1 Timing Analysis

It is the objective of the timing analysis to determine whether the timing properties of the given design are consistent with the requirements document. This analysis concentrates on predictable timing behavior under heavy-load conditions i.e., it is a worst-case analysis. It is assumed that all events occur with the specified minimum time between event occurrence. The performance of the communication system and the component is taken from past experience (e.g., a communication system can transport n messages per second, and the component m average messages per second, etc.).

4.2.2 Reliability Analysis

The reliability analysis is based on the self-checking property of the components. The reliability of all system functions is calculated given the MTTF of a component and the probability of message loss by the communication system. Estimates are made of error-latency times, repair times, state-restoration times, and the consequences on the availability of the system functions. The effects of failures of individual components and the loss of messages are analyzed and the reliability improvement which can be realized by the addition of redundant components and messages calculated.

4.3 Management Tools

4.3.1 Documentation

The generation of complete and consistent documentation is an integral part of all tools. Given the design database, different reports can be produced for different user groups. The documentation subsystem is based on "specimen documents" such that many of the detailed questions about structure, format and representation of the information in the reports are solved in a standard way.

4.3.2 Project Management

Project management data are predicted, collected and analyzed during the design period. These data can be used to estimate the project effort and adjust the project timetable. The development of a software tool for the estimation of the development of a complexity metric appropriate to our design methodology is in progress.

5. CONCLUSION

This design method has been tested on a number of "real life" control systems with very encouraging results to date. The method enforces the separation of concerns, e.g., separation of architecture design and analysis from detailed component design, separation of the deep semantic contents of a system description from the actual representation of the information, and so on. Particularly useful is the separation of the archival database from the real-time database and the representation of the real-time database by a set of periodic (distributed) state messages with unsynchronized access to the unidirectional information by all components of a cluster.

On the architectural level, even a complex control system can be decomposed in loosely coupled subsystems (i.e., clusters and components) with a recursive mechanism in the decomposition, i.e., every component can be decomposed into a new cluster consisting of a set of new components without changing the interface to the original component. The design environment for this methodology is not yet finished. A first version of the cluster-design tool has been completed, and work on some of the other systems, e.g., the requirements subsystem, the timing and reliability analysis and the project management subsystem is in progress. All tools for this development methodology will be implemented on a state-of-the-art UNIX workstation with a bit-mapped display and a laser printer as the output device.

REFERENCES

1.   Alford, M.W., "A Requirement Engineering Methodology for Real Time Processing Requirements," IEEE Trans. Software Eng., Vol. SE-3, 1977, pp. 60-69.

2.   Christian, F., Aghili, H., and Strong, R., "Atomic Broadcast: From Simple Message Diffusion to Byzantine Agreement," Proc. 15th Symposium on Fault-Tolerant Computing, Ann Arbor, Mich., June 1985, pp. 200-206.

3.   Henderson, P., "Finite State Modeling in Program Development," ACM Sigplan Notices, Vol. 10, No. 6, 1975, pp. 221-227.

4.   Kopetz, H., "Accuracy of Time Measurement in Distributed Real Time Systems," Mars Report 1985/4, Institut für Praktische Informatik, TU Vienna, Austria.

5.   Kopetz, H., and Merker, W., "The Architecture of Mars," Proc. 15th Symposium on Fault-Tolerant Computing, Ann Arbor, Mich., June 1985, IEEE Press, 1985.

6.   Liskov, B., "Primitives for Distributed Computing," Proc. 7th ACM SIGOPS Symposium on Operating Systems Principles, December 1979, pp. 33-42.

7.   Mascot Suppliers Association, The Official Handbook of Mascot, Malvern, U.K., 1980.

8.   Parnas, D., "On the Criteria to Be Used in Decomposing Systems into Modules," Commun. ACM, Vol. 15, No. 11, November 1972, pp. 1053-1058.

9.   Ross, D.T., "Structured Analysis, A Language for Communicating Ideas," IEEE Trans. Software Eng., Vol. SE-3, 1977, pp. 16-34.

10.   Saltzer, J.H., Reed, D.P., and Clark, D.D., "End-to-End Arguments in System Design," ACM Trans. Computer Syst., Vol. 2, No. 4, November 1984, pp. 277-288.

11.   Schlichting, R.D., and Schneider, F.B., "Fail-Stop Processors," ACM Trans. Computer Syst., Vol. 1, No. 3, August 1983, pp. 222-238.

12.   Wassermann, A.I., and Freeman, P., "ADA Methodologies, Concepts and Requirements," ACM Software Engineering Notes, Vol. 8, No. 1, January 1983, pp. 33-98.

DISCUSSION

Chairman: R. Karg (Brown Boveri, Mannheim, Fed. Rep. Germany)

D. Damsker (Damsker Consulting Services, New York, NY, USA)

In my view, there is a difference between the logical hierarchy of control functions and the network architecture of the control system. But even network architectures can be hierarchical. The communication architectures of control systems, as a matter of fact, are often hierarchical, because of response-time requirements. Time-critical tasks are solved locally in subnetworks, and the plant-wide transmission system conveys messages that are less time critical.

H. Kopetz

Considering the flow of information in a distributed system, I take the view of a set of nodes exchanging messages. There is no hierarchy in my model. In other models, of course, other entitites and relations between them could be identified.

J. Kramer (Imperial College, London, England)

I want to address the binding of software components to hardware components. As far as specification is concerned, I think there is no necessity to tie software components to the hardware. One can specify software components and the desired performance without yet deciding on hardware. When it comes to analysis of a particular design, you will have already performed the allocation of the software components to particular hardware. As a design method, I think, there is no necessity initially to tie software components to particular pieces of hardware.

H. Kopetz

Whenever you want to evaluate the design in terms of response time and reliability, you need this binding of software to hardware. I do not see any disagreement here.

J. Kramer

In my view it is a decision of whether we are doing the performance analysis to get a set of requirements, or whether we are doing it in order to see if the final system will meet them. I think we tend to do both.

A. Kündig (Federal Institute of Technology, Zurich, Switzerland)

Can you quantify the reliability of the unreliable datagram service you mentioned, for example, as a percentage of erroneous datagrams? To what layer of the OSI-model does the unreliable datagram service apply?

H. Kopetz

The reliability of the datagram service depends on the channel reliability. OSI supports the unreliable datagram at the transport level.

ARCHITECTURES FOR PROCESS CONTROL

H. KIRRMANN, T. LALIVE D'EPINAY and H.P. STOECKLER*

Brown Boveri, Baden, Switzerland / *Mannheim, Fed. Rep. Germany

ABSTRACT

The control of industrial processes by computers has brought not only new possibilities but also new challenges to engineers. The requirements in terms of response time, computing power, flexibility and fault tolerance are stricter than in the fields of commercial or scientific computation, since the work has to be carried out in real time. An additional difficulty is that most of these requirements are contradictory. A solution to the problems of complexity, flexibility and geographical separation is provided by a distributed architecture, which consists of (multiprocessor) nodes located at strategic points of the plant and interconnected by an industrial network. Brown Boveri have been developing such a distributed architecture in the form of building blocks and have tested it successfully in power plants and other control systems. Such a distributed architecture results in a high degree of freedom in the configuration and extension of a control system, and in a firm base for tolerance against faults. The geographical distribution of hardware finds its counterpart in the distribution of software in the form of functionality independent units which communicate over well-defined interfaces. The drawback of distribution is the communication bottleneck and consequent response-time problems. Several means are being used to reduce this bottleneck, including data reduction at the source and broadcast of process data to all interested receivers. The broadcasting of data permits considerable reduction in the communication traffic, also addresses the response time and flexibility problems, and provides at the same time the base for maintaining the redundant units in an actualized state for fault tolerance. One particular application has been developed at the Brown Boveri Research Center in the form of a fault-tolerance mechanism based on remote procedure calls.

109

1. INTRODUCTION

Brown Boveri have been manufacturing process-control systems for several decades in three principal fields:

- Load-dispatching systems for the distribution of electric power, water, heat, gas and oil
- Generation of electric power
- Automation of industrial processes, such as cement factories, food silos, continuous-casting plants, rolling mills, printing presses, etc.

The control system serves many purposes. It can be roughly divided into an "open-loop", or operator-controlled part, and a "closed-loop", or automatic part. The main "open-loop" function is called SCADA ("Supervisory Control And Data Acquisition") in telecontrol, but the same function exists under other names in industrial or power plants. SCADA displays the current process state to the operator and relays the operator's commands down to the controlled process. The SCADA system does not take decisions by itself. Additional open-loop functions, such as logging, simulation or application-oriented programs, assist the operator in making complex decisions. The "closed-loop" function consists of the autonomous control without human intervention of some parts of the plant in areas where the decisions are evident (protection and control). This function accounts for the largest volume in electronic plant equipment.

Operation of industrial plants without computers is today unthinkable. Just a few numbers may be cited to characterize the increase in importance of computer control in recent years: in a typical thermal power plant, the number of digital and analog process variables has increased from a few 100 to more than 10,000 in less than ten years. The number of microprocessors and microcontrollers in such a plant has risen from a handful to more than 1,000. The thermal power plant itself has about 8,000 measuring points, for temperature, pressure, speed and position, and about 1,500 command points for the actuators, e.g., motors, relays, solenoids and heaters. The number of different items of data, crude and processed, available to the operators adds up to 10,000. For a nuclear plant, about the triple this amount of data is considered. Systems capable of handling 100,000 different objects are in the planning stage for the supervision of power networks. Other industrial processes follow the same trend.

At the same time, the complexity of the control system has increased to such an extent that a plant would be unmanageable if computers were not there to reduce the information flow before it reaches the operator's console. The continuous information flow has been estimated to exceed 5kb/s, and without automation, 200 persons would be required to manage it.[8] The walls of the control room would be completely filled with instruments if every measuring point had its own display instrument.

The current trend is for the automatic part to increase and for the plant operator not to intervene in normal operation, but only to cope with exceptional situations. This evolution is similar to what is taking place in aviation: the roles of the former mechanical engineer, radio operator and navigator, as well as most chores of the pilot have been taken over by computers. Only the pilot and the copilot remain in modern aircraft and their role in normal flight is reduced to the supervision of automatic equipment, but their interaction is decisive in emergency situations.

2. HIERARCHICAL DIVISION

Control systems are divided hierarchically to cope with the high degree of complexity and to increase the modularity, as shown in Figure 1. The lowest level consists of the plant itself, with its sensing devices (e.g., thermocouples, microswitches, and photodiodes) and its motoric elements (e.g., motors, heaters, generators, amplifiers, and pumps).

Above the plant level, we find the Sensomotorial Control, consisting of the devices which directly control the physical actuators, such as high-power switches, power amplifiers, and valve-control pulses, and the sensors which read the physical state of the plant, such as temperature, pressure, flow, or position sensors, along with the attached transducers and A/D converters.

At the next level, the Unit Control uses the information provided at the sensor level to execute local closed-loop control and protection operations. The speed regulator of a motor belongs to that level, as well as the overcurrent, overspeed or overpressure protection. This level is normally not accessed directly by the human operator except for maintenance. The unit-control level is time-critical. It was traditionally executed by analog devices such as regulators and protection circuits, or by hard-wired digital logic or electromechanical relays. Today, the execution of the unit control is often done by dedicated, "front-end" digital processors, which are often built with specialized bit-slice processors to meet the speed requirements. These front-end processors are attached to the sensors and actuators either directly on the same backplane, or remotely by I/O buses.

At the next higher level, the Group Control coordinates several unit control, for instance, all functions belonging to the same unit, such as a furnace in a cement factory. At this level, some interlock functions are possible (e.g., preventing a motor from switching on if the silo material has not been loaded into a wagon). The group control used to be manual, and an operator console was attached to it. Today, this

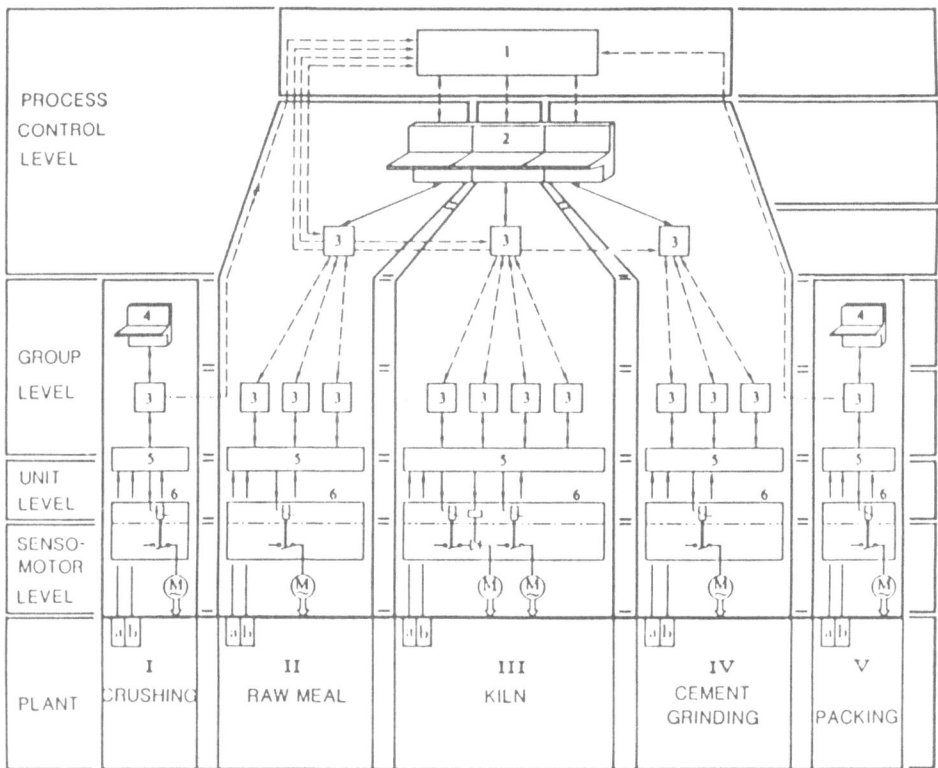

Fig. 1: Hierarchy of a Control System in a Cement Works

level is mostly carried out automatically, e.g., by a 16-bit microprocessor, or semi-automatically. As a special case, the group control is still manual in rolling mills, where the roll itself is controlled by an operator in a dedicated control room.

The Plant Control, at the highest level, operates on the different group controls. Since the groups represent different parts of the installation, this level is responsible for their coordination: control of the flow of material or energy between the groups, repartition of the workload on parallel groups, and so on. A human operator is always present at that level. There are today no plants whose plant control would be executed entirely in closed-loop by a computer, although some gas-turbine plants are close to it. One or more computers provide the operator with correct and timely information and assist him in his decisions. Additionally, the computer can relay commands to the actuators, but hard-wired direct controls which bypass the computer are still often found. Due to the volume of data involved, this level is executed by 32-bit computers.

One can also define a higher level outside of the plant itself, by considering several plants, for instance, in a power grid. Then, a <u>Network Control</u> is responsible for the distribution of load between the plants, for load/frequency regulation, for network-topology evaluation and for long-term planning of the operation.

The hierarchical division occurs quite naturally in a physical process. In fact, process-control systems have been so designed since their inception. This hierarchical model of a process-control system bears a certain similarity to the OSI layering of communication systems, especially with respect to the real-time requirements. The closer to the physical process, the stricter time constraints, and the greater the quantity of data. As the hierarchy is ascended, the data is processed and reduced, its quantity and flow diminish, the timing requirements become less critical, and human intervention is desirable and possible. At the same time, the complexity of the decisions increases as the levels are ascended. The lowest levels today are under complete computer control, except for maintenance or emergency bypasses. The trend is clearly toward increasing reliance on computers for the higher levels as well.

This neat hierarchy is breached by the presence of the plant operator. In normal operation, the operator intervenes only at the group level or above, but in exceptional situations, the operator needs a very clear picture of the situation if he is to take a critical decision. The operator must therefore have sufficient information and be able to focus on a specific part of the plant to study a problem. This means that the operator should intervene at all stages of the hierarchy, even down to the senso-motor level, which can be accessed for emergency manual operation. This vertical structure can also be found in the OSI model, whose manager structure extends in parallel to the protocol layers.

The operator must therefore have access not only to the reduced data, but also to the raw data, and this in real time. Of course, the operator's view of the system is always obsolete because of communication delays, but the values displayed in front of him should not be older than a few seconds, and should belong to a consistent set. Furthermore, the operator is not patient: he will not accept that the screen requires several seconds to fill with the data when he asks for a new view of the plant. These two requirements, data obsolescence and screen build-up time, prevent any solution which would permit the operator's computer to access the sensors directly to fetch the required data. Due to the communication delays, such an operation would take several dozens of seconds.

The solution is to keep a copy of the state of the plant in a process database close to the operator. This is similar to the "caches" used in high-speed computers. The data in this base are refreshed by the data sent over the communication links, either cyclically, on events, or upon demand. The operator's display computer has a fast

access to this database, and can build up a picture of the plant within a few seconds, modifying the displayed values as the new values are produced. Brown Boveri
have developed such a process database, called PRIMO, which is at the heart of
numerous telecontrol applications. The requirements here are becoming stricter: while
about 5 to 8 seconds for the picture build-up time and 2 seconds for the process-
state-refresh rate were acceptable, the current trend is to achieve times of 2 seconds
and 1 second respectively. Although this does not seem to be a large step, the effort
behind it is considerable since there exist physical limitations.

The process database is a key element in every SCADA system. The background
updating of the database imposes a heavy traffic load on the communication links,
because all data, not only reduced data, must be eventually transmitted to the database, in case the operator would like to look at them. This explains why, in the
typical power plant cited above, 10,000 data items are available to the operator, of
which only 8,000 are effectively read from the sensors.

The fact that the operator is separated from the process by the database has a positive side effect: instead of displaying real data, the database may submit simulated
data to the operator for training purposes. In this case, part of the computing power
is diverted to simulate the plant operation.

Another example, where a continuous flow of data is required, down to the lowest
level of the plant is the data logger. This unit has a similar role to the "black box"
of an airplane or train. It is used in case of trouble to reconstruct the situation
before and after a problem occurs. The data logger must be able to register the
events in real-time or, more precisely, in the exact order of occurrence. The logger's
output sometimes goes to a printer directly, or it can be transmitted to the operator
for analysis. At least two types of data logger are common: an analog data logger,
which monitors transient situations, and a digital data logger, which records all
operations in the plant, including the operator's commands.

3. ARCHITECTURE OF A PROCESS-CONTROL SYSTEM

3.1 Distribution

Load-dispatching systems for electric power may stretch over hundreds of kilometers.
Industrial processes are geographically distributed over a wide area. And, as illustrated in Figure 2, cement works, chemical or paper plants span some several hundred meters.

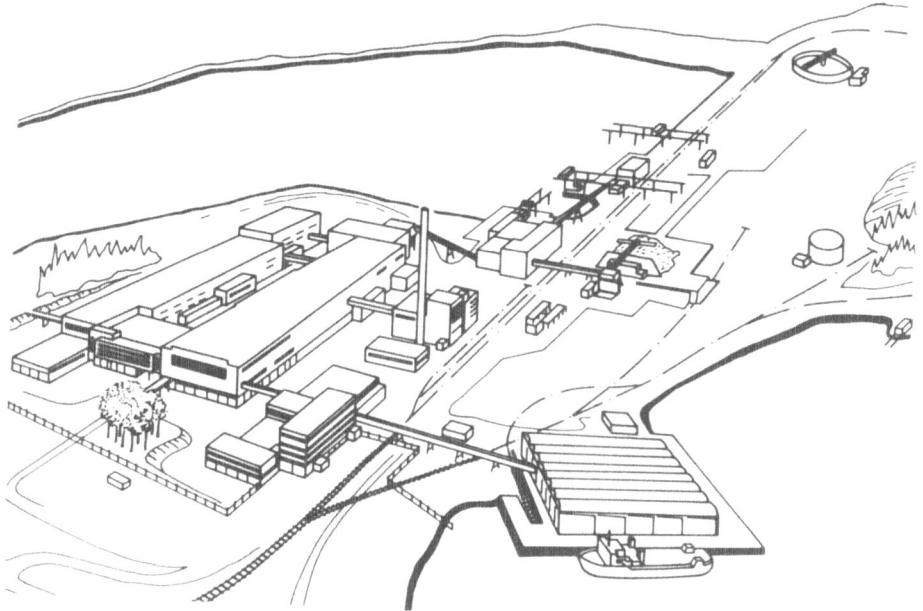

Fig. 2: Geographical Distribution of a Paper Plant

The inputs and outputs to the plant tend to cluster in a few, well-defined places. For instance, some typical industrial processes such as rolling mills are distributed linearly along a production line of several hundred meters. Most sensors and actuators are located around the main pieces of equipment, such as rolls, motors, furnaces and cutting tools. Other plants are scattered over a wide area, e.g., in harbor installations, but inputs and outputs to the process tend to cluster around a few objects such as cranes or conveyor belts. In power grids, the activity concentrates around the generation, transmission and distribution centers.

It is therefore quite natural that the computing power required to manage the plant is also distributed and concentrated where most work is required, to limit the data flow and achieve greater independence in case of failure of parts of the plant. A general rule is that the structure of the control system should match the structure of the plant it controls. A process-control system today is a distributed computing system, whether or not it is portrayed in that way. Only the smallest data-acquisition stations exhibit a concentrated structure.

The question is no longer whether the process-control system should be distributed, but whether the control should be centralized or decentralized. The arguments are about the same as those of federalists against centralists: a centralized system can respond faster, requires less interaction and the operator can control it better. A decentralized system requires more local intelligence and imposes a communication overhead, but is less sensitive to partial outages and can be more easily tested and expanded. In fact, the choice is quantitative: how much should be decentralized and how much centralized.

3.2 Hardware Building Blocks

One objective of a company like Brown Boveri, which implement control systems for a large number of applications, is to define a set of a few building blocks, which can be configured for a specific project, and be reused in another project with little development work.

The requirements placed on this family of building blocks are:

Universality: The modules can be used in both large- and small projects, but each
                   one should be sufficiently complex to reduce the number of blocks
                   required.

Extensibility: It must be possible to extend an existing plant either during the
                   planning or installation phase, or after years of operation, with little
                   or no down time due to the update operation. This also means that
                   the technologies of different vintages must coexist.

Dependability: It must be possible to meet specific requirements in terms of reliabi-
                   lity or availability with these modules, with few special elements and
                   as much transparency as possible.

Modularity: The building blocks must have two defined interfaces: one "horizon-
                   tal" interface in the OSI layering, which consists of the services
                   provided to the higher level and the services requested from the
                   lower level. Then a "vertical" interface which is the standardization
                   of the connection to other modules.

A distributed hardware architecture which fullfills these requirements has evolved in recent years, and is now practically stable. Today's plants are controlled by a network of computer nodes interconnected by several kinds of links, ranging from back-

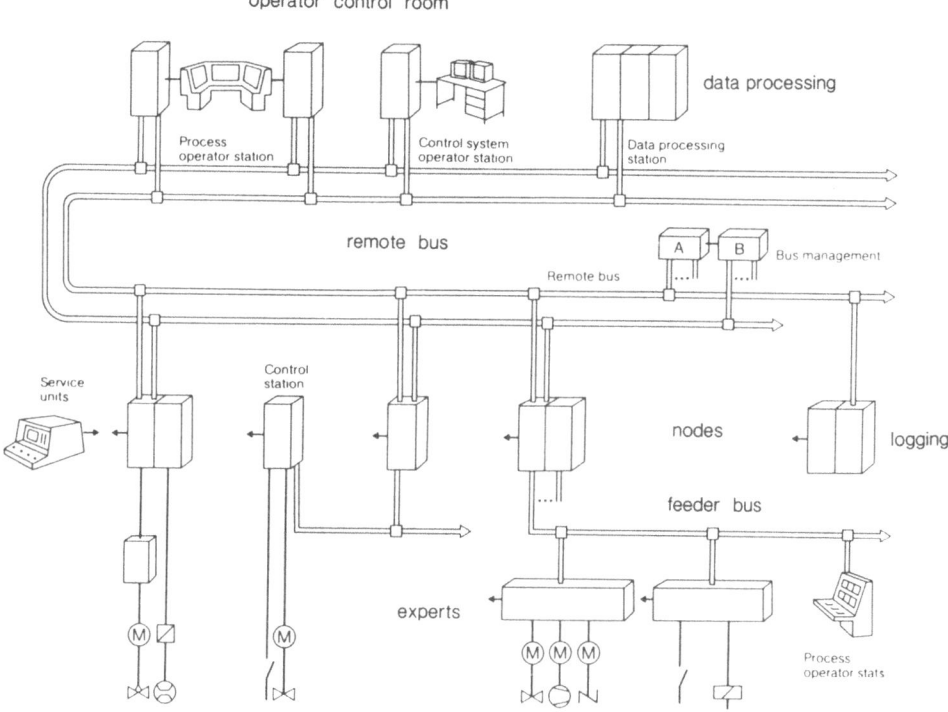

Fig. 3: A Distributed Control System for a Continuous-Casting Plant

plane buses to local networks, telephone lines, power-line carriers, microwave and radio links, according to the distance which must be spanned. Figure 3 shows a typical configuration in a continuous-casting plant.

3.3 Processors and Periphery

The nodes are dedicated to relatively complex functions, such as human interface, process database, logging station or group control. Each node may have one or several processors, and include peripheral devices, such as Winchester disks, display controllers, and consoles, all interconnected by the same backplane bus. Nodes for large applications use 16- or 32-bit minicomputers, whilst nodes for small data volumes use 16-bit microprocessors.

When the computing power of the nodes is insufficient to drive the process directly, or when the communication imposes a bottleneck over long distances, the control of the sensomotor level is delegated to one or several "experts" or "front ends", which are dedicated processors tailored for a specific application. Such experts are generally fast processors built with bit-slice processors which perform the basic analog and digital functions previously carried out by analog or digital circuitry. These front ends perform an information reduction, and their output data are transmitted by means of a feeder bus or a dualport memory to the higher level data-processing station.

Finally, for the acquisition of the plant data themselves, a broad range of modules with A/D converters, digital I/O, relays, synchroconverters, counters, and temperature transducers is available. These devices interface to a backplane bus or directly to the feeder buses.

3.4 Buses and Links

Data transfer through direct wires occurs only inside the command room and in the interface to the physical process. All other communication takes place over digital buses and links, the main advantage of which is to save on wiring and interface logic.

The nodes communicate with each other by means of serial buses, called the process highways, which are high-speed local networks such as BBC's PARTNERBUS or the IEC SC65 PROWAY, and which are often duplicated to achieve higher availability. When the distance between nodes exceeds a few kilometers, communication takes place over telephone lines, power-carrier or other long-haul channels, at a slower pace. The nodes are connected to the expert stations through the process highway or through direct point-to-point links, such as BBC's MAILBOX. The nodes or experts which access the process have interface boards on an I/O backplane bus, such as the ED1000. When the sensors are spread over a large distance, feeder buses such as BBC's P13 provide a better solution. Feeder buses have basically the same function as the backplane bus, but can extend to several hundred meters. Specialized I/O buses, such as BBC's RIO (Remote I/O) bus connect the transducers close to the process and reduce cabling costs. Peripheral devices, especially in the control room, are often linked by instrumentation buses such as the IEEE 488. The following table shows the main transmission media used in BBC products:

PRODUCT	FUNCTION	FEATURES	STATIONS	BIT RATE	DISTANCE
Power-Line Carrier	remote nodes	full-duplex event and cyclic modem transmission	2	50b/s-1200b/s	no limit
MAILBOX	expert to node	full-duplex dual-port memory	2	37.5 kb/s	1000 m
RIO	remote I/O bus	backplane extension	8	37.5 kb/s	2000 m
PARTNERBUS	highway	token bus multimaster reliable broadcast up to 512 bits	32	1Mb/s	500m
P14	remote feeder + highway	broadcast messages cyclic + event up to 8 buses in star	64/seg.	1Mb/s	1500 m (3000)
P13	remote feeder	central + cyclic 16-bit address 16-bit data	50	0.7 Mb/s	1000 m (2000)
4LB	local feeder of P13	cyclic 16-bit address 16-bit data	64	0.2 Mb/s	30 m
ED1000	backplane bus	read-write processor mainly I/O bus	21	2.6 Mb/s	40 cm
B448	backplane bus	multiprocessor read-write bus	127	16 Mb/s	3 m

3.5 Software Distribution

The geographical distribution of hardware finds its counterpart in the distribution of software. Although software has not yet attained the degree of modularity now taken for granted in the hardware, the tendency is for the different functions to be programmed in some high-level language as independent modules with a well-defined interface. The module concept incorporates the notions of abstract data types: a module manages a resource, such as a disk, a database, or an I/O controller. Modules provide services to other modules and request services from them. For instance, the human-interface module uses the services of the database module, which itself uses the services of a data-acquisition module, and so on.

The programmer of the modules is not concerned with the communication between modules. In fact, it is not necessary for the mode of communication to be known at the time a module is written. At run time, the modules may communicate over a network, over a shared memory or over a common database.

Furthermore, it is desirable to delay the question of where the modules will actually run until sufficient knowledge of the installation is available. Ideally the software should be written independently of the hardware on which it should run. This would allow the software to be written first, and then to see how it can be distributed among the nodes and the processors within a node.

This ideal is however problematic to achieve. First, some modules require specific resources, such as access to the plant signals, to a disk, or to a computer of sufficient power. The access of these resources over a network would be too costly in communication time. This means that a module may be tied from the outset to a specific site. Second, the ideal of transparent communication implies that the modules must be designed for the worst case, which is network communication, even if they can rely on a common memory within the same node. In fact, the network transmission time itself is negligible with respect to the overhead involved in message preparation and protocol. It is therefore meaningful to specify from the beginning that some modules will run on the same node and can therefore communicate over a common memory or even shared variables (which contradicts to some extent the concept of abstract data types).

A price must be paid for modularity. Even in a modern structured language such as MODULA-2, procedure calls will always be more costly at run time than threaded code, although all textbooks encourage programmers to structure their programs. Furthermore, if the application is split into several tasks, an overhead is bound with the task management. If the tasks must be written so that several can run in parallel, another overhead is required to manage their possible interaction.

Finally, the ideal of transparent communication also has its price. If, for the sake of modularity, the communication primitives are independent of the physical distribution, then the delays for communication within a node are about the same as for communication between nodes as a consequence of the communication management. However, communication within the same node using shared memory is an order of magnitude more efficient than network communication. Current operating systems require about 200 µs for a send operation via shared memory compared to 5 ms for a network message.

This is the key problem that all designers of layered architectures and communication protocols encounter: modularity imposes a high overhead, and there is a strong tendency to "punch" through the layers and bypass some to make the system efficient. Indeed, if the hierarchy is too heavy and the communication too slow, the programmer will be tempted to break the modularity. This could render the program comprehensible only to the original programmer and in any case make difficult software maintenance and system extensions.

The key to the use of modular software is an efficient "software bus", i.e., a communication structure which allows software modules to communicate efficiently. Such is the motivation behind BBC's Distributed Systems. The application is divided into well-defined modules called Functional Units (FUs). FUs can be assimilated to tasks in a conventional multitasking system, but they differ in the implementation. FUs may run as parallel processes in a multiprocessor.

The FUs are grouped in Clusters, a Cluster being an abstraction for a node. Within a Cluster, communication between the FUs takes place via common memory. All FUs belonging to the same Cluster must therefore run on the same node, either on the same processor or on different processors sharing a common memory.

FUs located in different Clusters communicate by messages over a high efficiency data channel, BBC's Channel Network, whether they reside in the same node (or even in the same processor) or in different nodes linked by a network. BBC's PARTNERBUS has been specially designed for this kind of application and supports a very efficient communication channel.

Figure 4 shows the way a process-control station can be configured. The distribution of functions around the data channel makes no assumption about the actual geographical distribution. The actual distribution of the modules is determined during system configuration. If the processing power within a node is insufficient to support multiple Clusters, the Clusters are distributed over several nodes in a network, according to the peripherals or other resources they need. Figure 5 shows how the FUs could be distributed in an actual system.

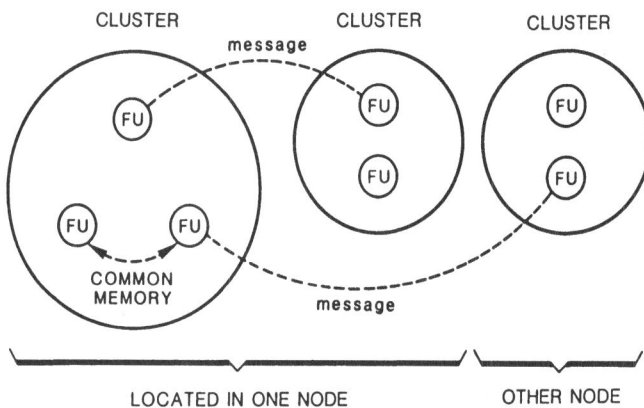

Fig. 4: Clustering of Functional Units

Extensions to this principle have been investigated at the Brown Boveri Research Center. If the processing power of a node is insufficient to support a full Cluster, then the different FUs may be distributed over several processors sharing a common memory. Each processor of such a closely coupled multiprocessor owns a private memory in which the code and most data of the FUs are loaded. The multiprocessor may be heterogeneous (the different processors are not interchangeable) or homogeneous.

In a heterogeneous multiprocessor, the FUs are always bound to the same processor which holds the resources they need (Figure 6). If the multiprocessor is homogeneous, it can be operated as a Pool of processors (Figure 7).

Subsequently, the FUs may be loaded in any of the processors and run there until the system is initialized again. This situation is termed a Static Pool. Alternatively, the operating system can be designed such that the FUs may change processor during their lifetime, for instance, at each interrupt or synchronization point. This is termed a Dynamic Pool, which exhibits a somewhat higher throughput than a Static Pool, at the expense of additional memory (each processor must own the code of each FU).

The computing power of a static or dynamic pool can be increased by adding processors and memory. There are, of course, limits to this extension, mainly because of the bottleneck which the common memory represents. However, if the different FUs are sufficiently autonomous, as is the case in process control, then multiprocessor systems with as many as 32 processors may be constructed.

Finally, promising extensions to the Channel Network concept are under development which combine the functions of intertask communication and state saving for error recovery.[1]

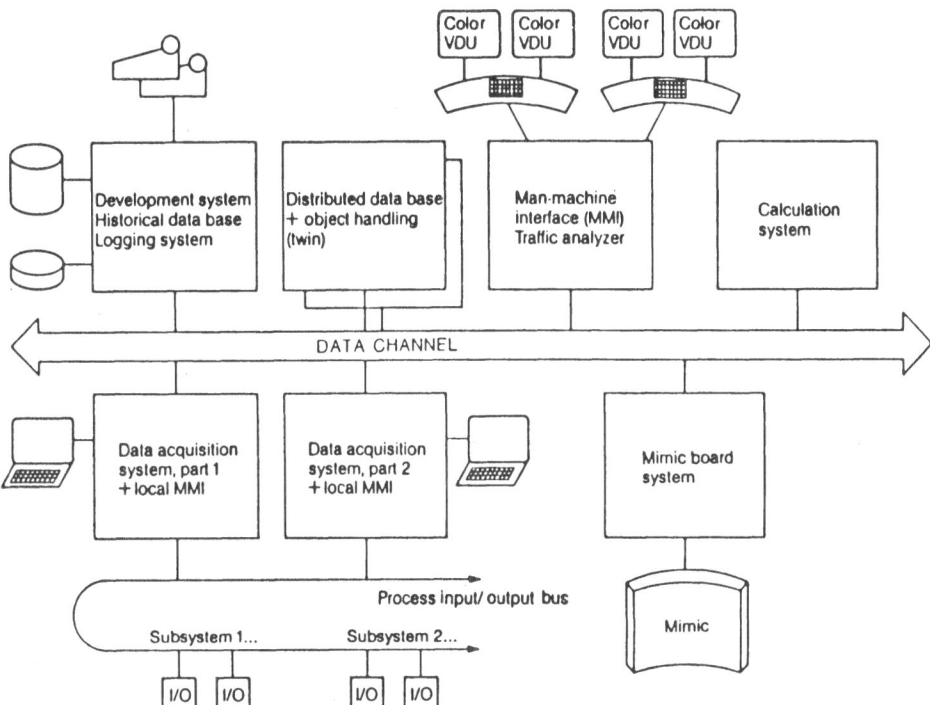

Fig. 5: Functional Unit Clustering and Distribution

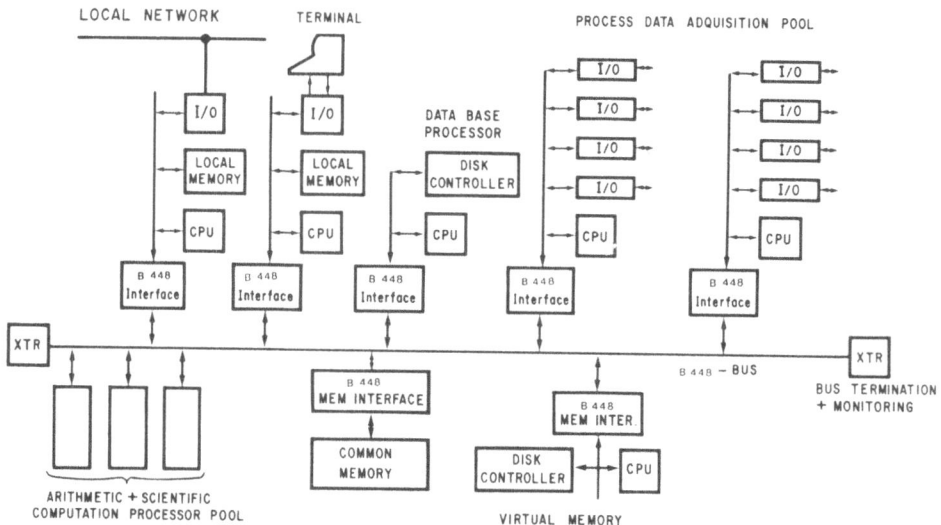

Fig. 6: A Heterogeneous Multiprocessor

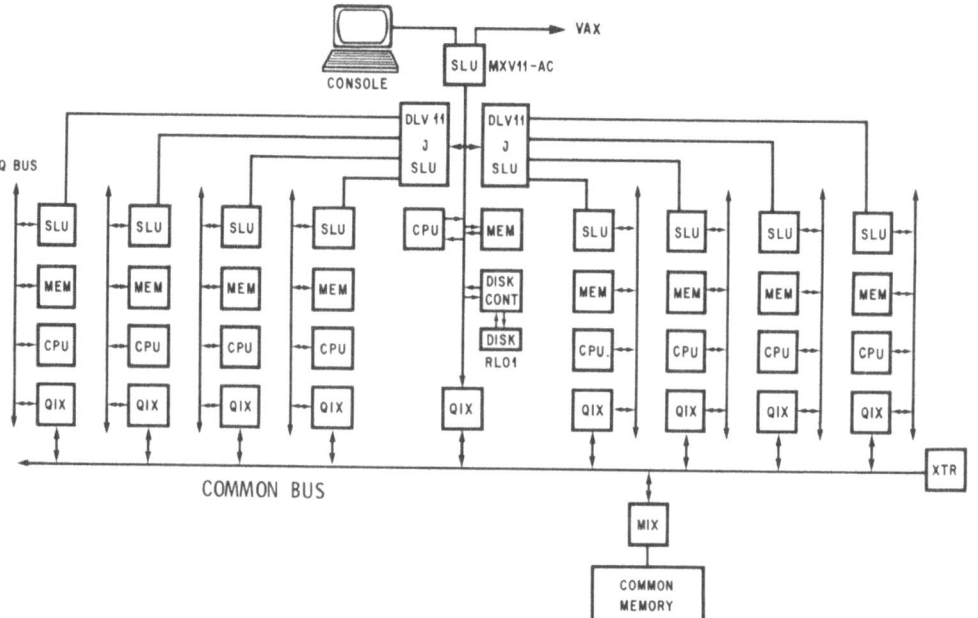

Fig. 7: A Homogeneous Multiprocessor

4. REAL-TIME CONSTRAINTS

4.1 Requirements

In contrast to users in the commercial and scientific computing fields, a technical process is inflexible with respect to time. Most processes involve continuous flows of material or energy, and buffering is either not possible or too costly. Concern for real-time behavior is present at all stages of the hierarchy, but especially at the lowest levels. Meeting the time constraints means two different things:

- Timely response: between a stimulus and a response, no more than a certain delay may elapse.

- Correct ordering: the sequence of events must be correctly reproduced. This is especially important for the data logger.

The strongest constraint appears in event ordering, which requires a time resolution of 1 ms in special electrical power-system telecontrol. Industrial applications allow some 10 ms for event resolution. Regulation and protection algorithms at the unit control level require reaction times below 20 ms, whilst regulation at the group level requires sampling rates in the range of 30 ms to 200 ms.

These constraints must be met in spite of delays in the system, which originate from:

- Communication Delays: These delays are divided into propagation delays, transfer delays and access delays. Propagation delays are due to the finite speed of electrical or optical signals down the link (which is lower than the speed of light in vacuum), and to repeater delays. This delay can be estimated as about 5 μs/km, depending on the loading of the cable. Transfer delays depend on the bit rate of the link, and are due to the finite bandwidth of the communication links. Access delays are due to the arbitration and waiting phases when the medium is shared. There is no access delay in a single master system.

- Computation Delays: These are due to the limited speed of the processor. For a given processor, the computation delay depends on the number of instructions and on the control structure (loops, branches) of the program. Therefore, the computation delay is a function of the state of the variables.

- Synchronization Delays: Although processing power or link capacity may be available, a task may be waiting for a resource to become free, or for another task to produce data.

The correct estimation of delays is fundamental to any real-time system. These delays limit the performance of the system, and measures have to be taken to reduce them and suppress bottlenecks, e.g., by using faster processors, better organization, or redundant communication. Further, the delays have an influence on the regulation algorithms which are usually based on a constant sampling frequency and they appear as (variable) dead times in the regulation loop. This is especially the case when the signals used in regulation loops pass through one or more communication media before being processed.

Basically there are two opposing schools of thought depending on whether or not the delays can be estimated, i.e., whether delays are deterministic or random. The first is the pessimistic school: if the delays can be given a worst-case upper bound, then the system should not make a distinction between the worst case and the normal case. This leads to cyclic or polling systems, which can guarantee real-time response under any circumstances. The optimistic school, on the other hand, assumes that the delays are random, that their upper bound is unknown, but that their mean value is small and their distribution known. The system should be designed for the average delay with a certain safety margin such that the probability that the timing constraints are violated is sufficiently low for all practical purposes. Ideally, the probability that the response time is not met should be lower than the probability of system failure in that interval. This leads to demand-driven systems, as discussed below.

In fact, whether a delay is deterministic or random is a question of how good is the model of the computing system. If adequate tools were available, then the upper execution bound for a program of arbitrary complexity could be predicted, provided there are no design errors. This, in turn, would allow reduction of the built-in safety margin of current systems and augmentation of the performance. However, the presence of basically nondeterministic elements makes this endeavor useless.

For instance, communication delays are reasonably well defined in terms of transfer and propagation time, but the access delay to a multimaster bus depends on the arbitration algorithm, and may be a random variable. For example, the access delay to a local area network of the Ethernet type is a random variable due the collision method. The probability of gaining access to such a bus within a given time may be high, but never equal to one. This is one of the arguments why a process highway with a nondeterministic arbitration such as Ethernet should not be used for time-critical process control, but only where it can be predicted that the bus load will never exceed a certain value.

Computation delays may appear to be deterministic, since digital computers are deterministic machines by nature. But programs may be so complicated and include so many loops and forks that the worst-case execution time cannot be determined but only estimated through test runs. How the program behaves under worst-case conditions is unknown, even if the program is correct. If programming bugs in seldom used parts of the program are taken into account, then the situation is beyond control.

Finally, synchronization delays are an order of magnitude more complicated to estimate than computing delays, since they involve the interaction between different programs, on the same node or on several nodes, and possibly the waiting for resources which comprise nondeterministic elements, such as disk drives. The access to a database, for instance, may introduce completely random delays, according to the strategy used for the locking policy.

4.1 Cyclic Transmission

Consider as a paradigm a simple station which collects data from sensors and processes them, for instance, to generate alarms or regulate the speed of a motor. The case of command transmission is neglected, since this function occurs less frequently. When the transducers of the sensor are located on the same backplane bus (Figure 8), the processor addresses them directly as I/O devices or memory locations. If they are remote, they can be accessed through a feeder bus.

Every sensor input must be read and processed within a fixed time. This is necessary, for example, to guarantee the sampling period for regulation algorithms or to take emergency action in the case of protection equipment. This can be achieved by letting the processor sample the process State at regular intervals. For instance, it

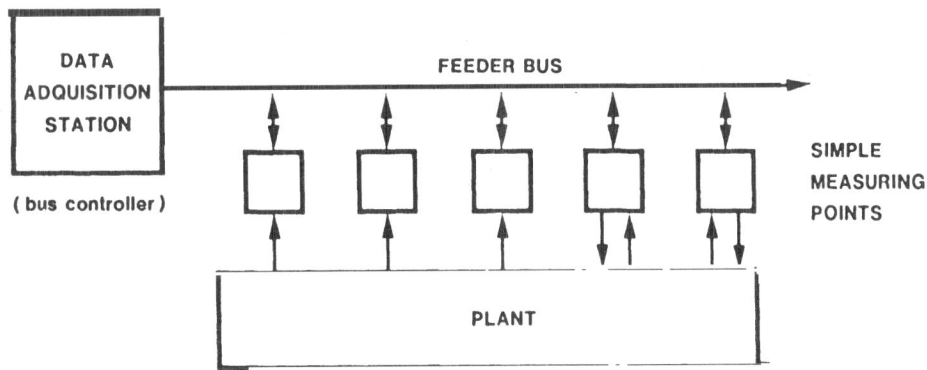

Fig. 8. Polled Data Acquisition

polls the sensors sequentially in a cyclical order. Each polling cycle takes a certain time, not only for the communication, but also for processing.

A cyclic operation guarantees the response time, provided the processing time for each input item has a fixed upper bound. As a weaker condition, it is sufficient for the sum of the processing times of all input items to lie below a fixed upper bound. This condition can only be met if the processing operations are simple enough to be evaluated. This is one of the reasons why the high-speed controllers at the unit-control level are programmed in a functional block language (FBL) which is executed cyclically rather than in a conventional high-level language such as MODULA-2. Since these languages have no program branches, and the execution time for each block is known, the worst-case execution time of the program can be determined precisely.

If the number of sensors is large, the cycle duration may well exceed the maximum time allocated for processing an item of data. For instance, if there are 1,000 sensors connected to a processor and 10 μs are needed to read and process each, the cycle time will exceed 10 ms.

This basic limitation of cyclic operation can be alleviated in several ways:

1) The sampling period need not be the same for all measuring points. Rapidly changing quantities such as currents are sampled more frequently than slowly changing ones, such as temperatures. The processor should use a multicyclical TDMA mode, where each variable is sampled at a convenient rate, independent of the bus load. This supposes that the processor has a knowledge of the sampling frequency of each variable, and requires a new configuration of the bus controller when the installation is expanded. The easiest way is to provide nested subcycles within a large cycle for the time-critical variables.

2) The feeder buses are optimized for a polled-operation mode. The buses carry short messages, typically 32 bits in length (analog sensor values are digitized to about 12-bit significant bits, digital sensor values are transmitted as 16-bit units).

3) Standard buses use two messages for each read operation (request/reply). This "split read" supposes that the bus controller at the transducer site has a certain intelligence. Since reading is the most common operation, the feeder buses should be capable of read transfers. This mode of operation is common in backplane buses, but practically unknown in local area networks. During a read operation, the master station sends out the address of the peripheral and receives the data during the same cycle.

4) In both cases, read or split read, the time delay between request and data exceeds the round-trip delay, i.e., it is higher than twice the bus propagation time. In a bus with a length of 1,000 m, the round-trip delay is about 10 µs. During this time, the processor could perform some processing. This calls for a separation of the functions of the processor and bus controller. The classic solution today at the expert-station level is to permit the processor to execute the data processing cyclically, taking the data from a dual-ported memory which is refreshed, also cyclically, by the feeder bus in the background.

5) The functions of data receiver and bus controller are separated. A controller on the feeder bus is responsible for the sampling of all variables. Its only function is to output the address of each input variable at regular intervals. To do this, the bus controller owns a system-configuration table which indicates the sampling frequency of each variable. In this case, the concept of source and destination becomes blurred. The bus controller sends out the address of a device, and the device sends out the data to other devices. The destination of the data considers the cycle as a write cycle, and the sending device as a read cycle. The data address is therefore to be considered as a source address.

4.3 Event-Driven Operation

In cyclic polling, the number of sensors which can be processed is limited, since each must be read once in each cycle, whether its value changed or not. In fact, the processor has no way of knowing that the sensor's input changed before it was read. This implies that a large amount of data is processed unnecessarily, and that the maximum cycle length also limits the number of connected stations.

Better use of the bus and processor is made if the processor reads data only if a variable changes significantly. The transducer board could then generate an interrupt request to signal that the data changed and ask for its reading. This requires that an Event be built at the transducer. An event is a change of a Boolean function of several variables (including time). For instance, an event could be that a binary variable changed state, that an analog variable changed by more than a certain amount since the last reading (typically 0.39%), or that the analog variable passed a certain threshold. Also, a certain combination of states can trigger an event, for example, if the product of current and voltage passes a certain value while the busbar switch is closed. To transmit the event, the transducer must have the means to access the station spontaneously. If communication takes place over a bus, this means that the transducer may become bus master, and the bus is necessarily a multimaster bus (Figure 9).

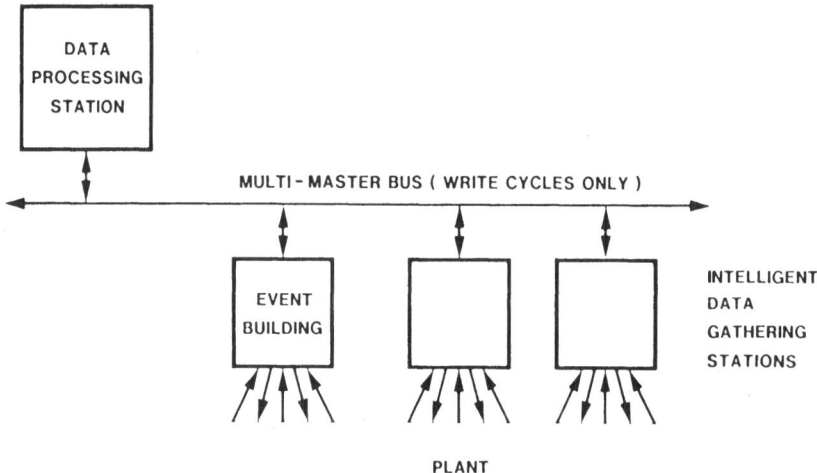

Fig. 9: Event-Driven Transmission

Event-driven operation must overcome four hurdles:

- The transducer logic must possess a certain intelligence to recognize the event. This function requires additional logic or a microcontroller. In any case, the transducer board will cost more, and it will be more economical to do the processing for several sensors at the same site (see Figure 9);

- The second problem is that in the worst case, all event messages come at the same time. Although this may seem unlikely, it is not, since the process variables are correlated: if the voltage changes, so does the current, the phase, the power, etc. This is why alarms tend to occur in bursts. The command station for a power grid can receive from 50 to 100 alarms a second. A case has been reported in which 1,500 alarms occurred in 53 s,[7] which would fill 60 screen pictures. Although the operator is not capable of looking at these alarm reports in real-time, the record must be kept for subsequent analysis, and no alarm message should be lost. Therefore, the program must deal with the worst case, that is, with the sequential handling of event messages. In this case, the overhead bound to the event handling makes the system slower than in the polled case.

- The third problem is defining which meaningful task the processor should perform when it is not responding to events. If the answer is "nothing", the data could just as well be polled cyclically. If some meaningful background tasks can be found, then the question is shifted to the definition of priorities.

- The fourth problem is caused by possible data losses. In a cyclic system, if an item of data is corrupted, it will be sent again at the next sampling period. In an event-driven system, a change of value may remain undetected because a transmission problem occurred. Therefore, a protocol must ensure that the event has been correctly received at the destination.

In fact, the operation of an event-driven system may well be so complicated that an upper bound for the execution time can no longer be guaranteed. On the other hand:

- An event-driven system can serve a far larger number of stations than a cyclical one. The net amount of resources tied, even including reserves, are lower than that needed for a cyclic system for all practical purposes.

- Fortunately, a physical upper bound generally exists to the number of variables which can change simultaneously, so that the worst case can be determined after careful analysis. Further, a delay can be introduced to prevent an event message being issued too often.[6]

It is important that the allocation mechanism, even in an event-driven system, be deterministic. Binary protocols, which change strategy under the bus load, have been proposed. These protocols run the bus in a random-access mode when lightly loaded, and go to a cyclic allocation under heavy load. Token buses such as BBC's PARTNER-BUS or GM's MAP show a very good behavior for this operation.

4.4 Comparison of State versus Event Transmission

The domain of validity of each solution is always changing. Generally, the time-critical lowest level will be operated cyclically, while the upper levels of the hierarchy are event-driven to operate more effectively on the huge quantity of messages.

Some buses mix the two modes of operation, for instance BBC's P14 has both a cyclic and an event-driven mode. There is at least one good reason why an event-driven system should also be operated cyclically: in an event-driven system, there is no control over the health of a station, while in a cyclic bus, each station is regularly polled, and is known to exist. Therefore, for error-detection purposes, it is necessary to run a cyclic interrogation of all stations as a background task. In both cases, there should be no volatile event messages, since there exists the probability that an event message may be lost. Either each event message is acknowledged, or the event messages are cyclically retransmitted.

We can compare the two modes in the following table:

CYCLIC TRANSMISSION	EVENT TRANSMISSION
Pessimistic policy	Optimistic policy
Deterministic	Stochastic
Polling	Spontaneous transmission
One bus controller for normal operation, dumb peripherals	Each station is an intelligent bus master
Feeder buses with read cycles	Bus has only write cycles
No protocol for retransmission	Error-handling protocol required
Limited throughput	Throughput limited only by congestion considerations
Short messages (~32 bits)	Long messages (<512 bits)
Adapted to functional languages (FB)	Adapted to procedural languages (PASCAL, MODULA-2)
Lower process hierarchy	Higher process hierarchy

Events should include the absolute value of the variable, not the increment since the last event. Incremental messages, such as an indication that a photodetector detected the passing of an object or that a variable increased by 0.10 %, should not be sent. An event should be transformed into a state as soon as possible. In the case of a photodetector, the pulse should be sent to a counter and the state of the counter transmitted.

4.5 Broadcast

The efficiency of both cyclic and event-driven links can be improved by considering that most data items are required in several places in the system, e.g., for regulation at the expert site, for the process database of the operator, for the service station, and for data logging. It has been estimated that the same input variables are used from 1.3 to 3 times in different places. This is why BBC's process-control buses support a broadcast mode. The logical development of the broadcast mode is that the messages, even in the event-driven mode, no longer bear the name of the destination, but carry a tag which identifies the source, or better, the data transmitted. Any station connected to the bus collects the data it requires as it passes by. To do this, every station has an associative memory to recognize the data to which it subscribes (Figure 10).

Broadcast operation allows very easy extension of the system. A broadcast bus is a kind of live database. Any device connected to the bus can update its own database from what it hears, and contribute to the database of the other devices by sending out its variables. The combination of broadcast transmission and cyclic operation has resulted in the P14 feeder bus. In addition to broadcast, this bus possesses a destination-addressed mode for service and diagnostics.

One key question is how broadcast protocols can be made resilient to transmission losses. The usual method of Positive Acknowledge and Retry (PAR) is only feasible if the identity of all destinations is known. This is not the case here, especially since the system should be transparently extensible. One possibility is to let all connected stations acknowledge, whether they use the data or not. This method is used in BBC's PARTNERBUS and P14. PARTNERBUS uses the token system to acknowledge, since the number of stations (32) is sufficiently low. P14 uses for that purpose a high-frequency noise channel with wired-OR behavior, which makes it a combination of baseband and broadband bus. Every station which received the data fault-free stops transmitting on the noise channel. The absence of noise indicates that all stations received. This detects transmission losses, but not a basic malfunction at the destination station, so it is necessary to test on a regular basis the proper functioning of the noise channel of each station.

The alternative is to have no acknowledge at all. Then, some care must be taken when transmitting event messages: if the event messages are retransmitted several times, a mechanism must be used to discard the duplicates, or a kind of high-level acknowledge must exist. Several broadcast protocols have been studied with the aim of providing a reliable broadcast.[4] There are numerous proponents of the "end-to-end argument" who suggest that the system should be trusted and only a high-level acknowledge be given. This is only feasible if no problems due to flow-control arise,

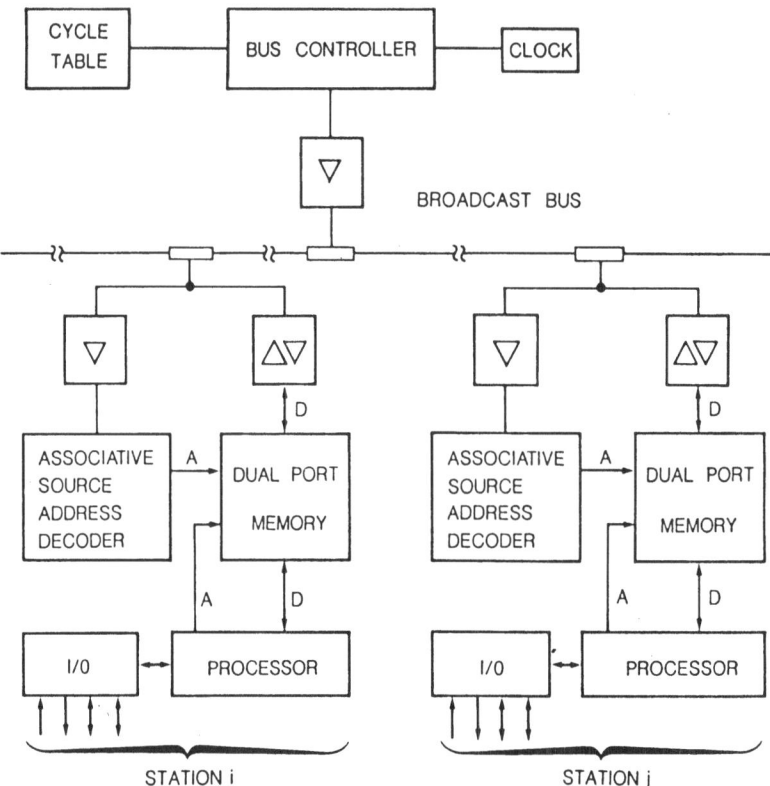

Fig. 10: Broadcast of Source-Addressed Data

and their avoidance requires either very large buffers or a method to predict exactly the data flow under the worst-case conditions. Thus we are brought back to the dispute of the optimists against the pessimists.

6. FAULT TOLERANCE

As dependence on computers increases, tolerance against their outage is becoming a mandatory feature of a process-control system. The question is not whether a completely fault-tolerant process-control system can be built. We know it can. But complete fault tolerance requires at least a threefold amount of resources: one resource to do the job, one to detect all errors (possibly by duplication and comparison), and a second redundancy as a spare to proceed in case of error.

The redundancy may come in the form of additional hardware, time or software. A full-blown fault-tolerant process-control system would cost about five times the cost of a nonredundant installation. Even if hardware costs are decreasing, installation, planning and software costs are not. Few clients would accept the cost involved in duplicating all sensors and actuators in their plants. Therefore, the art in fault-tolerant design is to include redundancy where it is needed to fullfill the dependability goals. Without such a clear goal, a cost-effective solution cannot be designed.

Different applications have quite varying requirements with respect to dependability. Most place strong emphasis on availability (relation of up time to total operation time). Others add to the availability figure a requirement of maximum allowed down time. For instance, modern power networks are specified to have an average failure rate at the various voltage levels of 1.5 interruptions/year, with a total down time of 60 to 90 minutes per year.[2] In a specific HVDC (High Voltage DC transmission) project, the availability is requested to be higher than 99.5 %. In other projects, data integrity is rated higher than availability: It is preferable to stop the process in the case of computer error rather than attempt recovery. The transmission media for telecontrol are subject to international norms regarding the maximum permitted residual error rate.[3]

On the other hand, ultrahigh reliability is rarely demanded: there are no plants today where an outage of the computer would cause a catastrophe, as is the case for fly-by-wire aircraft or the space shuttle. Plants generally have a safe shutdown state, and independent protection equipment is installed to protect critical parts. Only the protection circuits themselves have requirements of high reliability. Further, and in contrast to space probes, maintenance is possible in industrial plants, and one of the goals of fault tolerance is to delay maintenance until the scheduled date rather than to suppress it. The maintenance interval is therefore one key factor which fault tolerance can influence.

In many cases, the dependability goals can be met without specific fault-tolerance features. For instance, most outages do not take place in the process-control system itself, but in ancillary equipment, such as ventilators, power supplies, peripherals with mechanical parts, and sensors. Outages within an equipment cabinet are relatively seldom, and the reliability goals are better met by introducing an improved power supply than by duplicating the CPU. And even in those installations which lack an apparent redundancy, about one-third of the current software is there to cope with outage situations.

In most installations, some redundancy has become common. To meet the availability requirements, BBC has traditionally used twin computing systems, such as the BECOS 30 or BECOS 40 telecontrol configurations. The highway buses such as P14 or PART-NERBUS are duplicated, not because they would fail more often than a processor or a feeder bus, but because they are a key component whose outage cannot be tolerated.

There is however an increasing number of installations the dependability requirements of which cannot be met without a redundant control system. In that case, some extra effort is required. Let us suppose that the calculations show that a processing element, such as an expert station has insufficient reliability and should include some redundancy. The key parameter which dictates the architecture of that part is "How long can an outage be tolerated?".

If the time allowed is of the order of a minute, time is available to transfer the function from one node to a spare node which has basically no memory. The new node can then start up the programs, for example, by loading them from a disk, and reconstitute the previous state of the failed node by gathering the process state through a "general call", which is a polling of all peripheral devices. A general call takes about one minute in a power-control system. In that case, the process itself serves as a nonvolatile database, and the "cache" process database is actualized. The control is then switched to the new node.

If the allowed down time is shorter, the general call requires too much time. Since the transfer of data from the faulty node to the spare node should not be trusted once an error has been detected, it is mandatory to maintain duplicated databases in each of the nodes.

There are basically two methods of maintaining identical databases in each node. The first is replicated execution. Both nodes listen to the same buses and execute the same operations synchronously. Broadcast is a very convenient means of achieving this. The output of the spare is connected neither to the process nor to the bus. Parallel execution requires that the nodes be exactly synchronized, and that all inputs be routed to both nodes, in such a form that each receives exactly the same data at the same time. This rules out direct connection of process periphery to one of the redundant nodes. The other approach is updating. A direct link exists between the on-line node and the spare, by which the on-line node keeps its spare actualized at regular intervals. In case of failure, control is switched to the spare unit. The problem of updating is that any action which the main unit undertook since the last successful update is lost to the spare. The spare must therefore be able to reconstruct all data which the main unit received since the last update, and also to follow all outputs from the main unit since then, in order not to replicate them. The two solutions are shown in Figure 11.

The updating solution is appealing because it requires less computing power, and the spare node has capacity free to execute other operations. This solution has been retained for fault-tolerant computers in commercial applications. But the monitoring of the inputs and outputs of the main unit is a computation-intensive task which should

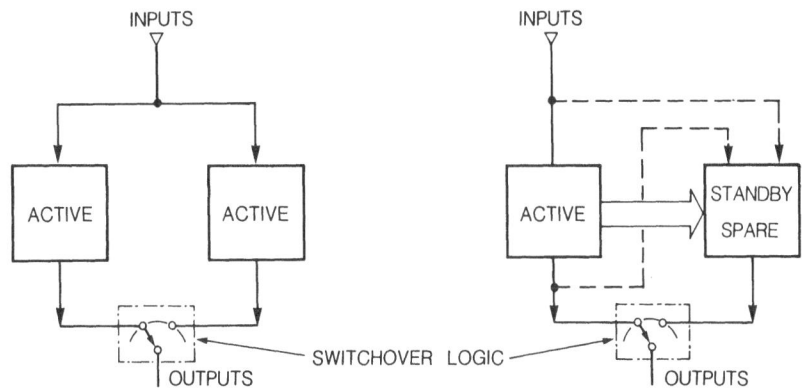

Fig. 11: Replicated Computing versus Spare Updating

be left to a communication controller. For process control, the question is what should be done with the excess computing power? Are there second priority functions which can be dropped in case of outage? In practice, these are not easy to find, and "graceful degradation" takes on a different meaning in process control. Process control is not expected to lose functionality in case of failure, only to lose redundancy margin.

The updating function must anyhow be present for the reintegration of a repaired unit. This aspect of fault tolerance has been often overlooked in the past, and attention focussed on the switchover phase, which turns out to be easier than the uplearning phase.

High-speed controllers such as those used in the expert stations have requirements of their own. The switchover time must not exceed some milliseconds. This requires that the spare be maintained continuously actualized by the main unit by some hardware without intervention of the processor, or that replicated execution be used. The classic solution of aerospace, the two-thirds voting, which lets three modules execute the same function and vote on the output, has little utility in process control. The probability of failure is about four times that of a nonredundant controller, and availability suffers accordingly.

Some applications require that no erroneous data ever be produced. This requirement can be met by using fail-stop components, or elements which switch to a passive state when an error in them is detected. This supposes that 100 % error detection is provided. Few fail-stop components are in use today. One of the reasons is that a fail-stop electronic circuit costs about three times the price of a normal element and is only capable of detecting hardware errors. Software errors, which account for about one-third of the errors in complex systems, are overlooked.

The principles of fault-tolerant computing are quite well known today, but the application of these principles to process control turns out to be a challenge, since the real-time interaction with the plant must be considered. In particular, the redundancy principle should be followed consistently not only in the control system, but in the plant itself. Replicating the sensors is relatively easy; replicating the control system is costly, but possible. Duplicating the output devices requires some form of switchover, and the reliability of this switchover device determines the reliability of the whole. The process-control engineer will therefore try to extend the redundancy as far as possible within the plant itself.

Fault tolerance is not a feature which can be added to a control system after the design is frozen. It must be planned from the outset. Fault tolerance has no general cost-effective solution. Each application will impose a different solution. Fault tolerance can be realized today with existing technology, and by adding to the existing building blocks specific fault-tolerant blocks, such as fail-safe error detectors and highly reliable switchover units.

7. CONCLUSION

Today, we are approaching the ideal of constructing sophisticated control systems from a large family of hardware and software building blocks, according to the needs in functionality, computing power and dependability of a specific application. Most theoretical problems are now recognized, solutions exist and have been installed. Distributed control systems, hardware building blocks, modular functionality, and integrated development tools have become common concepts in the literature. The question now is how efficiently these concepts can be implemented. Here, many problems related to the efficiency of communication in a modular system remain. A steady increase in computing power and memory size in recent years has become accepted, but theoretical limits exist. Fault tolerance only received solutions which are specific to a class of applications, but it is unknown whether a general solution would be cost-effective. The efficient implementation of these concepts will permit concentrating development effort on the optimization of the industrial process itself instead of optimizing the process-control system.

REFERENCES

1.   Aschmann, H.R., "Recovery in a Distributed Process Control System Based on a Local Area Network," Research Report KLR, 84-218C, December 1984.

2.   Frey,H., and Signer, K., "System Reliability Planning," <u>Brown Boveri Review</u>, Vol. 66, March 1979, pp. 216-224.

3.   Funk, G., "Data Integrity and Efficiency of Single Parity Check Product Codes," <u>ntz-Archiv</u>, Vol. 7, No. 4, 1985.

4.   Gueth, R., Zueger, S., and Kriz, J., "Broadcasting Source-Addressed Messages,"   <u>Proc. 5th International Conference on Distributed Computing Systems</u>, IEEE Press, May 1985, pp. 108-115.

5.   Krings, L., "Fast Rollback Techniques for Fault-Tolerant Computing," <u>BBC Research Report KLR 85-53B</u>, April 1985.

6.   Latzel, W., and Kahle, H., "Dynamik verteilter Prozessleitsysteme, insbesondere fuer Regelungen," <u>GMR-Fachberichte 1</u> "Prozessfuehrung und Datenuebermittlung",16 - 17 October 1984, Frankfurt VDE-Verlag-GmbH.

7.   Schaffer, G., Davis, M., and Sandmayr, H., "Supervisory Network Control Systems Reflect Users Requirements," <u>Brown Boveri Review</u>, Vol. 71, September/ October 1984, pp. 393-399.

8.   Stoeckler, H.P., "Kraftwerksleittechnik," BBC International Press Meeting, Baden, Switzerland, March 1982.

DISCUSSION

Chairman: R. Karg (Brown Boveri, Mannheim, Fed. Rep. Germany)

V. Haase (Technical University of Graz, Graz, Austria)

What is your approach to software redundancy? Do you use multiversion programming?

H. Kirrmann

To my knowledge, the concept of n-version programming has not as yet been widely applied to industrial process control. In the aerospace industry, this technique is being used, for instance for the Airbus. In many process-control systems, however, more than 30 % of the software is devoted to fault detection and error recovery, for example, for the detection of software errors and the data transfer and synchronization between the main computer and a backup computer.

U. Rembold (University of Karlsruhe, Karlsruhe, Fed. Rep. Germany)

When I look at fault-tolerant systems, it sometimes seems to me that the more computer hardware and software is involved, the less reliable the whole system becomes. Where is actually the limit where a complex fault-tolerant computer is less reliable than a single computer?

H. Kirrmann

It is a tradeoff. If you add complexity for the purpose of fault tolerance, you may reduce the reliability of your product and, at the same time, increase its availability. So it depends on what your goals are. You must have a clear understanding of the application for which the fault-tolerant computer is required.

H. Burkhart (Federal Institute of Technology, Zurich, Switzerland)

From the programmer's point of view, we need tools and languages with which we can configure software systems from modules and bind software modules to hardware resources. So far as I am aware, there are two approaches. One is a separate configuration language, like the Task Force Language of CM*. Another approach is found in recent programming languages. In Edison and Occam, for example, the configuration mechanism is incorporated in the programming language. Which kind of configuration technique do you have in mind?

H. Kirrmann

The configuration is to a large extent supported by the programming language, which is Modula-2. Utility programs for the configuration of distributed software modules and their binding to hardware still have to be developed.

PERFORMANCE MODELLING OF CONTROL SYSTEMS

M. VITINS and K. SIGNER
Brown Boveri, Baden, Switzerland

ABSTRACT

Control systems must satisfy stringent performance requirements that can be derived from the nature of the process controlled. The distinguishing characteristics of an electrical power system and their effects on the control system are described. A hybrid analytical/simulation model is developed that describes the control-system behavior for both nonsaturated and saturated resources. The model is incorporated in an application-oriented, user-friendly, interactive, performance-analysis tool that can be used to support the planning, design and tuning of a process-control system.

1. INTRODUCTION

The development of large process-control systems is continuously being influenced by the demand for increased functionality and better performance regarding throughput and response time, and by the requirement for higher reliability and safety. The fulfillment of these requirements should offer the benefit of better utilization of the controlled system.

The above trend has stimulated a growing interest in the use of quantitative techniques and tools for analyzing the tradeoff between the cost and the capabilities of the control system. The availability of such techniques and tools allows a consistent analysis of product variations and should thus support decision making during the planning, design, and tuning of a control system.

The paper describes a user-friendly, interactive, performance-analysis tool which has

141

been successfully developed, implemented and applied to practical process-control problems. The tool is based on a hybrid analytical/simulation model that describes the control-system behavior for both nonsaturated and saturated resources. Saturation typically occurs when the system is subjected to a large but temporary burst of information. An important feature of the model is the ability to represent directly both hardware and software components. The model is therefore well suited to the support of application-oriented studies. A typical application, which will be addressed in the paper, is the performance analysis for a large hierarchical multicenter/multi-computer process-control system for an electrical power system.[3]

2. THE PROCESS AND THE CONTROL SYSTEM

2.1 The Process

The process under consideration in this paper is an electrical power generation and transmission system for a large industrialized region. The electric utilities which are responsible for power systems invariably have the aim to supply electricity to con-sumers as reliably as required and as cost effectively as possible. The main process characteristics of an electrical power generation and transmission system are its distribution over a wide geographical area and the layered structure (its voltage levels). The operational philosophy of the power authorities is therefore normally based on a layered area approach, i.e., control responsibilities are hierarchically distributed according to geographical areas and groups of voltage levels. For example, responsibilities may be split into national, regional and district levels, corresponding to the highest, medium and lowest voltage levels, respectively.

The present state of electrical power systems is characterized by the following fea-tures:

- A steady increase in the demand for power
- Consumer expectations of continuous, stable supply
- Complex meshed transmission systems
- Environmental restrictions on development and expansion

each of which has a profound effect on the operating complexity with respect to a-chieving optimization of the system security and cost.

Power-supply optimization has thus two major components:

- Maximization of the use of existing power-system equipment
- Minimization of the cost of energy production

while ensuring acceptable security and quality of supply.

2.2 Control System

Electrical power systems are subject to rigorous demands of availability, security and quality of supply. These basic requirements on the power system are reflected in the rigorous availability and performance requirements concerning the supervisory functions, i.e., the control system.

As a consequence of the previously mentioned present state of electric power systems, control-system operators have to cope with:

- Narrow operating margins
- Accurate supervision of operating conditions, taking possible future events into account
- Complex switching operations, the consequences of which must be foreseen
- Continuity of supply without wastage of primary energy

The major task of the power-control system is to support the activity of the operator. Hence, a power-control system should:

- Present reliable and complete information about the state of the power system, especially an immediate overview of the system state following emergency situations
- Immediately notify abnormalities in the power system through alarming
- Indicate in advance the set of possible disturbances that would have major consequences for the system (e.g., by cyclic contingency analysis)
- Allow fast and reliable control of the power-system elements in order to guarantee the quality of the power supply
- Provide guidance to achieve secure and cost-effective generation and transmission of energy
- Facilitate record keeping through event logging and reporting

The power-control system should reflect the nature of the power system and also exploit the benefits of modular computer hardware and software structures. As mentioned above, an operational concept based on a layered area approach has been found to be the most appropriate. Normally, each of the operational areas within a layer has a control center which communicates with lower level remote terminal units and with centers of the same or higher levels. A typical hierarchically structured control system is shown in Figure 1.

Locally within each control center it is also beneficial to have a multicomputer system

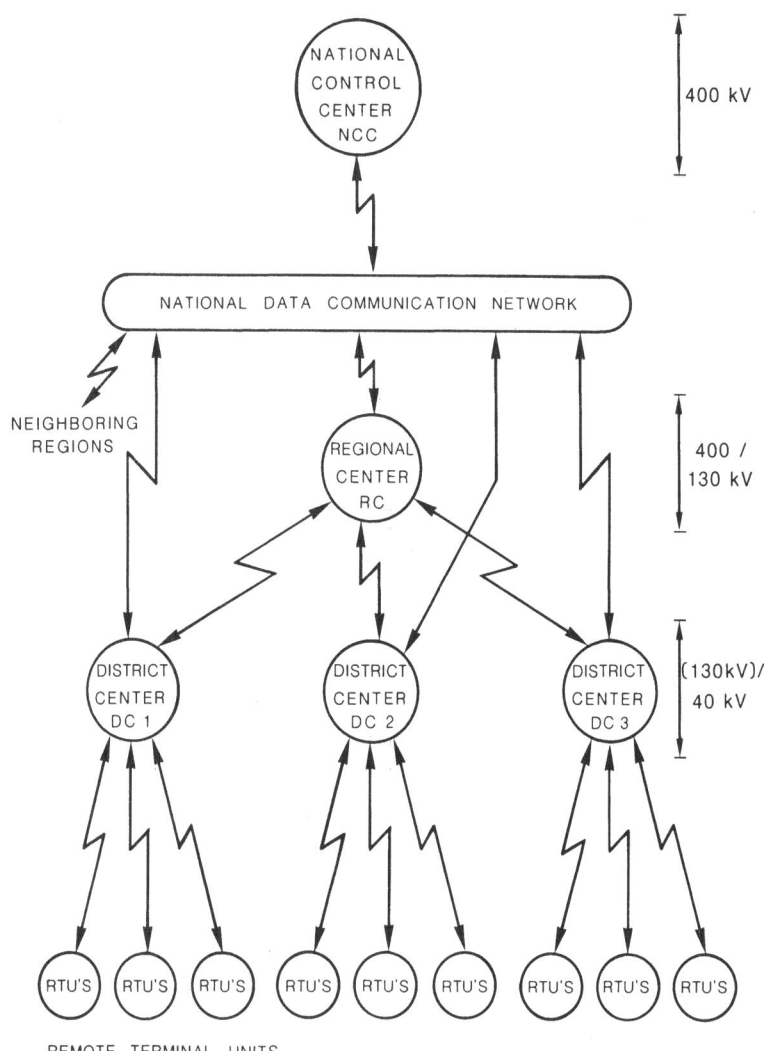

Fig. 1: Hierarchical Structure of a Power-Control System

with distributed functionality, as shown in Figure 2. Such a system has improved availability because it has a graduated range of performance and functionality. Throughput, response time and functionality can be selectively controlled by using computers of different performance capabilities and by varying the number of comput-ers in the center.

The center shown in Figure 2 consists of dual main computers, front-end computers, communication computers, and a MIMIC controller. All processors within the center are connected by a dual bus. The configuration can be extended by further front-end

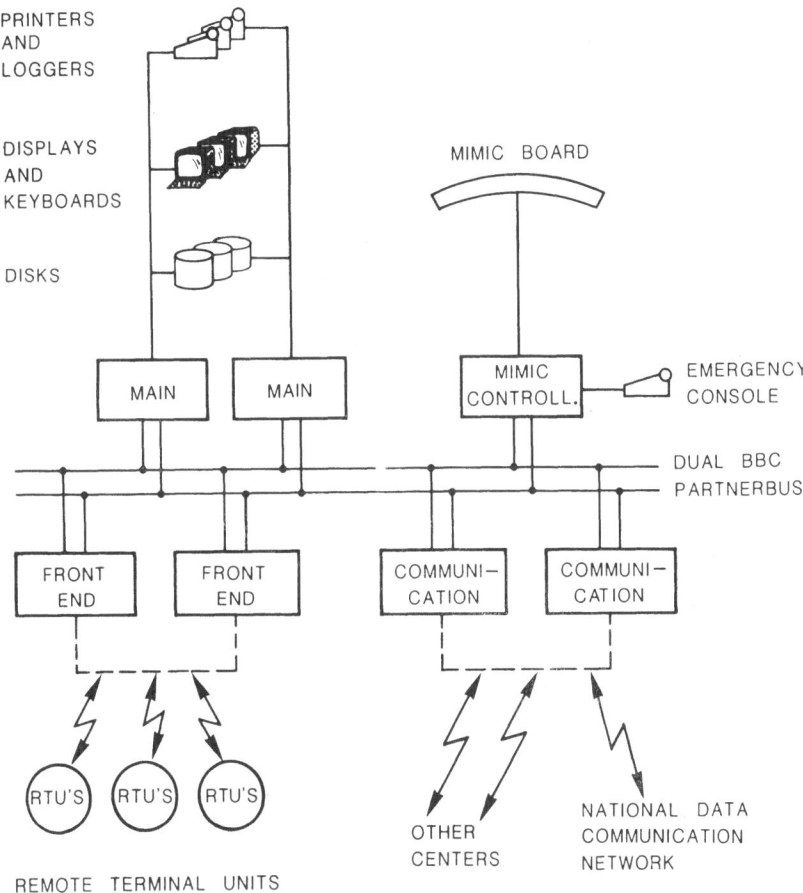

PRINTERS
AND
LOGGERS

DISPLAYS
AND
KEYBOARDS

DISKS

MIMIC   BOARD

MAIN   MAIN

MIMIC
CONTROLL.

EMERGENCY
CONSOLE

DUAL BBC
PARTNERBUS

FRONT
END

FRONT
END

COMMUNI-
CATION

COMMUNI-
CATION

RTU'S   RTU'S   RTU'S

REMOTE TERMINAL UNITS

OTHER
CENTERS

NATIONAL DATA
COMMUNICATION
NETWORK

Fig. 2: Hardware Configuration of a Center

and communication computers as demand arises. The main computers (DEC VAX type) perform supervisory and extended application functions, and support the man/machine communication. Interfaced to the main machines are powerful display controllers as well as numerous printers, loggers, etc. The front-end processors (DEC PDP type) perform all data-acquisition functions of telemetry management to the remote terminal units as well as numerous routine manipulations on the telemetered data. The communication processors (DEC PDP type) are assigned the task of managing all communication and data interchange between the center itself and any other interconnected control centers or communication networks. The MIMIC controller (microcomputer) drives the wall diagram, including recorders, and also supports an emergency console.

The telecommunication system, shown in Figure 1, is specifically designed to fulfill the following three data-transmission functions:

1) Information exchange between remote terminal units and district centers

2) Information exchange between centers

3) Information exchange with the data-communication network

The transmission media of the system in Figure 1 are radio links, each transmission function being based on a special protocol. The RTU's are grouped into subsystems (so-called "part" systems), each system having a transmission rate of typically 1,200 baud. The transmission speed of the other radio links (2 and 3) in the system is 4,800 baud. By contrast, the baud rate of the center's internal bus is 1 Mbit/s.

3. PERFORMANCE ANALYSIS

The supervision "quality" of a control system is largely reflected in the system avail-ability and system response times for both incoming and outgoing information and the guarantee that information is neither lost nor distorted. Performance analysis is important for both normal and emergency power system operating conditions.

In the normal mode, the power system is in a steady state. The rate of message traffic between the power system and the control system is relatively low. It is essen-tial to keep the resource usage by the Supervisory Control And Data Acquisition (SCADA) functions as low as possible in order to maximize the availability of re-sources to Power Application Software (PAS) which optimizes system security and operation cost.

The behavior of the control system during and following emergency situations in the power system must satisfy stringent requirements. Under such worst case operating conditions, the control system is subjected to a large but temporary burst of data. If system resources become saturated, an information-processing backlog will inevitably occur in one or more subsystems. Two questions are of utmost importance for the system performance:

1) How long will the backlog persist?

2) What is the maximum resulting message delay time?

3.1 Performance Analysis Goals

A versatile and comprehensive performance-analysis tool provides valuable support in the decision-making process during all phases of product/project development for a control system. Performance optimization in planning, design, and tuning of a control system is facilitated by analyzing product variants regarding throughput and response times.

The aim of <u>System Planning</u> is to investigate configuration and capacity alternatives in order to find the best tradeoff between the cost and capabilities of the control system. The configuration can be varied, for example, by choosing a different number of front ends in a control center. The capacity of a computer subsystem is to a large extent given by the CPU power, the disk-access speed, and the memory size. A major concern of a planning study is to ensure that sufficient of each of these resources is present, while avoiding the excess of any resource that produces diminishing returns. System variants must therefore be discussed on the basis of their resources utilization. For example, the performance of a CPU-bound system, in which the CPU is the resource that becomes overloaded, does not improve if the disks are replaced by faster (and more expensive) disks. However, the use of faster computers will improve the performance of a CPU-bound system. By varying the type of computer or disk, the system may change from being a CPU-bound system to a disk-bound system, or vice versa.

During <u>Design</u> or <u>Development</u> activity, the study of system variations helps to improve the understanding of the dynamic behavior of a complex system. Possible improvements in the software structure or the functionality of the software processes may result. A characteristic of a well-designed system is the absence of a situation in which a bottleneck component causes other (possibly expensive) components to become underutilized. The usual rule to maximize the utilization of a multicomponent system is to maintain a steady balance of the demands for each of the components.

In order to achieve balanced utilization, it is advantageous not to allow system behavior to rely on the abundance of just one resource. In a CPU-bound system, for example, in which there is a shortage of CPU, an effective technique to selectively distribute the CPU to the processes is to assign a CPU priority to each process. The sole reliance on CPU priorities is, however, totally inadequate if the system at a later stage becomes disk bound. An important factor influencing the behavior of a disk-bound system is the disk-access pattern of each process. In a disk-bound system, it is often advantageous to distribute the different program and data segments onto several disks with the aim of achieving a balanced and steady disk-access rate on each disk.

A common technique to ensure balanced resource usage for both light and heavy message traffic is the appropriate use of event and cyclicly driven software processes. An event-driven process offers the advantage that it can immediately respond to a triggering event, but suffers from the drawback that the resource demands grow and become large as the rate of the processed events increase. A cyclic process offers the advantage that its resource demands are constant, irrespective of the amount of actually useful data handling, but suffers from a possibly high cyclic load and long response time. The decision of whether a function should be implemented in an event,

SYSTEM DESCRIPTION

PERFORMANCE PARAMETERS

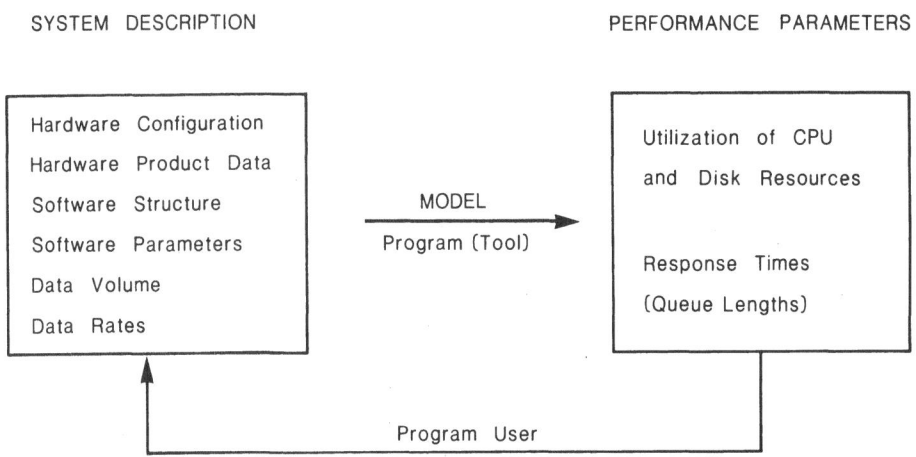

Fig. 3: A Performance-Analysis Tool

a cyclic, or even a combined switched event/cyclic driven manner depends on a number of criteria, such as the expected range of event rates and the overall significance of the process.

Tuning activities deal with the adjustment of parameters and with the local performance-oriented optimization of code (ensuring that it does not jeopardize the rest of the system). Code optimization must take into account the tradeoff between savings in resource usage and the cost of modifying the code, including the additional effort needed to maintain the modified code.

In all the types of investigations mentioned above, it is important to evaluate the effect of each system variation on both the immediate local-process or processor level as well as on the rest of the entire system. A local performance improvement can have the effect of shifting a bottleneck from one component to another, causing an overall degradation rather than improvement of the performance/cost tradeoff.

3.2 Performance Analysis of System Variants

The use of a performance-analysis tool to investigate system variations inherently leads to an iterative decision procedure, as shown in Figure 3.

In Figure 3, the system description consists of the following items:

- The hardware configuration, defining, for example, the number of front ends and the number of communication channels connected to each of the front ends

- The hardware-product data, including the type of computer (e.g., VAX 11/730, 11/750 or 11/780), disk and printer times

- The software structure, describing the software processes and their message-passing interactions

- The software parameters, which characterize the resource usage of the software processes

- The data volume, indicating the types and the total quantity of the data items to be handled by the system

- The data rates of the messages sent to the control system, indicative of a power system in both normal and emergency situations

The performance parameters calculated on the basis of a system model are

- CPU and disk loads for all computer subsystems

- Response times (for single hardware or software components and for the over-all system)

- Queue lengths

The tool illustrated in Figure 3 can be applied to addressing, in a uniform way, a wide spectrum of planning, design and tuning activites that are the expertise of individuals with quite different backgrounds, such as sales, design and on-site engineers. Each of these individuals is typically interested in the impact of a small number of specific parameters on the overall system performance. Based on the outcome of the calculated performance parameters, the program user decides whether and which further system variants need to be considered to achieve a specific goal.

In order to provide effective support, it is essential that the tool can be used inter-actively and that the dialog interface supports the application-oriented view of the program user (rather than forcing the user to become deeply involved with the under-lying mathematical performance modelling technique and program implementation).

4. PERFORMANCE MODELLING

4.1 Modelling Techniques

A wide spectrum of techniques has been used for computer-system performance modelling. The techniques have been arranged in order of complexity and cost of development effort in Figure 4. The easiest performance-modelling technique to apply is the
use of simple rules of thumb, which are generally based on widely recognized observations and safety measures. Linear projection of load and utilization offers a more
organized approach. It establishes growth rates based on measurements of various
resource utilizations, which are then used for extrapolation purposes. The benchmarking approach consists in taking a representative sample of the workload, running it
on the proposed hardware system, and measuring the performance. Unfortunately,
benchmarking provides answers to a rather limited scope of problems, since it is very
difficult, in practice, to carry out the benchmarking in a fully controlled manner for
all relevant situations. Simulation has become a popular form of modelling, since it
enables the analyst to model the system with the desired degree of accuracy before
actually building the computer system. Simulation models are, however, difficult and
costly to construct, validate and run. Although the techniques mentioned are very
popular, they are not always the most cost effective. Analytical and hybrid modelling
techniques are becoming increasingly powerful, and can provide answers to many
performance issues at moderate cost.

4.2 Analytical and Hybrid Modelling

The mathematical technique for solving performance problems is queuing theory.
Queuing theory is, in general, concerned with techniques for analysing the flow of
objects through a network that contains restrictions. Much of the early work on queu-

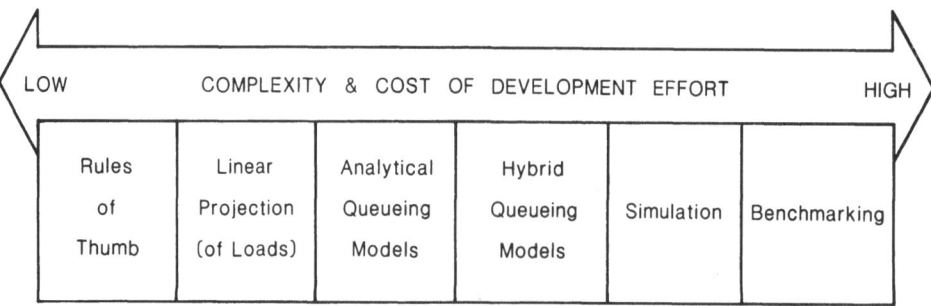

Fig. 4: Modelling Techniques for Computer Systems

ing theory, particularly in the late 1950s, was strongly influenced by the traditional aims of mathematics and pure science. Emphasis was put mainly on the generality and power of the underlying mathematical theory (making full use of probability and stochastics theory) and less on satisfying the needs of end users (e.g., engineers). As a consequence, queuing theory, as a method for the analysis of practical problems, has remained in a rather primitive state. In addition, it turns out that many practical engineering problems are easy to solve approximately but very difficult to solve exactly. This situation is reflected by the observation that much of the queuing literature is concerned with exact solutions or with lower and upper bounds for the steady-state behavior (e.g., the average length of a queue). When applicable, analytical techniques are more attractive than other techniques because of their relatively low cost, general applicability, contribution to understanding, and ease and flexibility of use.

In many practical problems it is not sufficient to restrict the performance analysis to steady-state behavior. Rather, "rush-hour" situations, during which objects arrive at the queuing network at a rate that temporarily exceeds the capacity of the system, must be considered as well. It is during rush hours that the system resources are temporarily overloaded and large queues develop, such that problems of real concern arise. The proper compromise between the cost of a high-capacity system and the inconvenience of a large delay can only be found if both steady-state and foreseeable exceptional rush-hour situations are taken into account.

Analytical techniques for dealing with the rush-hour problem include fluid approximations (based on graphical methods) and diffusion approximations (based on partial differential equations).[6] In complex queuing applications, in which the stimulus is time dependent and the maximum processing capacity depends on a multitude of parameters, the analytical solution often turns out to be too difficult. Hence, discrete-event simulation models were favored initially.

The combination of discrete-event simulation and mathematical modelling, to form a hybrid overall model, offers a cost-effective technique for modelling complex systems. The objective is to produce a technique that combines the economy of analytical models with the accuracy and real-time characteristics of discrete-event simulation. A hybrid model is particularly suitable for studying computer systems over long periods of time, for which analytical models would not be sufficiently accurate and for which discrete-event simulation models would be too expensive to operate, particularly when rapidly occurring events must be included. A number of papers have been published over the past ten years on the development of hybrid techniques for the modelling of computer performance.[8,1,5,4] In hybrid models, the simulation technique is typically used to model long-term behavior, while an analytical model is used to model short-term behavior, characterized by intense bursts of events.

Hybrid models share with simulation models the advantage of being particularly well suited for modelling real-time control systems because of their ability to incorporate complex and time-dependent job arrival patterns. There exists no simple way to represent complex arrival patterns in a mathematical model.

Schwetman[8] developed a hybrid model of a computer system, in which the resources of the computer system have been partitioned into two mutually exclusive sets, denoted by long-term resources and short-term resources, respectively, creating a two-phase model. Simulation is used to model the arrival of new jobs and the allocation of long-term resources to the currently active jobs (e.g., core allocated for a duration of the order of seconds or minutes). This data is then fed to the analytical model of the short-term resource consumption (e.g., CPU allocated for a duration of the order of milliseconds). The analytical model predicts the completion time of the first (the shortest) job to be processed. If one of the active jobs completes its workload or if a new job arrives before an active job completes, the long-term resources are redetermined and the loop is repeated. The model yields acceptable results when the consumption of short-time resources occurs at a rate which is much greater than changes in the use of the long-term resources.

Foxley[1] has developed a hybrid model for a virtual computer system, using analytical models of page rates and of steady-state computer-system performance, and discrete-event techniques to model real-time aspects such as loading and scheduling of jobs. The use of an analytical model of page rates eliminates the need for time consuming simulation at the level of page faults.

Gomaa[2] has also developed a hybrid model that combines regression and simulation methods. The regression model is used to determine the elapsed time of a job, and a numerical technique is used to predict the delay time experienced by each job due to competition from other tasks.

In this paper, we will apply the hybrid modelling technique to model both the steady-state and the rush-hour behavior of a process-control system. The proposed control-system model consists of two queuing submodels that capture the major data-processing activities of the system:

1) The Software Submodel: This model is given in terms of an open queuing network of software processes which receive and send messages. This network describes how messages entering the control system propagate from process to process as they are processed by the software. The network is open because messages can enter (and leave) the network. When saturation occurs, e.g. due to rush hour conditions, a message backlog will exist in the network and nonstationary behavior results.

2) The <u>Hardware Submodel</u>: This model is given in terms of a <u>closed queuing network of hardware resources</u> which serve the software processes. The network describes how software processes access the hardware resources of the computer system. The network is assumed to be cyclic and is hence closed.

A hybrid technique is developed to determine the combined behavior of the first and second queuing networks for a given time-dependent system stimulus. Analytical methods are used in both the first and second models, whenever appropriate. However, numerical techniques are also applied in order to facilitate the incorporation of the following features:

1) The system stimulus can consist of stepwise-constant functions of time.

2) The hardware model is not entirely predefined. The resource demands of the software processes not only depend on a fixed scheduling algorithm, but are also influenced by the structure and current state of the software model. For example, a software process which is capable of handling the received messages without significant delay does not need to compete as strongly for hardware resources as a process whose message input queue has a backlog.

3) The software processes can possess different CPU priorities.

5. THE SOFTWARE MODEL

5.1 The Queuing Network

The software model describes how messages entering the system are processed by the software. Each message received by the system is considered to be a job that requires processing by a predefined sequence of software processes. Different types of messages may initiate different chains of processing. Each software process is assumed to possess one message-input queue that receives all messages to be handled by that process. The messages in a queue are assumed to progress through the queue according to the First-In-First-Out (FIFO) discipline.

The processes and their message-passing interconnections constitute an open queuing network as illustrated in Figure 5.

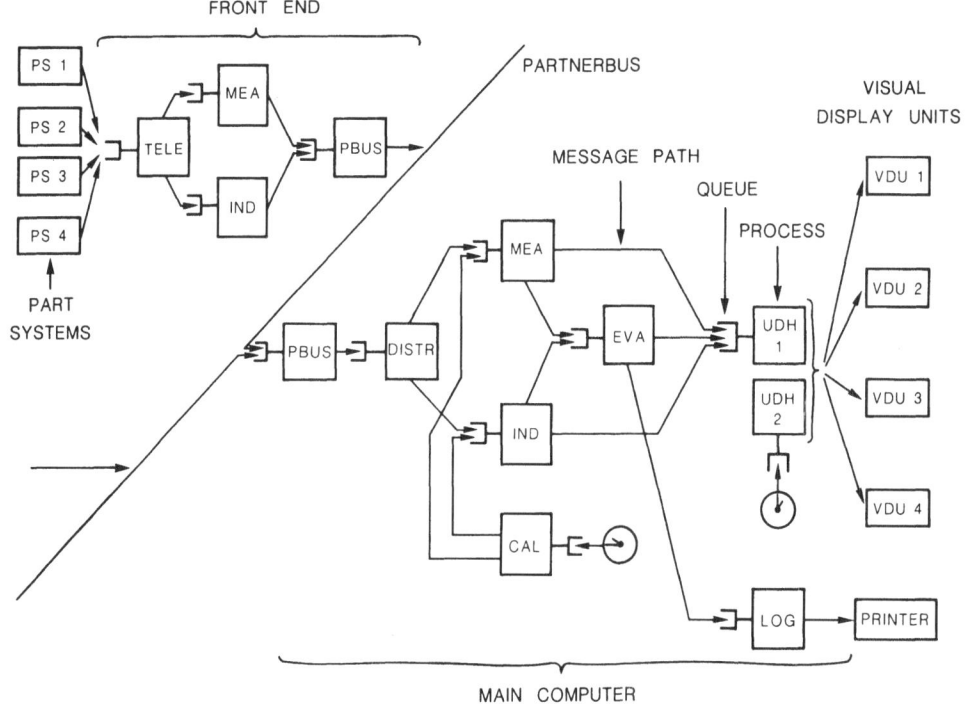

Fig. 5: Open Queuing Network of Message Queues

List of Software-Process Names

TELE	Telemetry Handler
MEA	Measurand Handler
IND	Indication Handler
PBUS	Partner Bus Driver
DISTR	Distribution Handler
CAL	Calculation Handler
EVA	Event/Alarm Handler
UDH-1	Update Handler (Event Driven)
UDH-2	Update Handler (Cyclic Driven)
LOG	Logger

The software model shown in Figure 5 illustrates the type of data processing that be-comes predominant and critical in the front ends and the main computer of a power-control center under peak-load conditions, i.e., when an abnormally large burst of messages enters the control system from the power system. The model can be used to determine the message throughput and the response time between the arrival of a stimulating message at a front end and the update of a corresponding object on the screen via the main computer.

The software processes shown in Figure 5 represent only a small, though highly relevant, subset of the entire Brown Boveri Energy COntrol Software (BECOS)[3] and turn out to be quite sufficient to describe the major resource-consumption patterns under peak-load conditions (cf. Hardware Model). The model is thus suitable for analyzing the effect of a "scarce" resource on message throughput and response time. Note, however, that the model ignores many aspects, such as the influence of power-application software, operator commands directed to the power system, communication computers, mimic-board controllers, and standby computers, all of which are not primarily important for the problem under consideration.

In Figure 5, two types of messages are assumed to arrive at the control system:

- Measurand-type messages which contain analog information (e.g., voltage or power) and are typically transmitted to the control center via a cyclic protocol,

- Indication-type messages which contain discrete information (e.g., switch or tap positions) and are typically transmitted to the control center when the state of a physical device changes.

Measurand- and indication-type messages are received by the Measurand and Indication Handlers, respectively, which update the database and trigger further processing. The Event/Alarm Handler determines how a new value of an indication or measurand is to be brought to the attention of the operator. Indication- and measurand-type data can also be created indirectly via the Calculation Handler rather than via a front end.

Processes that receive messages from other processes or from the external system are called event-driven processes, in contrast to cyclically driven processes which are triggered cyclically by internal clocks. In the model shown in Figure 5, the screens are updated by both event-driven and cyclic Update Handlers.

The arrival of external messages is defined as a stepwise continuous Fluid Flow (i.e., in terms of messages per second), rather than as a discrete flow. In the model, both the interarrival times and the service times of the messages will be considered to be deterministic. This view enables us to study the nonstationary behavior of the system in an approximate way and free us from stochastic intricacies.

Figure 6 illustrates the time behavior of system-input messages that reach the software system during an emergency situation in the controlled system. The average rate is initially relatively high but declines suddenly at times t1 and t2.

M. Vitins and K. Signer

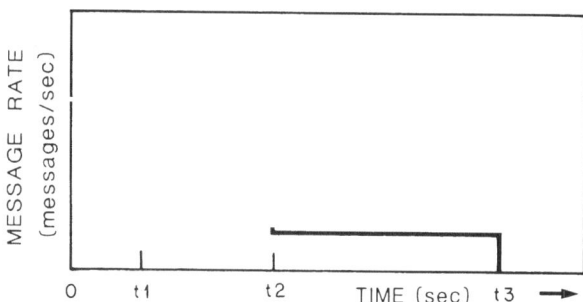

Fig. 6: Typical Burst of Messages Entering a Node from the External System

The time behavior of the queuing network is defined as a sequence of system events. In the software model, an event is defined as a change in the rates of messages being received or sent from any of the processes. During the time between two events, the message rates remain constant and the system will be said to be in a state of equilibrium.

5.2 Fluid Approximation of a Software Process

During an equilibrium interval, the behavior of a process, say $P[i]$, and its message queue can be fully characterized by the following state variables:

- The average incoming rate $z[i]$ of messages arriving at the message queue of the process

- The average input rate $x[i]$ of messages departing the queue to be handled by the process

- The queue length $q[i](ts)$ at the start of the equilibrium interval (denoted by ts)

Given the above state variables, the following dependent variables can be derived:

- The average output rate $y[i,j]$ of output port j is proportional to the average input rate $x[i]$, i.e.,

$$y[i,j] = k[i,j] \ x[i],$$

where $k[i,j]$ is a nonnegative, real-number amplification factor.

The output message(s) of the process are assumed to be generated instanta-
neously as the input message is processed. Note that the conservation principle applies to the message queues but not to the processes: Any number of
output messages may be generated upon the reception of one input message.
Thus, the sum of all $k[i,j]$'s for all j's need not be equal to 1.

- The queue length $q[i](t)$ varies linearly (for $t > t_s$ during the equilibrium
  interval) and is given by the integral of the difference between the incoming
  and input rates, i.e., $z[i] - x[i]$. In particular, if the incoming rate exceeds
  the input rate, the queue length increases, thus augmenting the backlog of
  messages in the queue.

Figure 7 illustrates graphically the state variables and derived quantities of a process.
All state variables and derived quantities are considered to be nonnegative real numbers, representing average values. This is a macroscopic model of a process, the
so-called fluid approximation, that ignores the stochastic properties of the message
arrivals and departures. On a more detailed microscopic level, the queue length can
be considered to be a nonnegative integer number, since microscopic events are
generated whenever individual messages enter or leave the queue. An important
consequence of the macroscopic point of view is that a queue remains empty when the
arrival and departure rates of that queue are equal (i.e., $z[i] = x[i]$), assuming the
queue was originally empty.

5.3 Maximum Response Time

We assume, for the moment, that the state variables of all the software processes are
known for all equilibrium intervals during the period of investigation. The state
variables will be determined in the next section, which incorporates the hardware

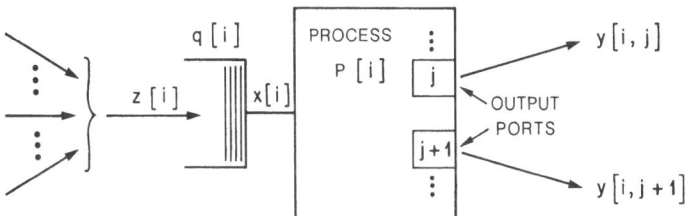

Fig. 7: Characteristic Variables of a Process P[i]

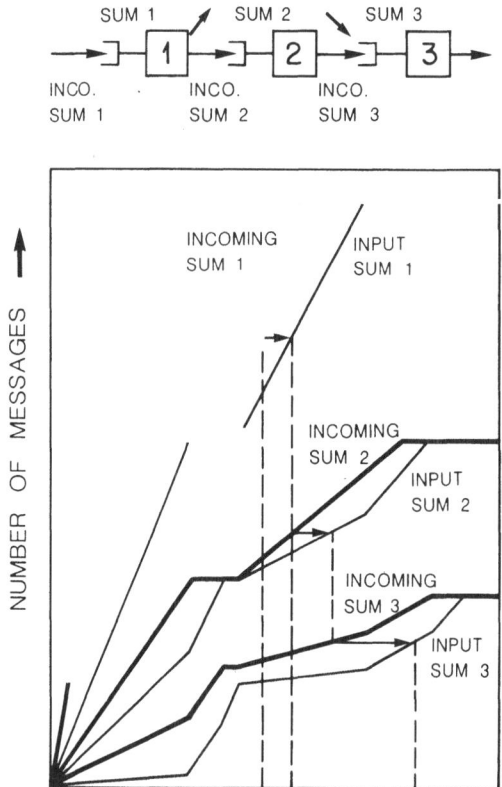

Fig. 8: Graphical Determination of Message Response Time

model. In the following, we describe a well-known[6] graphical method for determining the response time of a nonstationary queuing system. Figure 8 illustrates the technique.

The method is based on analysing the message totals. The total number of all incoming and input messages for each process can readily be determined by integrating the message rates over time, assuming that the totals are initially zero (at $t = 0$). The length of a message queue at some time t_i is given by the difference between the incoming and input totals at that particular time. Note that the response time of a particular incoming message arriving at a queue, say at time t_i, is obtained by determining the horizontal distance between t_i and the time at which the message is input to the process (denoted by t_i' in Figure 8). Clearly, the response time of a single queue can vary linearly as a function of time within an equilibrium interval. Viewed over the whole time scale, the response time for messages to progress through a given queue varies in a piecewise-linear fashion.

The response time required for a message to progress through an entire path in the network can be obtained by chaining the response times of the individual queues along that path. As soon as a message is input to a process, the process immediately generates an output message which is then deposited in the queue of the succeeding process(es). The subsequent queues are then traversed in a similar manner until, finally, the last message queue along the path has been traversed. Again, the response time for messages to traverse a given chain of queues varies in a piecewise-linear fashion. The maximum response time can thus easily be found.

6. THE HARDWARE MODEL

In the software model considered in the previous section, a system event was defined as the moment of a change in the rates of messages being received or generated by any of the software processes. System events mark a sudden change in the resource demands of one or more of the processes, an effect not modelled in the software model so far.

The purpose of the hardware model is to determine the influence of the hardware resources on the input rates of the software processes. In the absence of hardware saturation, the message-input rate of each software process is simply equal to the incoming rate. All queues have zero length and no message backlog occurs. As soon as hardware saturation occurs, however, at least one of the software processes will no longer be able to cope with the messages sent to it and the input rate of that process will drop below the incoming rate.

In order to establish a relation between the message flow and the hardware demands of a software process, we adopt the following assumptions:

- The usage of hardware resources of a process $P[i]$ is proportional to the message input rate $x[i]$.

- A process accesses the hardware resources cyclically. The message-input rate of a process is proportional to the cycle rate of that process.

The hardware model presented in this section is solved by means of the hybrid approach: the simulation technique is used to track the sequence of events and to store the state variables of the processes for each of the equilibrium intervals. An analytical model is used to determine the state variables during an equilibrium interval. After completing the simulation, the results are fed to the graphical determination of the message response times as described in the previous section.

In order to construct a mathematical model, we introduce the following distinction:

- If the message queue of a process is empty, the process is termed a <u>Noncompeting Process</u>.

- If the message queue of a process is not empty, the process is termed a <u>Competing Process</u>.

The message-input rate of a noncompeting process is equal to the message-incoming rate, and depends directly on the preceeding processes and the external stimuli (but is not directly related to the availability of the hardware resources).

Competing processes occur when there is a shortage of (at least) one of the resources. The message-input rate of a competing process is independent of the message-incoming rate and depends only on the rate at which the process is scheduled to access the hardware resources. If the incoming rate to the queue is larger than the rate input to the process, the message queue becomes longer, otherwise the queue becomes shorter.

Figure 9 illustrates the resource-consumption patterns of competing and noncompeting processes in a CPU/disk-bound system. Both competing and noncompeting processes use the hardware resources cyclically, but noncompeting processes also spend some time in a waiting state. The waiting time (called "think time" in conventional queuing models) depends on the rate at which messages are received as compared to the time spent using resources.

An iterative technique has been proposed[9] to determine the competing and noncompeting processes and to determine the cycle rates of the processes. The cycle rates can be applied to determine the message-input rates of the processes and, in particular, the time when a queue is expected to become empty (thus causing the associated process to change from the competing to the noncompeting set). The next significant event is given by either the first queue that empties or the next change in the external stimulus rate.

Note that the cycle time of a software process in the hardware model is primarily given by the resource-usage times and to a lesser extent by the times to progress through the queues. By contrast, we emphasize that the message-delay times in the software model are primarily caused by the queue-traversing times and to a negligible extent by the actual processing times.

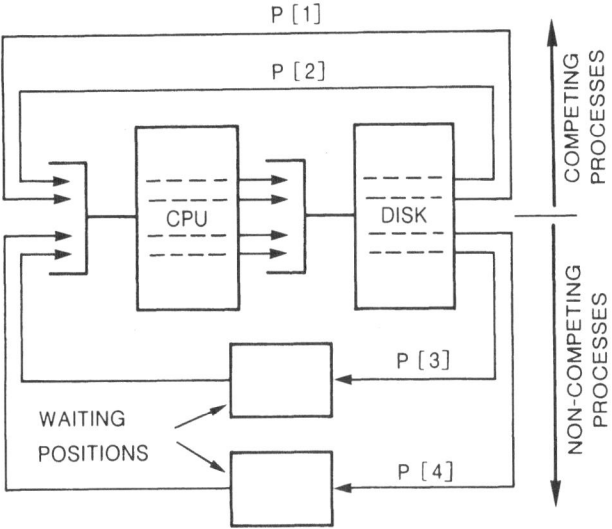

Fig. 9: The Hardware Model

7. A PERFORMANCE EVALUATION AND PREDICTION TOOL

The theory described in this paper has been implemented in the form of an interactive software tool.

7.1 User Interface

An important feature of the tool is that the user need not be concerned with algorithmic- and data-managing details of the program and performance-analysis techniques. These aspects are largely hidden from the user allowing him to concentrate on the analysis of the performance aspect.

Another important feature of the program is that the user can at any time ask the system

   - What he is dealing with (Where am I?)

   - Which commands he can enter (What can I do?)

In order to present this information to the user at all times, the screen is subdivided into four separate areas, as shown in Figure 10:

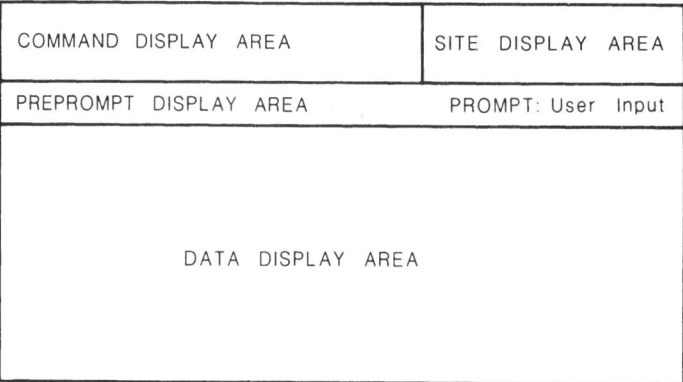

Fig. 10: The Screen Layout

- The <u>Site</u> display area, indicating the computer subsystem currently being considered (in the case of a multicomputer control system).

- The <u>Command</u> display area, indicating the top-level commands that the system accepts.

- The user <u>Interaction</u> area, consisting of <u>Preprompt</u> and <u>Prompt</u> areas. The preprompt area is used to signal user input errors and to request further input from the user if the top-level command requires additional parameters.

- The <u>Data</u> area, in which the problem data and program results are displayed.

Note that only the fourth display area is actually used for displaying performance-related information and results, while the first three areas are dedicated to displaying dialog-state information. The user interface can be viewed as being a special, simpli-fied, VT-100-oriented implementation of the XS-2 dialog system[7] whereby several general features have been discarded (such as the use of a mouse, full graphics, general tree manipulations, a hierarchy of commands, and the trail facility).

The program user has the following commands at his disposal:

- A set of performance-related commands. There is a command that computes and displays the maximum response times and queue lengths. Furthermore, the resource loads and process behavior can be displayed as a function of time.

- A set of paging commands. The user can look at the results pagewise in segments that fit onto the screen, when the program is operating in the semigraphics mode.

- A set of parameter display and modification commands. All data (except the
software structure) that is read in from the data files can be displayed on the
screen and modified interactively.

- A set of miscellaneous commands, including a help and a quit command. Furthermore, the user can flip the display from the semigraphics to the scroll
mode and vice versa. A save command causes the current values of the parameters and all the program results to be written into a user-defined file that
can be directed to a hard copy device for documentation purposes.

7.2 Implementation

The tool is written in the programming language Modula-2^{10} and executes on a VAX
11/780. The main program controls the user dialog and calls appropriate internal
procedures that must be executed upon a user request. The internal procedures are
defined within different so-called library modules. The entire program currently
consists of forty such modules. Modules have been designed for:

- Reading a general network of software processes from a data file and setting
up an internal-program data structure,

- Determining the process behavior as a function of time, using the internal data
structure and given system-input rates,

- Determining the maximum response time for a message to follow a given message path for given process behavior,

- Displaying the message rates, resource loads and response times on the screen
or on a result data file. The results can be displayed on the screen either in
the normal scroll mode or in a no-scroll (i.e., semigraphics) mode that supports tabular and bar graph output.

Each library module exports procedures and data types for use in other library
modules or for use in the main program. The major advantage of this highly decomposed program design is that program modifications can often be carried out locally
within only one module, leaving other modules unchanged.

8. CONCLUSION

A hybrid model has been developed to analyze the performance of a process-control system, such as a power-system control center. The model successfully achieves its objective of predicting the performance of system variants under realistic operating conditions. Important features of the model are the ability to vary the intensity of message arrivals at the control system, ranging from steady state to large transient bursts of messages, and to predict the effect of a possible temporary saturation of system resources.

The model combines a network of software processes with a network of hardware resources to produce an easily understandable overall model of the system. Due to the computational efficiency of the model, it can be used as the basis for an interactive decision-making support tool for addressing performance issues during system planning, design and tuning. An interactive tool has been developed that provides a comprehensive set of user commands and performance-related output.

REFERENCES

1.   Foxley, E., and Ali, M., "Hybrid Performance Modelling Method for Virtual Environments," Computer Performance, Vol. 5, No. 3, September 1984, pp. 159-168.

2.   Gomaa, H., "A Hybrid Simulation/Regression Modelling Approach for Evaluating Multi-Programming Computer Systems," in Computer Performance, K.M. Chandy and M. Reiser, eds., North-Holland Physics Publishing, Amsterdam, August 1977, pp. 209-234.

3.   Goudie, D.B., Davis, M., and Spatz A., "The BECONTROL Family of Supervisory Network Control Systems," Brown Boveri Review, Vol. 71, September/October 1984, pp. 406-415.

4.   Kimbleton, S., "A Heuristic Approach to Computer Systems Performance Improvement I - A Fast Performance Prediction Tool," Proc. AFIPS 1975 NCC, Vol. 44, AFIPS Press Montvale, N.J., 1975, pp. 839-846.

5.   Lasser, D., "Productivity of Multiprogrammed Computers - Progress in Developing an Analytic Prediction Method," Commun. ACM, Vol. 12, No. 12, December 1969, pp. 678-684.

6.   Newell, G.F., Applications of Queuing Theory, 2nd ed., Chapman and Hall, London, 1982.

7.   Nievergelt, J., and Weydert, J., "Sites, Modes, and Trails: Telling the User of an Interactive System Where He Is, What He Can Do, and How to Get to Places," Technical Reports, Institute of Informatics, Federal Institute of Technology, Zurich, Switzerland, No. 28, January 1979.

8.   Schwetman, H., "Hybrid Simulation Models of Computer Systems," Commun. ACM, Vol. 21, No. 9, September 1978, pp. 205-209.

9.   Vitins, M., "A Performance Model for Process Control Systems," <u>Proc. Interna-</u>
     <u>tional Conference Modelling Techniques and Tools for Performance Analysis</u>,   5-7
     June 1985, France, pp. 245-261.

10. Wirth, N., <u>Programming in Modula-2</u>, Springer-Verlag, Berlin 1982.

DISCUSSION
Chairman: R. Karg (Brown Boveri, Mannheim, Germany)

J. Magee (Imperial College, London, England)

Have you validated the model in practice? What were the results when you compared the behavior of actual systems with values the model predicted?

M. Vitins

All the parameters you need in this model have in fact been taken from measurements on the control system. When the measured parameters were put into the model, it showed the expected overall behavior. It reflected what the designers of the control system expected of the model. In practice it is very difficult to create a stimulus of the complex type we assume in the model and to measure the response times and the queue lengths in the real system. It has, nevertheless, been our experience that the model is sufficiently accurate and well suited for analyzing the overall effect of system variations on performance.

T. Lalive d'Epinay (Brown Boveri, Baden, Switzerland)

How flexible is your model when you change the basic layout of the control system? When you change to a different application, what would be the effort to change the model?

M. Vitins

The network structure of the software processes within each of the computer nodes is read from a file. Hence, changing the application means just changing an input file for the program.

At this point I would like to add that the model can be applied to any network of software processes that has no cycles in it. The presence of a cycle means that there could be a message cycling around between the software processes, impeding the throughput of the entire system. In the case of real-time applications such cycles are not important under worst case operating conditions.

R. Karg

You mentioned that you assume a deterministic disturbance of the control system. How do you deal with a stochastic stimulus?

M. Vitins

Most of the queuing theory in the literature is concerned with the problem of queues that occur because of the stochastic nature of arrivals. In industrial applications, performance specifications are generally defined in a deterministic fashion from the very outset. The negotiations between the customer and the vendor lead to specifications of maximum response times and maximum throughput and resource consumptions for predefined deterministic scenarios. Average values for response times and queue lengths based on stochastic arrivals are much less important. Thus, the absence of stochastics in the model is not really a drawback, but rathermore an advantage, since the thinking of the people that are involved is based on deterministic assumptions.

SESSION II
HUMAN-COMPUTER INTERACTION
Chairman: H. Sandmayr

DIALOG DESIGN: PRINCIPLES AND EXPERIMENTS

J. NIEVERGELT

Federal Institute of Technology, Zurich, Switzerland

C. MULLER and H. SUGAYA

Brown Boveri, Baden, Switzerland

ABSTRACT

Programming methodology has emphasized different criteria for judging the quality of software during the three decades of its existence. In the early days of scarce computing resources the most important aspect of a program was its functionality - what it can do, and how efficiently it works. In the second phase of structured programming the realization that programs have a long life and need continuous adaptation to changing demands led to the conclusion that functionality is not enough - a program must be understandable, so that its continued utility is not tied to its inventor. Today, in the age of interactive computer usage and the growing number of casual users, we begin to realize that functionality and good structure are also not enough - good behavior towards the user is just as important. Interactive use of computers by users with a wide variety of backgrounds, in different applications, is ever increasing.

Novel styles of interfaces that exploit bitmap graphics have emerged after a long history of interactive command languages based on alphanumeric terminals. This development is repeating, with a twenty-year delay, the history of programming languages: large collections of unrelated operations are being replaced by systematic structures that satisfy general principles of consistency. We do not attempt to give a comprehensive or balanced survey of the many approaches to human-computer interface design being investigated today. We summarize a few general issues that designers of human-computer interfaces encounter, and illustrate them by means of examples from our research activities, some of them performed jointly with industry.

A survey and classification of design errors common in today's command languages leads to concepts that any human-computer interface should provide. In particular, answers are needed to the following questions that characterize common difficulties experienced by users: <u>Where am I</u>? <u>What can I do here</u>? <u>How did I get here</u>? <u>Where else can I go and how do I get there</u>? Such questions reveal the fact that the fundamental problem of human-computer communication is how to present to the user what he cannot see directly, namely, the state of the system, in such a manner that he can comprehend it at a glance. If the output language of a system, what the user sees or can get onto the screen easily, is rich and well structured, the command language proper, or input language, can be simple - ideally the minimum set of selecting the active data, selecting the active operation, and saying "do it", all by just pointing at the proper places on the screen. The <u>Sites</u>, <u>Modes</u> and <u>Trails</u> model of interaction answers the questions above by displaying to the user on demand his current data environment (site), his current command environment (mode) and the past dialog (trail).

This model has been implemented in several experimental interactive systems: we describe XS-2, a comprehensive operating system that supports application-independent dialog over graphics workstations, and EASY, a reduced, portable version that makes minimal assumptions about the system on which it runs. Several applications have been developed to run under these dialog packages, and the interaction realized on them has influenced the style of dialog programmed into applications on conventional operating systems.

Two experiments use Prolog as a tool for rapid prototyping of interactive applications - the ease with which Prolog programs can be modified is the main attraction of this style of programming. First, a general-purpose command interpreter is described which simulates the dialog behavior of arbitrary command languages described by Prolog rules and facts. Second, an experiment in system-driven deductive user guidance is reported.

Although most of the problems of human-computer interaction are software problems, the search for better hardware tools is also rewarding. Whereas the commercial development of mice has gone into the direction of devices that provide only one or two buttons to the user, we describe a prototype "mighty mouse" designed to make optimal use of the user's dexterity.

The behavior of interactive programs will improve when the current generation of commercial software has outgrown its usefulness, and the next generation of programmers is free to make a fresh start. Powerful tools for automating the syntactic aspect of programming human-computer interfaces are becoming available and will improve the consistency of dialogs.

1. DESIGN OF HUMAN-COMPUTER INTERFACES:
THE MEDIA ASPECT OF SOFTWARE ENGINEERING

During the three or four decades of its existence, the craft of programming, dignified by the name of software engineering during the sixties, has emphasized entirely different characteristics of a program.

In the fifties and early sixties the functionality of a program was the key issue: What your program can do for you. What problems can it solve, at what cost in time, memory and dollars? Programmers were proud of their optimization skills, and the field of algorithms and data structures emerged in response to this concern for efficiency.

During the sixties the data-processing and computer-science community became aware that commercial programs have a long life and need to be "maintained", i.e., adapted to changing hardware and specifications. They had to be passed from one programmer to another, thus their legibility, as determined by the structure of the source text, became the key issue. How does the program look to a programmer? The discipline of structured programming emerged to guarantee professional software standards.

During the eighties the spread of personal computers and workstations has changed the meaning of the term "computer user": no longer does it denote a programmer, but rather an end-user of an interactive application program. He never sees a source program, but he experiences its behavior. How your program behaves towards the user has become the issue that determines not only productivity, but whether a program will be used at all. Human factors is the discipline many people hope will do for program behavior what structured programming did for program legibility, and what algorithms and data structures did for program efficiency.

This hope may be misplaced. Program efficiency has easily measured yardsticks of time and memory, and the field of algorithms and data structures called upon finely honed tools of programming lore and mathematics. Program structure, a less easily quantified aspect, yielded in part to established organization tools such as hierarchic decomposition, and in part to formal logic to express rigorous specifications, but remained partly a matter of taste and never developed as objective a theory as did algorithms and data structures. Program behavior towards the user is a softer field still. We look to psychology, in particular cognitive science and the technology of controlled experiments to provide a scientific backbone to what is largely a subjective judgment of user preference.

The computer-driven graphics screen is a medium for communication - the only two-way mass communications medium we have. By looking at how the skillful use of other media is taught we will discover what we may realistically expect, and what not, in terms of guidelines for the programmer of human-computer dialogs: A catalog of frequent errors to avoid, some important concepts to understand, and a few rules of thumb the observance of which avoids the most obvious errors. We discover these exemplary errors, concepts and principles not by deduction from any axioms, but the hard way: by trial, error and observation with many systems and users.

In this paper we summarize the experience our research group has gathered over a decade of experiments with implementing interactive systems. We present examples of a half-dozen projects and attempt to draw lessons from their strong and weak points. We cannot attempt to cover the activities of the many large centers working in the field of interactive systems and human factors, but we believe the projects presented below are typical of the issues that arise in the design of human-computer interfaces anywhere.

2. PROBLEMS, CONCEPTS, PRINCIPLES, AND EXPERIMENTS

The User's Questions: It is useful to begin the study of human-computer interfaces by analyzing common errors in today's systems, but here we will only refer to literature on this topic.[6,11] Observation of novice users without preconceived notions provides particularly valuable insight into the fundamental design question of how a machine should present itself. Most of the recurring difficulties they encounter are characterized well by the questions:[8]

- Where am I? (when the screen looks different from what he expected)
- What can I do here? (when he is unsure about what commands are active)
- How did I get here? (when he suspects having pressed some wrong keys)
- Where else can I go and how do I get there? (when he wants to explore the system's capabilities)

We are beginning to learn that the logical design of an interactive system must allow the user to obtain convenient answers to the questions above at all times. In other words, the human-computer interface must include queries about the state of the system (without changing this state), about the history of the dialog, and about possible futures. This principle is much more important for today's computerized machines, black boxes that show the user only as much about their inner working as the programmer decided to show, than it is for mechanical machines of the previous generation, which by visible parts, motion and noise continuously give the user a lot of state information.

Concepts: In order to answer the user's basic questions in a systematic way, the designer of an interactive system must also design a simple user's model of the system: a structure that explains the main concepts the user needs to understand and relates them to each other. The major concepts will certainly include many from the following list: the types of objects that the user has to deal with, such as files, records, pictures, lines, and characters; referencing mechanisms, such as naming or pointing; organizational structures used to relate objects to each other, such as sets, sequences, hierarchies, and networks; operations available on various objects; views defined on various types of objects, such as formatted or unformatted text; commands used to invoke operations; the mapping of logical commands onto physical I/O devices; the past dialog, how to store, edit and replay it; the future dialog, or help facility.

The user's model of the system is a state-machine: it has an internal state and an input-output behavior. Components of the state must include:

- The user's data environment (data currently accessible)
- The user's command environment (commands currently active)

The questions Where am I? and What can I do here? are then answered by displaying the current data and command environments, respectively. In our experimental systems these concepts have been incorporated into a Sites, Modes and Trails model of interaction, in which the user can invoke this information at any time in system display windows, regardless of which application program is running.

Principles: Most considerations about man-machine dialogs relate to one of two fundamental concepts: the state of the system and the input/output behavior. The state is concerned with everything that influences the system's reaction to user inputs. The input/output behavior is concerned with interaction: what the user sees on the screen, what he hears, how he inputs commands on the keyboard, mouse or joy stick. Today's fad is to engineer sophisticated-appearing interfaces: fast animation in color, sound output and voice input. This is desirable for some applications, unnecessary for others, but in any case it does not attack the main problem of man-machine communication.

> The fundamental problem of man-machine communication is how to present to the user what he cannot see directly, namely the state of the system, in such a manner that he can comprehend it at a glance.

Insufficient state information is responsible for most of the difficulties users encounter. Imagine that you leave your terminal in the middle of an editing session because of an urgent phone call. When you return ten minutes later, the screen looks exactly as you left it. Even if the system state is unchanged, you may well be unable to resume work where you left off:

D: "What file was I editing?", "Is there anything useful in the text buffer?";
C: "Am I now in search mode?", "What is the syntax of the FIND command?";
T: "Has this file been updated on disk?", "Has it been compiled?".

Such questions indicate that part of the state information necessary to operate this system has to be kept in the user's short-term memory. When the latter is erased by a minor distraction, the user needs to query the system to determine its state. Hardly any system lets him do this systematically at all times. Today's systems provide state information sporadically, whenever the programmer happened to think about it. Examples abound. The file directory can be seen in the file server mode, but not in the editor; in order to inspect it you must exit from the editor, an operation that may have irreversible consequences. In order to see the content of the text buffer you may have to insert it into the main text. To find out whether you are in search mode you may have to press a few keys and observe the system's response - not always a harmless experiment. A designer who follows the following principle avoids all the problems above:

> The user must be able at all times to conveniently determine the entire state of the system, without changing this state.

Realizing this principle has two consequences. First, it implies that the system designer must clearly distinguish inspecting the state from operating on the state. He must define a separate class of viewing commands for displaying representations of selected state components onto the screen, which do not change the user's active data and command environment. Second, there must be universal commands that are active at all times and processed directly by the operating system, not by any application program that may be running at the time. At least the viewing or state inquiry commands postulated above must be universal, but general dialog control needed in every interactive utility are also best made universal. By taking them out of the text editor, the diagram editor, the database query language and incorporating them as universal commands into the system, the user obtains the impression that all these utilities "talk the same language".

Experiments: The craft of designing human-computer interfaces is highly experimental and experiments in systems development are costly - prototype development and testing tends to take a few years, and if a project is not brought to completion, then there is nothing to show and the effort is wasted. What kind of experiments can a university research group, with modest resources and students who are available for a limited length of time, undertake to test its hypotheses and hope to contribute to the state of the art?

We have solved this goal-setting problem by focusing on a few general research directions that have inspired us over a decade and by identifying within these trends autonomous projects that can be realized, used, and judged on their own. Such a subproject is often championed by a Ph.D. student who stays with it for a couple of years, who can draw on results produced by his predecessors and who is supported by younger students who work on term projects or Master's theses. If things go well, the software produced by a subproject combines with other autonomous components to produce a large package, if not, it provides a lesson in software (in)compatibility. An important characteristic we look for in defining a research project is the possibility to find a user community to test the product. This we do by producing software that can be used for instruction at the university, or by seeking cooperation with industry, as was the case for several of the projects to be described.

One of the broad guidelines that has guided our selection of projects is the concept of how a machine should present itself to the user, as embodied in the Sites, Modes and Trails model of interaction described above. Another guideline is that hardly any of the software in use today fully exploits the power of interaction, and that the difficult question of which tasks should be programmed and which ones should be left to people is rarely given adequate attention. We are not in agreement with much of the work that sails under the flag of artificial intelligence, because it often aims at complete automation of a task. For us an automated expert system is not nearly as useful a paradigm for computer use as that of a human expert aided by an interactive tool.

We now survey a half dozen experiments we have been involved in over the past few years, and try to extract some lessons from them. Section 3 describes the creation and use of teaching and demonstration programs we have used at the Federal Institute of Technology in Zurich. Section 4 presents experimental integrated systems based on the Sites, Modes and Trails model of interaction and an industrial application. Since the system is complex, the application needs to be embedded in it. Section 5 describes the opposite approach: a portable man-machine interface that adapts to the application, in this case illustrated by a deductive database package. Sections 6 and 7 provide examples of the uses of Prolog: for rapid prototyping of command languages and for user guidance in a knowledge-based system. Section 8 presents an attempt to standardize dialog-control functions in hardware.

3. INTERACTIVE DEMONSTRATION PROGRAMS FOR EDUCATION

Small computers are being increasingly used as communications media: As stand-alone devices that replace a video projector or an automatic slide show, or supporting a speaker as does an overhead projector. Our experience is primarily with educational uses of such demonstrations: simulations such as animated displays of algorithms in execution, and tutorial programs that include exercises. Commercially published courseware is not yet available, so the only way for a teacher to use the computer-driven graphics screen as a medium is to write his own teaching programs.

An interested teacher can learn to program an animation in his subject matter, say, the display of curves in calculus. However, few teachers today have the interest or ability to program the dialog-control part of an instructional program well - resulting in courseware that crashes on unanticipated inputs, where the user can get stuck in dead ends. CAI author languages facilitate courseware production by noncomputer specialists, but they are usually too restrictive to allow animation - still under the spell of the programmed instruction movement of the fifties, they tend to force all interaction into the straightjacket of multiple-choice questions.

An approach to courseware production has been described recently[10] which separates subject matter content, whose presentation is left unrestricted, from dialog control, which is standardized to the extent that it can be automated. An interactive program presents itself as a network of dynamic pages. At any moment the user is at a page (where am I?) activating page-specific commands (what can I do here?). A page is called dynamic because, unlike a book page, animation may evolve; but the page does not fade or scroll away, the user remains at this page until he issues a motion command that brings him to another page. The motion commands are standardized to match the structure of the class of dialog networks. We have chosen series-parallel-repetition networks as a class that allows exploration of the network by means of the set of motion commands next, select, skip and repeat or back, as shown in Figure 1.[16] A program generator translates a series-parallel-repetition network, presented as a short expression in a formal notation, automatically into a corresponding frame program shown at right. This frame program can be executed to test its dialog behavior, with pages that announce their name but have no other content. The author/programmer now fills in the dummy procedure generated for each page with code specific to the desired content of the page. Thus he programs the creative component of his demonstration program, to the best of his ability, but the routine and error-prone dialog-control component has been automated away.

4. INTEGRATED INTERACTIVE SYSTEMS

4.1 XS-2: Sites, Modes and Trails Laid Out on a Screen

In order to test the practicality of the Sites, Modes and Trails model of interaction, we have developed two operating systems that support human-computer dialog, the eXperimental Systems XS-1[2] and XS-2.[4,14] These systems are intended as experiments to test the extent to which unrelated interactive applications programs can be in-

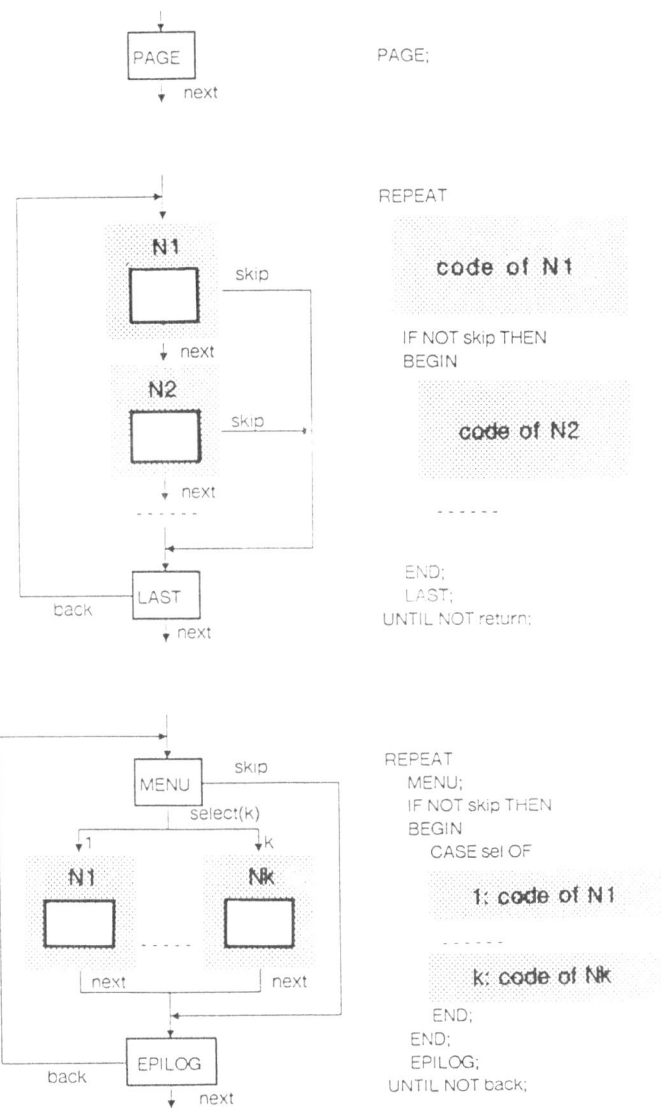

Fig. 1: Series-Parallel Dialog Networks and Corresponding Frame Programs

tegrated into the operating system, in the sense that they all share the same data access and dialog-control facilities. In particular, we were interested to learn what benefits and constraints the applications programmer can expect when working with a system that supports design and execution of application-independent dialog control. Such integrated interactive systems have since become commercially available for various personal computers and graphics workstations. They all impose a common model for interaction on all application programs, often based on window management. We believe XS-2 goes further than other systems in the systematic identification of application-independent dialog control.

The Model of Interaction: The screen shown in Figure 2 illustrates the interaction model embedded in XS-2. The application window is at the bottom of the screen, in this example executing a graphics editor being used to produce slides. The remaining part of the screen is allocated to three system windows. In the middle of the screen we see the sites window at left, a directory of all the data in the system with the active data (the current site) highlighted, and the command or modes window at right, with the active command highlighted. The trails window is at the top, showing from left to right the history of data activated, application commands and universal commands issued. The pop-up menu of universal commands is shown in the figure as having been activated in the commands window.

System State: The state of XS-2 is visible at all times in the sites and modes windows. A site is a tool for structuring data and a command for structuring application functions. They are organized into the same tree structures. Each site and command has a name and is classified by its type. A site represents the data attached to it and relates it to its neighbors, e.g., sons and brothers (accessible data set). Similarly, a command represents the function of the application attached to it and relates it to its neighbors (accessible command set). A point of interest in a dialog is set by selecting a site (currently active data), selecting a command (currently active command), and executing the selected function. The history of site-command pairs forms a trail.

Universal Commands: Because the universal commands are defined as motions and operations on a tree, they have a well-defined meaning in each of the system windows. The "universal commands" serve to explore the system state and facilitate learning the system.[5] The viewing command Displace changes the appearance of the screen without changing the state, whereas the motion commands Select or ↑ Return change the current site or the current command and thus activate new data or commands.

Command Language: The behavior of an interactive program is described in terms of the sequence of user actions that can be accepted during program execution. The XS-2 command language[14] is used to specify command structures of interactive programs in the form of a tree based on regular expressions. The command tree is used for input (selection of a node) as well as output (prompting).

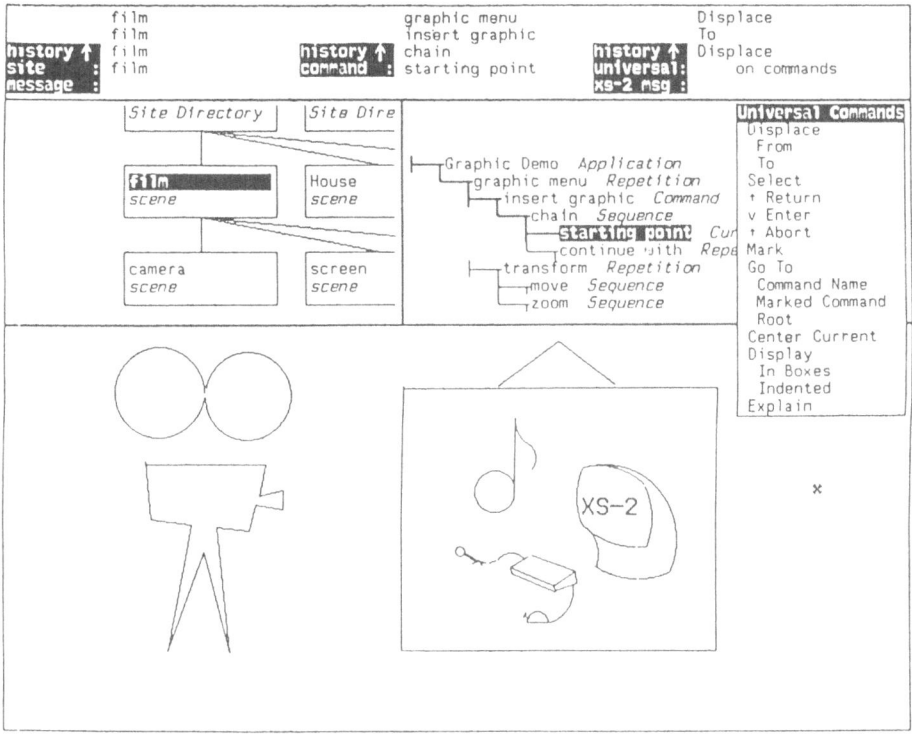

Fig. 2:   The XS-2 Screen with Application and System Windows, and the Pop-up
           Menu of Universal Commands

4.2. An Example of Rapid Prototyping in XS-2:
The Performance-Model BECOS

The BECOS simulation program is a performance model of a distributed process-moni-
toring and control system for electrical energy.[17] The hierarchical system architecture
of BECOS is parameterized through a small set of configuration rules: a main computer
is connected to several front-end systems, which in turn are interfaced to the real
processes. A cost-effective solution can be found by choosing a minimal configuration
depending on certain system loads. The BECOS program had been implemented in
Modula-2 on a DEC VAX-11 computer using an alphanumeric terminal. It was rewritten
to run as an interactive application program under XS-2 in order to test the feasibility of integrating a sizable industrial application program into the XS-2 user interface.

Rapid Prototyping: Developing an application under XS-2 involves three phases:
designing a command tree using the syntax-driven tree editor; testing this command

language with the aid of an automatically generated skeletal action module; and programming the action module in Modula-2. The first two phases are repeated until the
command language specification becomes final. The syntax of commands is displayed on
the screen in the form of a tree, as shown in the following sample language with
three commands: show the parameters of BECOS, set new parameter values, and
compute the performance of this configuration.

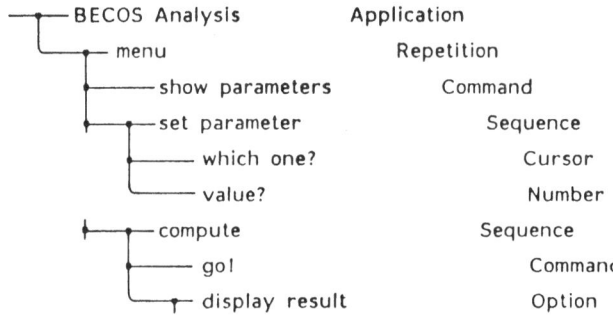

In the third phase of application development the commands need to be tied to action
modules that implement the intended application functions. The application service of
XS-2 provides a first assistance by automatically generating a dummy prototype application module based on the command tree. This executable Modula-2 program must now
be filled in with action specific code, but the definition and activation of commands
is not part of the application code, it is handled by the XS-2 kernel. This presents a
striking contrast to conventional interactive programs, where the command handling is
interspersed with the code for execution of the functions of the application.

Experience: The designer of BECOS and an XS-2 expert spent several hours of prototyping to finalize the representation of the configuration rules and the command
tree. The existing BECOS application program consisting of about ten Modula-2 modules was then adapted to the XS-2 interface in about a week. As a result of XS-2
handling the user dialog, the size of BECOS was reduced by one third while improving the user interface to allow flexible access to all parts of the model: to the configuration components and to the commands, which was impossible in the previous
version. Since XS-2 is much larger than BECOS and thus constitutes a significant
overhead for running this single application, a simplified user interface based on
XS-2 notions but specific to BECOS was developed.[16] Thus the integrated software-
tools environment XS-2 served as a development system that shaped the user behavior
of a new version of BECOS.

5. A PORTABLE HUMAN-COMPUTER INTERFACE TO THE
SMART DATA INTERACTION PACKAGE

User interfaces are mostly developed on a case-by-case basis, in that each application program has its own independent user interface, partly because of a lack of general-purpose and machine-independent user interfaces. EASY is a user-interface package that can be used by application programs to simplify, standardize and improve their user interface. It is application and machine independent and provides support for many common types of input, generalized as commands and parameters and for many common output requirements, such as screen refresh. Existing programs can be adapted to use EASY with little effort, since EASY implements most user interface requirements. New programs can best be designed with EASY as a user interface, thus relieving the implementor of this otherwise nontrivial task.

EASY is in many ways related to the interactive systems XS-1 and XS-2. Just as with these systems, EASY divides the screen into several nonoverlapping windows, one of which is reserved for exclusive use by the application. EASY displays the state of the system in the form of trees, and displays one tree in each system window (as distinguished from the application window). The site tree, in the site window, displays the data available to the user as well as to the application. The site window is always visible to the user, but can be shrunk or enlarged any time if a larger application window is desired or if a larger part of the data environment should be displayed. A number of universal commands is defined in the site window: these commands are generally classified into viewing commands, which affect the status display, selection commands, which affect the data available to the application, and editing commands, which are used to actually change the available data. The site window constantly reminds the user of where he is, i.e., of the current data environment.

Typically, a user of an application with an EASY front end proceeds by first selecting a site as a data environment, then entering all the parameters for a given command and finally selecting the command itself. In general, for different applications, most of the action will be in either the site or the operation window, rather than in both; in either case, all universal commands are available in both windows.

The application calls EASY whenever it is ready to accept a command, EASY returns control every time a command or a new data environment is selected. Otherwise, control rests with EASY, i.e., with the user. This arrangement ensures that EASY and the application do not interfere with each other and helps keep the interface between the two as simple as possible. Control returns to the application whenever the user enters a command, so EASY always returns a record identifying the command entered by the user and the values of all the parameters that belong with that command, whether strings, numbers, options (menus) or site-tree identifications (files).

The application need not concern itself with how the parameters were entered, and EASY does not concern itself with the actions appropriate to each command.

EASY is the man-machine interface of the Smart Data Interaction package,[3] a general-purpose interactive data manipulation package which integrates the grid file data structure that provides multikey access and Prolog as a deductive query and manipulation language. EASY allows the end user to communicate with the Prolog interpreter by means of data- and command trees, and Prolog provides full access to the features of the Grid File System. User interaction with EASY-Prolog is substantially different from user interaction with a conventional Prolog interpreter. The site tree of EASY-Prolog is available to the end user for storing Prolog facts, rules and queries as subtrees, so that they can be edited using EASY's built-in tree editor. A database or query stored in a subtree has to respect the following scheme:

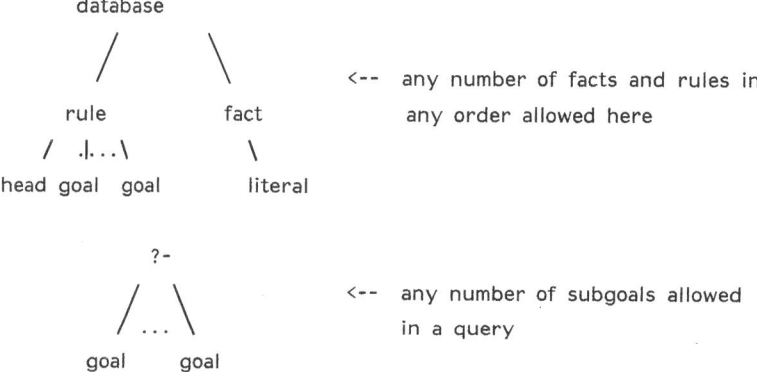

Head, goal and literal are Prolog literals represented as trees. The following tree (indented) shows an extract of a Prolog database that contains facts of the form city (City, Country), followed by two queries over this database:

```
* Site Root
    database
      fact
        city
          zurich
          switzerland
      ...
      fact
        city
          rome
          italy
```

```
queries                      <-- structural nodes allowed anywhere
  ?-
    city
      City                   /* ?- city(City,switzerland). */
      switzerland
  ?-
    city
      City
      Country
      write(City)            /* ?- city(City,Country), write(City),
      write(' in ')                write(' in '), write(Country), nl, fail. */
      Write(Country)
      nl
      fail
```

Prolog itself does not directly access the tree: when the user indicates that a subtree should be used for a query or added to the database, EASY-Prolog parses it, builds a corresponding Prolog term and calls the appropriate procedure from Prolog. Having a Prolog database accessible in structured form and being able to access queries as data, instead of having to type them in every time, represents a considerable improvement over conventional Prolog implementations.

The machine- and application-independent user interface EASY brings substantial advantages to both users and application programmers. The user is in full control of the data and command input, can refer to the system state at all times, and thus answer the questions Where, When, How? The application need not know about any of the user's specific actions, instead it obtains its inputs in a compact and well-structured form, and can call the command interpreter whenever it is ready to execute a new command. The machine-independence of EASY facilitates porting applications to machines other than the one on which they were developed and shows that a good user interface does not require elaborate hardware. The hardware requirements of EASY are limited to a video terminal or screen with character-positioning capability and a keyboard or pointing device capable of moving a cursor on this screen (arrow or function keys are satisfactory).

6. PROLOG AS A TOOL FOR THE EXPERT

The first step in the design of an interactive application is often the formulation of a model that describes the communication of the end user with the application. At this early stage of the design, the model of interaction is only present in a vague form in the designer's mind. The rapid realization of a prototype demands a flexible tool that supports the formalization of the model and allows an incremental design with executable intermediate results.

We used Prolog to define a general-purpose command interpreter to simulate the dialog behavior of an arbitrary command language. The interpreter is driven by a set of Prolog facts which specify the command language and can be modified interactively at any time. The fact that the Prolog parser can be user-tailored by defining new operators makes it easy to design a formalism in which the command language can be expressed. As a case study we chose a language similar to the command language of the UCSD p-System, based on menus, repetitions of commands, selections of alternatives and sequences of parameter specifications. The syntax for this specific command language is as follows:

```
command     = name ':' type actionlist.
name        = <Prolog literal>.
type        = 'rep' | 'seq' | 'alt' | 'exec'.
actionlist  = '[' action {',' action} ']'.
action      = command.
```

The following declarations tell the parser that seq is a unary operator with precedence 150, that ':' is a binary operator with precedence 175, etc., whereafter the parser accepts the command language as normal Prolog expressions:

```
?- op(15Ø, fx, seq).    ?- op(15Ø, fx, alt).    ?- op(175, xfx, ':').
?- op(15Ø, fx, rep).    ?- op(15Ø, fx, exec).
```

Command definitions in Prolog concisely express the intended meaning: for example, at the top level of the UCSD p-System the user chooses repeatedly (rep) among filer, editor, compiler, execute.

```
ucsd            : rep   [filer,editor,compiler,execute].
filer           : rep   [transfer,change,volumes,date,quit].
change          : seq   [param(from), param(to)].
param(PROMPT)   : exec  [write(PROMPT), write(' : '), read(PARVALUE)].
        ...                         ...
```

The interpreter for command languages of this kind consists of about 20 Prolog clauses (facts and rules) and can easily be adapted to more general command languages. New concepts can be introduced incrementally and the resulting Prolog program represents, at each stage of the design, an executable prototype simulating the dialog behavior of the underlying model of interaction. The pure simulation can be turned into a functional prototype of an interactive application by combining the application-independent command interpreter with application-specific Prolog predicates, as it has been realized in the 'exec' command above.

The interactive capabilities of Prolog and the flexibility of the parser make Prolog particularly interesting in the field of rapid prototyping. Due to the declarative character of Prolog, the resulting programs can be considered as executable specifications of the prototyped application.

7. USER GUIDANCE: USER-DRIVEN EXPLORATION
VERSUS SYSTEM-DRIVEN DEDUCTION

User-Driven Exploration in XS-2: The user guidance offered by the concept of trail in XS-2 is of a purely syntactical nature. The answer to the question How did I get here? is obtained by the user inspecting a graphical representation of the stored trail. Where else can I go and how do I get there? is answered by extrapolating the trail and displaying a menu of all legal next steps. It is worth considering how this user-driven guidance could be made more effective. The past dialog can be used as a source of knowledge which may help the system to understand a present dialog, providing an interpretation to incomplete or erroneous commands, as provided by the DWIM (Do What I Mean) facility in the InterLisp programming environment.[15] A future dialog can be made available by using stored trails for common problem-solving tasks as command macros.

Experience with XS-2 shows that user-driven guidance in a complex interactive system is suitable for expert users, but not for beginners. The main difficulty of providing useful user guidance in a user-driven dialog stems from the fact that the inferential knowledge for problem solving lies in the brain of the user, not in the computer to be used.

System-Driven Deduction: A knowledge-based system may contain a network of solution steps for problem solving, in the form of an and-or graph, represented in a form suitable for automatic inference. The system's inference engine can then make decisions based on factual knowledge and deduced by matching user inputs against the solution network. We have developed a rule-based expert system and used it to investigate user guidance in knowledge-based systems by providing an explanation facility that answers "why" and "how" questions.

In this experimental system the solution network is constructed by production rules of the form "if Premise then Conclusion". When used for diagnostic applications, the production rule can be interpreted as "if Causes then Effect". Multiple possible effects for the same cause can be expressed in separate production rules. The production rule corresponds to a node (conclusion) and links to it (premises) in the inference network. A solution (possible causes) can be found by proving a list of hypotheses (possible effects).

As an example, a library classification scheme might contain associations such as: the joint presence of the two index terms window and mouse imply (and-node) that the term human-computer interaction also applies to the document in question. Another and-node may combine this newly deduced term human-computer interaction with another term (original fact or deduced), such as menu to yield the conclusion that the term command language also applies.

In such a deductive system the user is automatically guided by the system, which asks him questions to acquire needed factual knowledge (query-the-user[13]). Deduced facts are stored in the inference network as a history of reasoning. Reasoning steps and deduced facts can be used for explanation. The system's answer to the user question "Why do you ask me this?" may be "Because your answer will help me determine whether rule A is applicable or not."

8. THE "MIGHTY MOUSE": FREQUENT DIALOG CONTROL COMMANDS AND THEIR REALIZATION IN HARDWARE

In cooperation with the Yokosuka Electrical Communication Laboratory of NTT we have designed a five-key mouse[12] and are experimenting with it to determine its suitability for standardizing dialog control on different personal computers.[1] The motivation for developing this "mighty mouse" requires explanation, since it points in the opposite direction of commercial "mouse architecture". The mouse popularized by the Xerox Alto computer had three keys, but its successor on the commercial Xerox Star has only two and the mouse of the Apple Macintosh has a single key, according to the "theory" that the user is confused by many keys. However, the software developed for these computers tells a different story: often, different logical commands are multiplexed onto these one or two keys and coded by such gimmicks as a "fast double click" (different from two slow clicks), or a click modified by a key simultaneously depressed on the keyboard. These examples lead to the conclusion that it is not the number of keys that confuses the user, but rather the changing semantics assigned to the keys! If a physical input (key, button, etc.) is always assigned the same function, then we can remember the meaning of more than one such device.

The idea is to freeze a half dozen of the most important dialog-control commands: universal commands interpreted by the operating system rather than by the application program, into hardware, in particular the mouse. These commands include motion and selection (changing data and command environment), viewing (control over how

much of an object is to be seen), and operations on the history (undo). With such a
modeless input device, restricted to a few universal dialog-control commands, the user
can browse through an entire system while keeping a hand on the mouse, touching
the keyboard only when he wants to change from browsing to working with an application.

According to these considerations we have designed the mighty mouse shown in Figure 3 to have five keys and a thumbwheel, tuned to the agility of a person's hand.
Thumb and index finger are agile, so they control analog inputs. The other three
fingers control 0-1 inputs. The thumb moves with greater freedom than any other
finger, so its key can be switched into two different settings; in addition, the thumb
glides upwards to turn the thumbwheel.

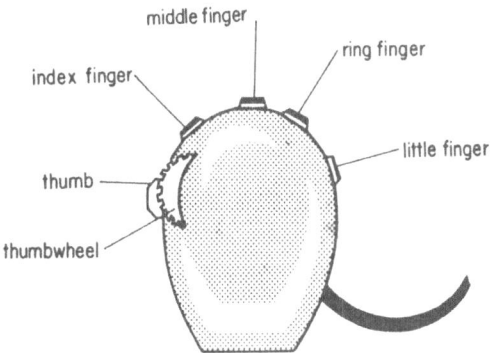

Fig. 3: Illustration of Mighty Mouse

The following functions have tentatively been assigned to these keys. The thumb key
realizes a linear motion command - gas pedal and gear shift in one. Varying pressure
moves the cursor forward or backward with variable speed, depending on the setting
of the key. A hard click means forward or backward by a unit, for example scrolling
to the next or previous page on the screen. The thumb also turns a thumbwheel that
sets the angle of an oriented cursor. An object is identified by moving the cursor to
it, but for graphical objects such as intersecting lines, an oriented cursor (with x,y
and an angle coordinate) is useful to identify one line out of a cluster. The index
finger controls a "show me" key. Increasing pressure asks for increasing depth of
detail, for example through zooming. A hard click selects this object, that is, turns
it into the active data environment to be operated upon by subsequently issued commands. The remaining keys are conventional. The middle finger activates a pop-up
menu that may offer additional operations, the remaining two fingers of little agility
are assigned important but relatively infrequent functions: the ring finger activates

undo, the little finger exits into the hierarchically next higher data or command environment.

A user must memorize these meanings once and for all. Thereafter, he should be able to explore an entire system that implements a standard "mighty mouse dialog control", even if unfamiliar with the rest of the hardware. In order to master an application program, he will undoubtedly need to learn commands specific to the application, but in order to browse and determine whether he really wants to use it, the universal commands of the mighty mouse may suffice.

9. CONCLUSION

People's awareness of human-computer interfaces has increased drastically since the beginning of the decade and design practice has changed significantly - we now balk at machine behavior that was standard only a few years ago. Whereas five years ago only a minority of programmers had first-hand experience in writing or using highly interactive programs and most considered dialog programming unworthy of study, the proliferation of personal computers and workstations with bitmap graphics screens has changed everything - now everybody is concerned about human-computer interfaces. This may be a passing fad, and we suspect that the frequency of conferences devoted to human factors will diminish - but the subject will certainly not fade away. We suspect interest will develop in much the same way as the publicity wave that accompanied electronic pocket calculators - once you can take them for granted because everybody has several calculators, there is no point in discussing them any longer. And as soon as professionally produced human-computer dialogs are consistently well-behaved, most users and programmers will take good interaction programming for granted.

The reason for believing that the state of the art of dialog programming will improve rapidly is that dialog control can be automated, in the same sense that machine-code generation has been automated with the help of compilers. In order to avoid misunderstanding, we should recall the distinction between semantics and syntax of a human-computer interface. Semantics is concerned with the functionality of an interactive program, what services it offers, and will certainly remain in the domain of human creation: Exactly what functions should we build into a document editor? is best answered by trying to improve on existing editors. Syntax is merely concerned with the formal presentation of agreed-upon concepts. A rich set of dialog concepts is now available: windows, menus, selection, "what you see is what you get", screen masks, etc. Operating systems, program libraries, program generators and programming languages that support such concepts are becoming available. A programmer will have to decide which of these software environments he will work with, much as he decides what programming language to use. A given dialog programming environment will support certain paradigms and interaction techniques and not others, much as a programming language is selective in the programming techniques it supports. If the dialog programmer adheres to the recommended style, he will not have to worry about the details of interaction, but can concentrate on the creative semantic design.

The analogy with programming languages, which went through a similarly rapid evolutionary phase a quarter of a century ago, leads to a prediction about the development of human-computer interaction. Just as no programming language gained supremacy over all others, but at least a dozen attracted large subcultures of devotees, so no single style of human-computer interaction will become standard, but different styles, based on different concepts, expectations and user communities, will coexist. In this early stage of development, there is room for many experiments in human-computer interaction, so that there will be a sufficiently rich population of ideas from which successful future products will evolve.

10. ACKNOWLEDGEMENTS

Among the many people who have contributed to our research on interactive systems at the Federal Institute of Technology, Zurich over the past decade, we would like to express our thanks in particular to M. Arnoldi, G. Beretta, E.S. Biagioni, H. Burkhart, P. Fink, K. Hinrichs, H. Hinterberger, A. Kierulf, F. Kronseder, B. Plattner, J. Stelovsky, A. Ventura, and J. Weydert.

REFERENCES

1.   Ackermann, D., and Nievergelt, J., "Die Fünffinger-Maus: Eine Fallstudie zur Synthese von Hardware, Software und Psychologie," Berichte ACM Tagung Software-Ergonomie '85, Stuttgart, September 1985, pp. 376-385.

2.   Beretta, G., Burkhart, H., Fink, P., Nievergelt, J., Stelovsky, J., Sugaya, H., Weydert, J., and Ventura, A., "XS-1: An Integrated Interactive System and Its Kernel," Proc. 6th Internat. Conf. on Software Eng., Tokyo, 1982, IEEE Computer Society Press, pp. 340-349.

3.   Biagioni, E.S., Hinrichs, K., Muller, C., and Nievergelt, J., "Interactive Deductive Data Management - the Smart Data Interaction Package," Proc. GI-Kongress '85, Wissensbasierte Systeme, Munich, October 1985, pp. 208-220.

4.   Biagioni, E.S., Nievergelt, J., Stelovsky, J., and Sugaya, H., "Can an Operating System Support Consistent User Dialogs? Experience with the Prototype XS-2," Proc. ACM Annual Conf., Denver, CO, 1985, pp. 476-483.

5.   Darlington, J., Dzida, W., and Herda, S.,"The Role of Excursions in Interactive Systems," Int. J. Man-Machine-Studies, Vol. 18, 1983, pp. 101-112.

6.   Hayes, P., Ball, E., and Reddy, R., "Breaking the Man-Machine Communication Barrier," IEEE Comput., Vol. 14, No. 3, March 1981, pp. 19-30.

7.   Kitahara, Y., "Information Network System - Telecommunications in the 21st Century," The Telecommunications Association, Tokyo, 1982.

8.   Nievergelt, J., "Errors in Dialog Design and How to Avoid Them," in Document Preparation Systems, J. Nievergelt et al., eds., North-Holland, Amsterdam, 1982, pp. 265-273.

9.   Nievergelt, J., and Ventura, A., Die Gestaltung Interaktiver Programme, B.G. Teubner, Stuttgart, 1983.

10. Nievergelt, J., Ventura, A., and Hinterberger, H., Interactive Computer Programs for Education - Philosophy, Techniques, Examples, Addison-Wesley, Reading, Mass., to appear 1985.

11. Norman, D.A., "The Trouble with UNIX," Datamation, November 1981, pp. 139-150.

12. Ohno, K., Fukaya, K., and Nievergelt, J., "A Five-Key Mouse with Built-in Dialog Control," ACM SIGCHI Bulletin, Vol. 17, No. 1, July 1985, pp. 29-34.

13. Sergot, M., "A Query-the-User Facility for Logic Programming," Integrated Interactive Computing Systems, P. Degano and E. Sandewall, eds., North-Holland, Amsterdam, 1983, pp. 27-41.

14. Stelovsky, J., "The User Interface of an Interactive System," Doctoral Dissertation 7425, Federal Institute of Technology, Zurich, 1984.

15. Teitelmann, W., "InterLisp Reference Manual," Xerox Palo Alto Research Center, 1978.

16. Ventura, A., "Einsatz und Programmierung des Computers als Werkzeug für den Unterricht," Doctoral Dissertation 7752, Federal Institute of Technology, Zurich, 1985.

17. Vitins, M., and Signer, K., "Performance Modeling of Control Systems," Proc. Symposium on Computer Systems for Process Control, R. Güth, ed., Plenum Press, New York - London, 1986.

DISCUSSION

Chairman: H. Sandmayr (Brown Boveri, Baden, Switzerland)

J. Ludewig (Brown Boveri, Baden, Switzerland)

One comment on your remark that the input language will result as a byproduct of the output language. In my view the distinction you are making is that of having semantics as an abstract concept but syntax as a means of human-computer interaction, which I think is very true. What you demonstrated seems to be extremely useful for the casual user. Do you think the same principles should apply for professional users, let us say for programmers who are using the computer every day?

J. Nievergelt

The issue of whether an interaction is input or output is often a matter of choice. At the semantical level I want to trigger an operation. Whether I input that by typing "please do that operation for me", or whether a menu is displayed and I pick an entry, is perhaps a superficial issue, but much of man-machine communication is surface. So I am simply arguing, whatever the semantics of the operation is that you wish to trigger, use our tremendous pattern recognition ability so that the system displays to you all it can do, and you simply say, "that". So I stick to my claim that essentially the input language is not relevant if you have a well-structured output. Now, there is the question of whether a novice or an expert user would like to use the same type of interface. I have a subjective opinion which I cannot prove, but I would say, yes, the same concepts apply to novices as well as to experts. The difference is that an expert likes to say everything in abbreviated form and the novice likes to see the same thing spelled out. So my advice is, design precisely the same type of interface and have a switch for selecting abbreviation or long form. The words will be different, but the concepts are the same.

G. Koch (Biomatik, Freiburg, Fed. Rep. Germany)

How can the system analyst separate the concerns of dialog and application? Is there a standard for the data interface between the dialog system and the application program?

J. Nievergelt

De facto standard user interfaces are beginning to emerge in commercial products, for example, the multi-window screen based on the desk-top metaphor that came out of research at Xerox PARC. Within five years we will see formal standards for user interfaces defined by standards organizations.

ENGINEERING GRAPHICS: OVERVIEW

J. ENCARNACAO

Technical University Darmstadt, Darmstadt, Fed. Rep. Germany

ABSTRACT

CAD Systems are of increasing importance both as a basic technology and as a tool for engineering applications. This paper surveys different types of CAD workstations (PC-based, high-performance) and the possibilities of interconnecting them with special reference to the subject of graphics and geometry standards. The CAD workstation market and the world revenue in engineering graphics are briefly surveyed. The user benefits of CAD and the expected trends for CAD system technology are listed. The bibliography lists references to related work and further technical details.

1. INTRODUCTION

The structure of a CAD System[7] is discussed first. This is followed by consideration of CAD on PCs, on high-performance CAD workstation and on related interfaces. The growth of the workstation market and the corresponding worldwide revenue is then discussed. The paper ends with a brief discussion of the most important CAD benefits and the expected technological trends.

2. STRUCTURE OF A CAD SYSTEM

As shown in Figure 1, an industrial system may be characterized as having administrative and geometric needs as applications to be served by user and data services.[11] The user services give access to remote modes and to engineering workstations; the

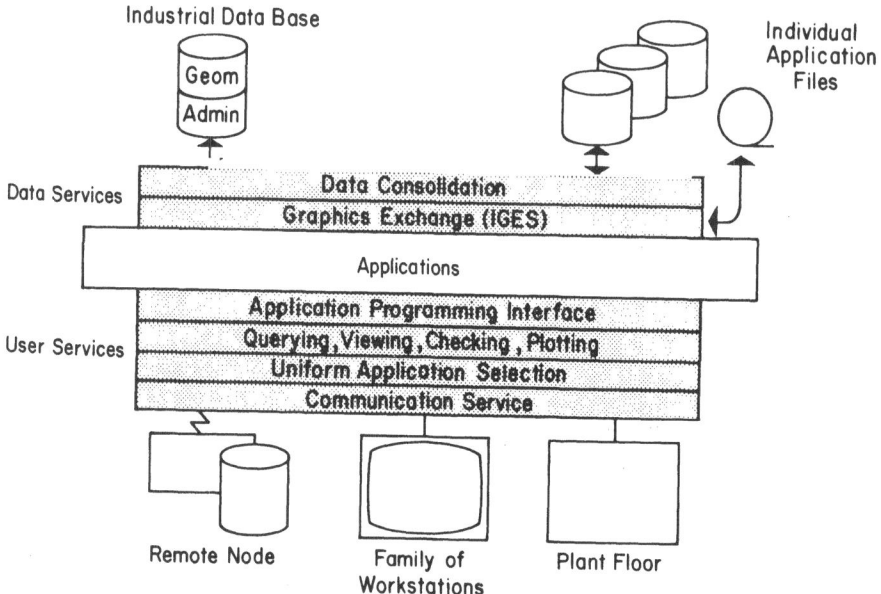

Fig. 1: Structure of an Industrial System[11]

data services give access to the individual application files and to the engineering database.

Examples for data services are:

- Data consolidation
- Graphics data exchange

The expected user services include:

- Application programming interface
- Querying, viewing, checking and plotting
- Uniform application selection
- Communication services

Both the CAD/CAM application profiles and also user profiles are dependent upon the functionality expected or needed. Typical application phases, with increasing levels of functionality in the mechanical engineering area, are

- 2D Drafting
- 2D/3D Design
- Analysis, simulation

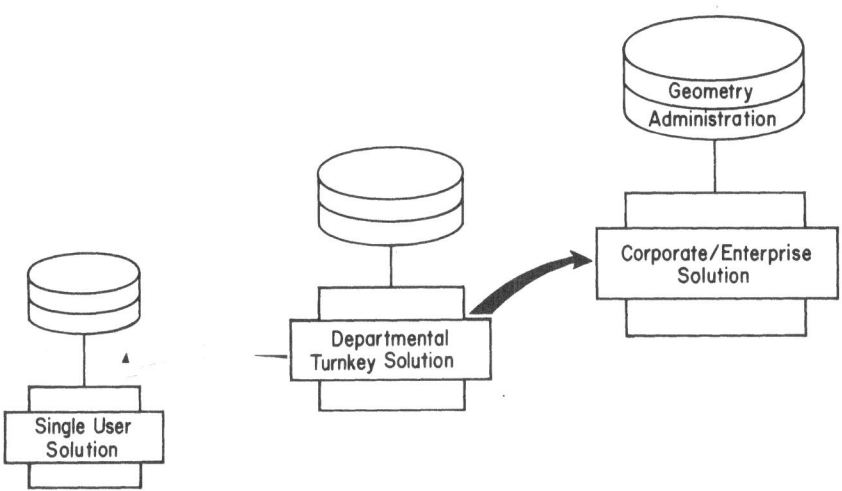

Fig. 2: Different Solutions for a CAD/CAM System[11]

- Numerical control
- Integrated administration
- Solids design, modelling

As the user progresses from 2D drafting to solids design and modelling, the system functionality will increase from a departmental to an integrated solution as shown in Figure 2.

There are basically three classes of workstations to serve these purposes:

1) PC-based workstations
2) Standard engineering workstations
3) Intelligent, high-performance, engineering workstations

3. CAD ON PC'S

PCs for CAD purposes[7] today have 16- or 32-bit processors and cost in the range of $ 5,000 to $ 10,000.* Software packages are available for $ 500 to $ 3,000. Enhance-

---

* U.S. dollars

ment of the standard display resolution of 720 × 300 costs an additional $ 200 to $ 500. The principal I/O devices now used are mice ($ 100 to $ 300) and plotters ($ 1,500 to $ 18,000). To increase system speed, PC-based CAD systems usually include RAMs with at least 512 kb matched coprocessors (e.g., 8087 or 80827) and hard-disk storage for local support of the data handling. There is a trend towards the increasing use of floating-point processors in such systems.

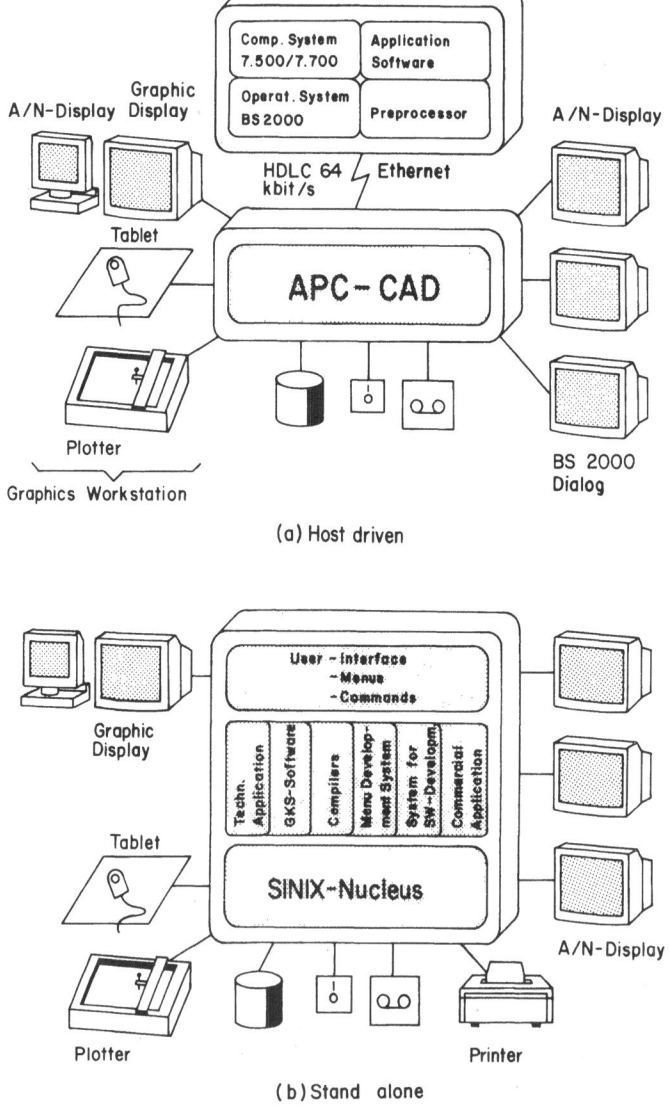

Fig. 3:   Example of a PC-Based Graphics System (Siemens APC-CAD Based on Siemens PC-MX2)[1]

A typical system architecture for PCbased CAD is shown in Figure 3.[1] The architecture may be basically stand-alone or host-driven.[6]

The strongest supplier of PCs for CAD is IBM with 3270 PC/G, 3270 PC/GT, PC/XT and PC/AT. Compatibility is an issue of great importance for which these are two aspects: the one based on the IBM operating system MS-DOS, and the other based on standardized environments such as UNIX, C and GKS. The first form of compatibility differentiates between

> a) MS-DOS compatibility only
> b) MS-DOS compatibility with additional read/write of PC-DOS diskettes
> c) as (b) with additional support of PCs ROM BIOS extension to MS-DOS
> d) a direct hardware compatibility based on an IBM PC-like configuration or emulation of the IBM PC-hardware

The second form of compatibility is based on an operating system of the UNIX-family,[2] a widely used programming language (today such a language is usually C[14]), and standard graphics interfaces and packages (the first ISO international standard for this is GKS[7]).

Application or support software running on PCs is available for

> - Presentation graphics
> - Image enhancement
> - Drafting and design
> - Analysis, simulation and statistics
> - Painting

and also for general-purpose application programming and for post- and preprocessing (e.g., for NC-interfacing).

Based on the application profiles and the user needs there are the following PC-based system configurations:

> - Local interfacing based on V24
> - Local networking (e.g., based on Ethernet-LAN)
> - Networking of PC-pools (e.g., based on Ethernet)
> - (Ethernet) networking based on the ISO-protocols

4. INTELLIGENT, HIGH-PERFORMANCE ENGINEERING WORKSTATIONS

Today's traditional engineering workstations are based on a microprocessor supported by various special processors to implement special functional needs. Basic components of such a workstation include the memory, the display, the keyboard and the disk. The more advanced architecture shown in Figure 4[5] includes, among other components,

- PROMs and RAMs
- Storage handling and error processing
- Bit-mapped graphics controller
- Input controller
- Interface processor
- LAN controller

Another concept aiming at a very high performance for an engineering workstation is based on three different processors (application processor, graphics processor and image processor) served by a special-purpose bus interconnecting them. This architecture is additionally supported by so-called "accelerators" either in silicon or at least in firmware, which implement such things as geometric operations, special algorithms for viewing and shading, matrix operations, etc. By these means the workstation architecture gains a high degree of optimization in its performance and functional capabilities. An example of such an architecture is shown in Figure 5.

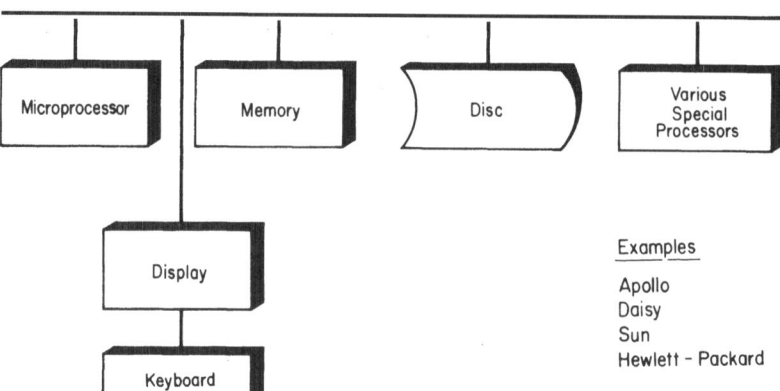

Fig. 4: The Traditional Workstation[5]

HARDWARE

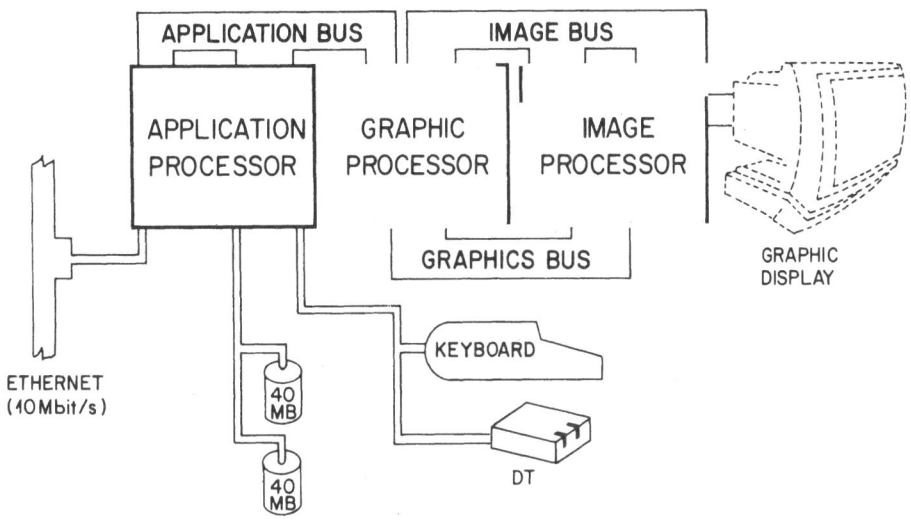

- APPLICATIONS PROCESSOR
- GRAPHICS PROCESSOR
- IMAGE PROCESSOR
- DISPLAY
- BUS STRUCTURE
- PERIPHERALS

Fig. 5: High-Performance Engineering Workstation[5]

5. INTERFACES AND STANDARDS

In the context of the system software for CAD systems, there are two basic types of interfaces

a) Graphics interfaces
b) Geometry interfaces (product-data interfaces)

The graphics interfaces define the following communication levels:

- Application program / graphics package (e.g., GKS and 3D-GKS[9,10])
- Graphics package / archive system (e.g., GKSM and CGM[9,4])
- Graphics package / graphics device driver (e.g., CGI[3])

These standards and/or standards proposals define the system interfaces at the level of the graphics data used by the graphics system to produce, manipulate and/or store pictures which are defined by a set of graphics primitives (e.g., polylines, poly-markers, text, and arrays).

The geometry interfaces are standardized transfer files for communication between different CAD systems (see Figure 6[16,8]). These interfaces are defined based on different performance levels:

- Geometry functionality (A, B, C,...) (e.g., solids, surfaces, wire frames)
- Application area (a, b, c,...) (e.g., mechanical engineering, electrical engineering, civil engineering, plant design)
- Application phase (1...n) (e.g., design, drafting, analysis, production, assembly, quality control)

On the basis of this framework we may have, for example, a performance level A1a defining the data-transfer format for solids in the design phase of a mechanical engineering application.

The CAD data-transfer formats are conceptually structured in levels based on the differentiation between high-and low-level systems, with a system-dependent and a system-independent part (Figure 7).

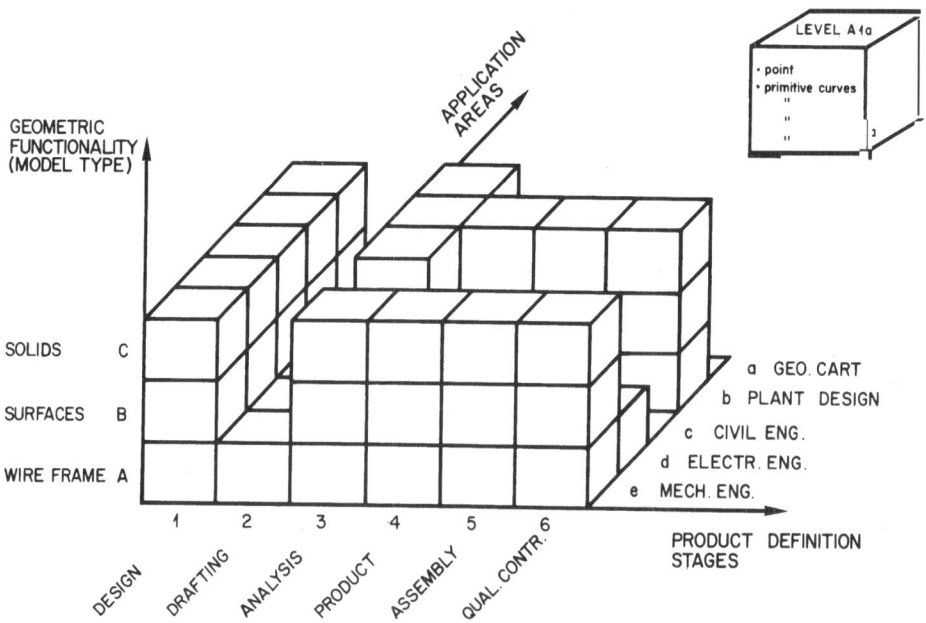

Fig. 6: Performance Levels for Geometry Interfaces[8]

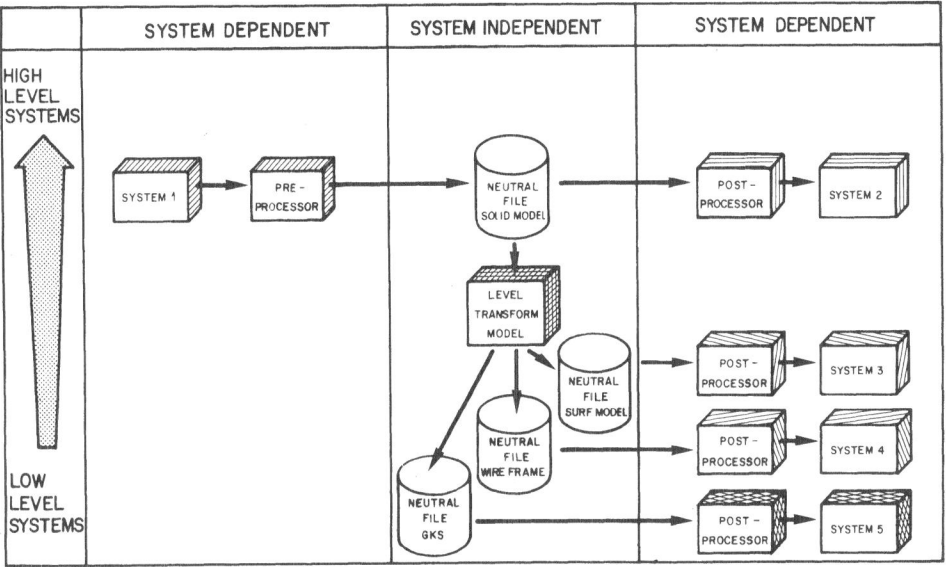

Fig. 7: Level Definition Concept for Geometry Standards[8]

The implementation of the geometry interfaces is based on the following layers:

- Application-oriented layer (defined by specific entity packages and the semantics, which are related to the application area, the product definition stage
  and the geometric functionality)
- Basic logical layer (defined by entity classes, which are application independent)
- Physical layer (defined by "tools" for the physical representation of the entities)

These layers are exemplified in Figure 8.

These interfaces are of great importance for achieving the following goals:

- Portability and device independency (Graphics and CAD-software should run
  without too much additional effort on different computer environments using
  different graphics-I/O-peripherals).
- System integration (Based on well-defined interfaces and specifications, graphics and CAD systems should be an intrinsic part of the overall application
  environment; this implies the need for the possibility of data transfer and
  communication between the different system parts and for the integration of all
  these parts, including the CAD, into an integrated system architecture).

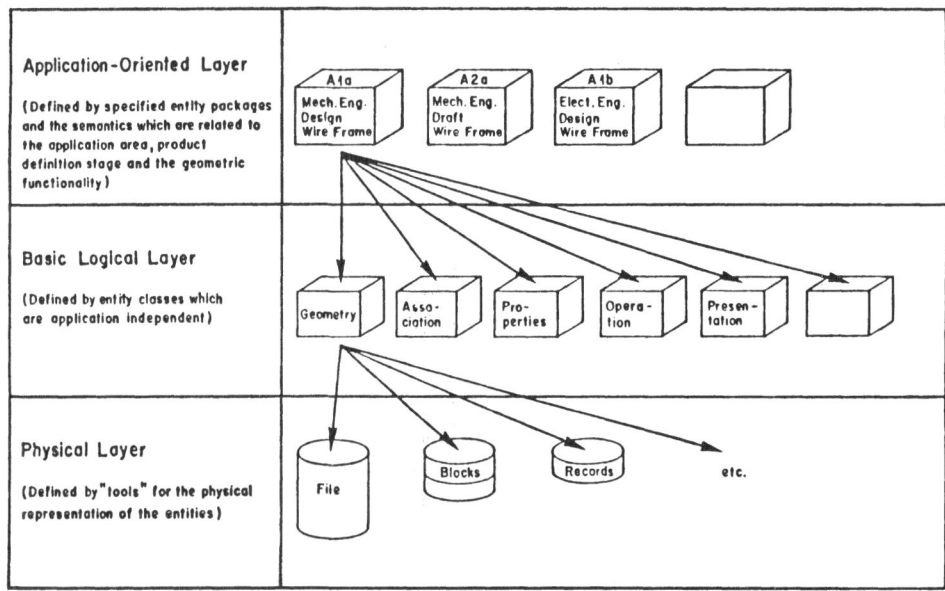

Fig. 8: Examples of the Level Concept for CAD/CAM Data Transfer[8]

6. GROWTH OF THE CAD-WORKSTATION MARKET

Carl Machover has estimated the following growth of the workstation market[15] as shown in the following examples for a representative application area (all revenue figures are millions of U.S. dollars).

    a) Integrated circuit workstations
        - Revenues: from $ 370 (1985) to $ 1,150 (1988)
        - Installations: from 5,120 (1985) to 13,530 (1988)

    b) Printed circuit board workstation market
        - Revenues: from $ 230 (1985) to $ 870 (1988)
        - Installations: from 2,790 (1985) to 12,960 (1988)

If CAD systems are classified according to their architecture into three classes

    - Turnkey and host-based systems
    - Workstations and low-cost systems
    - Personal/desktop systems (computers),

then based on the same source[15] we have the figures shown in Figure 9.

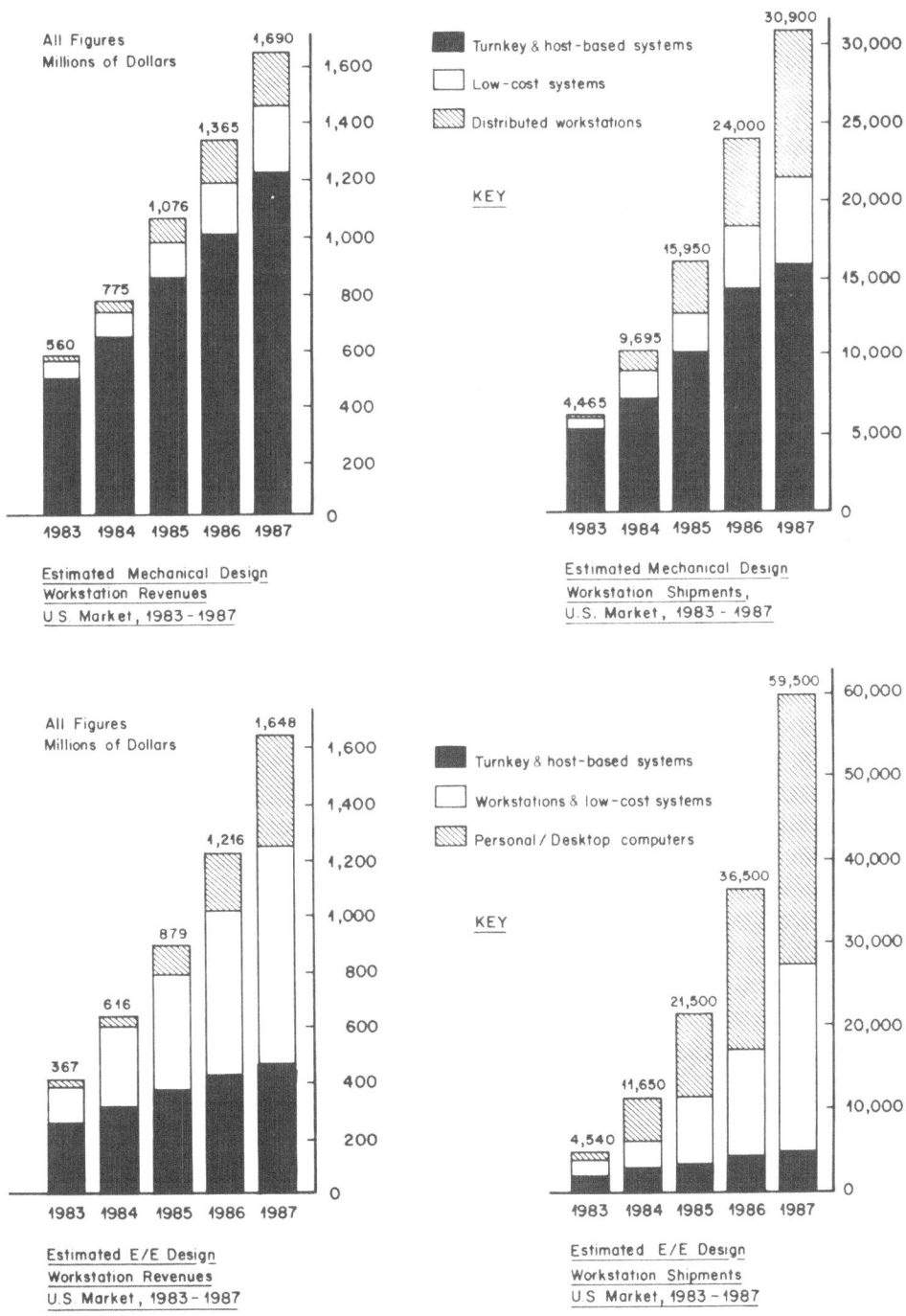

Fig. 9: Growth of the Workstation Market[15]

From Figure 9 we can deduce the following:

1) Mechanical design workstation market

   Although the bulk of the revenue is still based on turnkey and host-based systems, the share of the other two types of systems is becoming increasingly more important ($ 160 million in 1983 to $ 450 million in 1987). This fact becomes even more evident if we consider the number of installations: 1,500 in 1983 to 15,000 in 1987, i.e., 50 % of all workstations installed for applications in this area by 1987.

2) Electrical/electronic design workstation market

   Here the impact of workstations and low-cost systems as well as of personal/desktop systems (computers) is much stronger; the revenue is expected to increase from 30 % (1983) to 75 % (1987). Workstations and low-cost systems have a share of about 50 %. The largest number of installations here is based on personal/desktop systems (computers).

From the data shown in Figure 9 we can clearly identify the following trends:

- Revenue is increasingly based on workstations and low-cost systems, as well as on personal/desktop systems (computers), and decreasingly on turnkey and host-based systems.

- The number of installations based on personal/desktop systems (computers) is increasing dramatically, with the result that in the medium and long term the demand for networking of these systems and for the integration of the application software running on them will become increasingly important.

7. ENGINEERING GRAPHICS WORLD-WIDE REVENUE

According to an IBM Study[11] the expected revenue in 1987 will be of the order of $ 9.1 billion, growing from $ 4.2 billion in 1984 to $ 18.6 billion in 1990 (Figure 10). The distribution within these $ 9.1 billion by application areas is expected to be: (in billions of U.S. dollars):

Mechanical	4.73
Electrical	2.73
A/E/C	1.18
Other	0.46

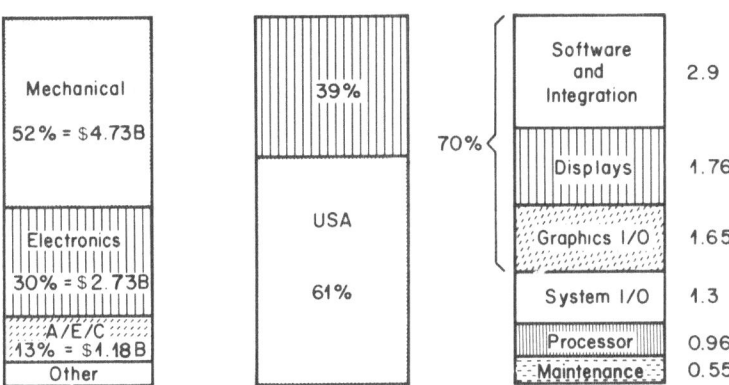

Fig. 10: Engineering Graphics World-Wide Opportunity in 1987 ($ 9.1 Billion)[11]

It is also important to see that 32 % of the $ 9.1 billion (i.e., $ 2.9 billion), will be invested in application software and in system integration. This reveals the relevance of standards and of well-specified system interfaces.

The development of the different types of workstations (display; graphics I/O) also represents a very large part of the revenue (37 %), confirming the trend that in future engineering graphics and CAD will be largely based on workstations, rather than on turnkey or host-based systems.

8. CONCLUSION

CAD systems are of great importance for all engineering applications. A user survey published by IEEE[12] reports on the following average of user answers to questions addressing the benefits of using CAD in their application environment:

- Improved accuracy	90 %
- Reduced errors	78 %
- Shorter design cycle time	76 %
- Increased productivity	75 %
- Reduced costs	70 %

Based on this evaluation, these are the most frequently cited reasons for using CAD and for justifying its price/performance value. CAD is a rapidly developing market in which the impact of technological development is very strong. The expected trends may be summarized as follows:

1) Hardware
   - Lower-priced entry level systems (PCs)
   - Lower cost for engineering workstations
   - More stand-alone and LAN-connected systems
   - 32-bit mini- or microcomputers as basic CPU
   - Sharing of information between low-cost, micro-based systems and higher cost, mini-based systems (data sharing; software migration)
   - Integration of CAD/CAM/CAE (need for interfaces and standards; also a database issue)

2) I/O-Devices
   - RISC architectures (Reduced Instruction Set Computers) will be used to achieve more flexibility
   - Increase in number of numerical-control devices (integration with CAM)
   - Interfaces to robots
   - Lower-priced plotters, printers and digitizers
   - Xerography for hard copy

3) Software
   - Easier-to-use graphics languages
   - Incorporation of DBMSs
   - Integration of graphics and CAD-software with estimating, material invoices and accounting programs
   - Added interaction and improved quality of the man-machine interface.

4) Communications
   - Increased networking (need for graphics and geometry protocols)

REFERENCES

1. "Arbeitsplatzcomputer für CAD," Siemens AG, München, Doc. No. U 2238-J-Z91-1, 1985.

2. Bach, F., Domann, P., and Remmele, W., _Unix Transparent_, Springer-Verlag, Berlin, 1985, (ISBN 3-540-15150-8).

3. CG-VDI Baseline Document, ISO/TC97/SC21/WG 5-2 N-1511.

4. Computer Graphics Metafile, ISO/TC97/SC21/WG 5-2, ISO DIS 8632/Part 1, 1985.

5. Encarnaçao, J., "Möglichkeiten der VLSI-Technologie für die Entwicklung zukünftiger graphisch-interaktiver Arbeitsplatzrechner," Proc. GI'84, Springer-Verlag, Berlin, 1984, pp. 16-28.

6. Encarnaçao, J., "PC-Based Computer Graphics," Proceedings China-Graphics'85, October 1985.

7. Encarnaçao, J., and Schlechtendahl, E.G., Computer Aided Design, Springer-Verlag, Berlin, 1983, (ISBN 3-540-11526-9).

8. Encarnaçao, J., Schuster, R., and Vöge, E., Product Data Interfaces, Springer-Verlag, Berlin, 1986, (ISBN 3-540-15118-4).

9. Enderle, G., Kansy, K., and Pfaff, G., Computer Graphics Programming, Springer-Verlag, Berlin, 1984, (ISBN 3-540-11525-0).

10. GKS-3D Version 3.4, ISO/TC97/SC21/WG5.2 N 277 Rev., 1985.

11. IBM-Seminar - Seminarunterlagen - Stadthalle Sindelfingen, Juni 1985.

12. IEEE Computer Graphics and Applications, Vol. 5, No. 8, 1985.

13. Information Processing Systems, Computer Graphics, Graphical Kernel System (GKS), ISO 7942.

14. Kernighan, B.W., and Ritchie, D.M., The C Programming Language, Prentice-Hall, Englewood Cliffs, N.J., 1978.

15. Machover, C., "Tutorial Notes for EUROGRAPHICS'84," Graphic Workstations.

16. Schuster, R., Tutorial T4.2, "CAD/CAM Product Data Transfer," Tutorial Notes, CAMP'85, September 1985.

DISCUSSION
Chairman: H. Sandmayr (Brown Boveri, Baden, Switzerland)

L. Krings (Brown Boveri, Baden, Switzerland)

Standards for the application interface of basic graphical functions have been established, for example, in GKS or PHIGS. In order to enable the exchange of graphics software in a local area network of workstations, standards for device interfaces and metafiles ought to be defined as well. What developments do you expect in this area?

J. Encarnaçao

I expect two different levels of development for graphics interfaces in a local area network of workstations:

- one based on graphics standards (GKS, CGI, CGM etc.),
- another based on model-exchange standards (IGES, EDIF, ODIF, PDES, MAP, etc.).

In Germany we are pursuing research work on both levels in the frame of the DFN-project, a project aimed at setting up a German research network. My group at the Technical University Darmstadt is, in particular, involved in the development of communication interfaces which are based on the GKS-functionality.

H. Sugaya (Brown Boveri, Baden, Switzerland)

The availability of a high-resolution bitmapped display with a pointing device (e.g., mouse) has opened up a way to explore new techniques for input/output operations such as windows, menus, and masks. Microsoft "windows", for example, are available for PCs. Many nongraphics applications such as document editors, programming tools, or spread sheets, however, are based on such techniques. Will these window- or menu systems be standardized for that class of applications separately from the complex workstation concept in GKS?

J. Encarnaçao

Concerning the implementation of window managers on PCs, standardization activities utilizing CGI and a subset of GKS have been started. There is also ongoing work to build window managers supported by GKS under UNIX. My group in Darmstadt participates in the development of window managers which are able to run on PCs and workstations.

SESSION III
SOFTWARE ENGINEERING
Chairmen: H. Sandmayr and O. Klammer

ISSUES IN PROCESS AND COMMUNICATION STRUCTURE
FOR DISTRIBUTED PROGRAMS*

B. LISKOV and M. HERLIHY

Massachusetts Institute of Technology, Massachusetts, USA

ABSTRACT

Many proposals have been made for structuring distributed programs. This paper
looks at one such proposal, the one embedded in the Argus programming language
and system. The paper provides a discussion of decisions made in the two major areas
of process structure and communication, and compares the chosen structures with
alternatives. The paper emphasizes the rationale for decisions and the issues that
must be considered in making such decisions.

1. INTRODUCTION

A major issue in languages and systems today is the structure of distributed programs: programs that run at one or more nodes connected by a communication network. Examples of questions that arise in this regard are the kind of modular structure that is suitable for distributed programs, and how the modules that make up a
distributed program should communicate.

Argus is a language/system designed to support the construction and execution of
distributed programs. Argus is intended to support only a subset of the applications

---

213

that could benefit from being implemented by a distributed program. Two properties distinguish these applications: they make use of on-line data that must remain consistent in spite of concurrency and hardware failures, and, although they are real-time programs, their real-time constraints are not severe. Examples of such applications are office automation systems and banking systems. Argus is a high-level language and is not intended to support the implementation of operating system kernels; indeed, the implementation of Argus requires such a kernel.

Like any such system, Argus supports distributed programs by providing a model of distributed computation. This model is based on a number of decisions about distributed program structure and contains some unusual features. The purpose of this paper is to describe the Argus model, and to explain the basis for the decisions that were made in designing the model. In the course of the discussion, alternative positions are discussed and compared with the selected features. The discussion is limited to two major areas: communication and process structure.

The remainder of the paper is organized as follows. In Section 2 we give a brief overview of Argus, providing the background for a subsequent, more detailed discussion of our decisions concerning communication and process structure. In Section 3, we discuss the major issues that arise in communication and process structure and reach some conclusions that constrain the form of solutions that we consider acceptable. We also discuss some of the design goals of Argus. In the remainder of the paper we describe communication and process structure in Argus; in each case we compare the selected feature with alternatives and analyze the tradeoffs inherent in the decision.

2. OVERVIEW OF ARGUS

This section presents an overview of the basic components of Argus. Argus is an extension of the CLU[18,16] programming language. A more detailed description of Argus has been given elsewhere.[17]

2.1 Guardians

Applications in Argus are implemented by one or more modules called guardians. A guardian in Argus resides at a single physical node, although a node may support several guardians. A guardian encapsulates and controls access to one or more resources, such as databases or devices. Access to the protected resource is provided by a set of operations called handlers that can be called from other guardians. Internally, a guardian consists of a collection of data objects and processes to manipulate those objects. The processes within a guardian can share the objects directly, but sharing of objects between guardians is not permitted. Arguments to handler calls are passed by value;[10] references to objects may not be passed in handler calls, ensuring that objects local to a guardian remain local, and that a guardian retains control of its own objects.

A guardian can survive crashes of its nodes (with as high probability as needed[14]). A guardian's state consists of <u>stable</u> and <u>volatile</u> objects. When a guardian's node crashes, all processes running at the time of the crash are lost, along with the guardian's volatile objects. The guardian's stable state remains intact; upon recovery, the guardian runs a special process to reconstruct its volatile state.

2.2 Atomic Actions

In systems where on-line data are read and modified by on-going activities, there are consistency constraints that must be maintained in the presence of failures and concurrency. Such constraints apply not only to individual pieces of data, but also to distributed sets of data. For example, when an attempt is made to transfer funds from one account to another in a banking system, the net gain over the two accounts must be zero, regardless of failures and concurrency.

Perhaps the single most important conclusion we have drawn from our experience designing Argus is that such consistency properties are very difficult to maintain using only the techniques provided by traditional languages. Our approach is to provide activities called <u>actions</u> as the primary method of carrying out computations in Argus. Actions are <u>atomic</u>: they are indivisible and recoverable. By <u>indivisible</u>, we mean the execution of one action never appears to overlap (or contain) the execution of another.* The effect of executing multiple concurrent actions is <u>serializable</u>:[7,23] actions are scheduled in such a way that their overall effect is as if they had been executed in some sequential order. If each action individually preserves the system's consistency, then indivisibility guarantees that any concurrently executing collection of actions will also preserve consistency. By <u>recoverable</u>, we mean the overall effect of the action is all-or-nothing: either all changes made to the data by the action happen, or none of these changes happen. An action that completes all its changes successfully <u>commits</u>; otherwise it <u>aborts</u>, and objects it modified are restored to their previous states. When an action completes it either commits at all guardians or aborts at all guardians.

---

* Nested actions, described below, are a partial exception to this rule.

Atomicity is actually achieved via the data objects shared among actions; these objects must be implemented so that using actions appear to be atomic. Objects that support atomicity are referred to as <u>atomic objects</u>. Thus actions are atomic provided they share only atomic objects. Argus provides a collection of built-in atomic data types; users may also define their own atomic types.[26]

Examples of built-in atomic types are atomic arrays and atomic records, which are similar to ordinary arrays and records except that they provide synchronization and recovery. Strict two-phase locking[7] is used to provide serializability for actions using the built-in atomic types. The lock on an object of built-in atomic type is acquired automatically when a primitive operation of the type is called by an action, and is held until the calling action completes (commits or aborts). Recoverability is provided by copying the object the first time an action executes an operation that modifies it. All changes are made to this copy (called a <u>version</u>). The new version replaces the original if the action commits; otherwise the new version is discarded.

The original (committed) version of a stable object resides on stable storage,[14] and the tentative versions reside in volatile storage. When an action is about to commit, a two-phase commit protocol[9] ensures that none of the tentative versions it has created has been lost as a result of crashes. In the first phase of the protocol, the guardians prepare to commit by copying the tentative versions of stable objects to stable storage without discarding the old versions. If it can be determined that all guardians have prepared successfully, the action commits and the old versions are discarded; otherwise the action aborts, and the old versions are retained.

Argus also provides nonatomic data types, and a semaphore-like primitive to synchronize concurrent access to nonatomic data. Nonatomic data is more efficient than atomic data, since synchronization and recovery need not be implemented. It is used for two purposes: for private data of actions, and for implementing user-defined atomic data types.[26]

Actions may be nested. Nested actions, or subactions, are a mechanism for limiting the scope of failures, as well as for introducing concurrency within an action.[25,21] An action may contain any number of subactions, some of which may be executed sequentially, some concurrently. This structure cannot be observed from outside; i.e., the overall action still satisfies the atomicity properties. Subactions are atomic with respect to other subactions of the same action. Subactions can commit and abort independently, and a subaction can abort without forcing its parent action to abort. The commit of a subaction is conditional: even if all subactions commit, aborting the parent action will abort all of the subactions. Further, the two-phase commit protocol described above is undertaken only at the completion of top-level actions; changes to stable objects become permanent only when top-level actions commit.

3. BASIC DECISIONS AND GOALS

In this section we discuss the two major goals of the Argus design: safety and simplicity. We also discuss the issues that arise in two main areas: the form of communication between guardians, and the process structure of guardians. Our approach to these issues is based on a model in which guardians act in two different capacities, as clients and as servers. Clients ask other guardians to do things for them by making handler calls; for example, G is a client of H if G calls one of H's handlers. Servers provide services for other guardians, e.g., H is a server for G. Of course, a guardian may be a server for some group of (client) guardians, as well as a client of some group of (server) guardians.

3.1 Goals

The first major goal of the Argus design concerns safety for atomic actions. If actions communicate only via atomic objects then it is guaranteed that they are atomic. A simple way to provide this guarantee would have been to provide only atomic data in Argus. This approach was rejected, however, to avoid the expense of atomic data when non-atomic data is perfectly safe, as within a single action.

Therefore, we felt it was important to structure Argus to minimize the likelihood that nonatomic data will be shared erroneously. This goal is achieved primarily by keeping each action's local variables hidden from other actions, as will be discussed in subsequent sections.

Another aspect of safety is to avoid confusion about the action on whose behalf a process is executing. Our approach here has been to keep the correspondence between actions and processes implicit: each process executes on behalf of a single action, and the identity of this action is established automatically when the process is created.

The second major goal was simplicity. Although we believe that atomic actions and atomic objects must be provided by Argus, their presence raises the possibility that the language might become too difficult to use. To avoid this danger, we tried to make the rest of the language as simple as possible, primarily by shaping the rest of the language to conform to the action structure. In addition, we have refrained from including in Argus features that might be useful, but whose full implications are not yet well understood. Our position on expressive power is that the need for a feature will become clearer when programmers are forced to cope with its absence. We expect that some of the omitted features will be added later.

3.2 Form of Communication

Communication in Argus is high level; users communicate in terms of ordinary data objects and are shielded from low level details such as packet management. Our present form of communication is based on some major decisions arrived at by studying applications. These decisions and their effect on expressive power are discussed below.

Early in the Argus design,[15] we inclined toward a communication mechanism in which sending messages was decoupled from receiving replies. As we looked at more and more applications of the type Argus was intended to support, we became convinced that clients almost always need responses to requests. For example, an answer is clearly required when a client of a printer spooler server inquires whether its previously spooled file has been printed. Even if the client simply asks the server to spool a file for printing, the client may require confirmation that its request has been carried out. If the request has failed, the client could then communicate with another spooler.

The desire to receive a reply in the latter case actually stems from perceptions about reliability. If hardware and software were highly reliable, then it would not be necessary to check back (via the reply) to make sure the request had been acted upon. Neither hardware nor software is this reliable today, and we doubt they ever will be, although improvements can be expected. Thus, clients really will want replies, at the very latest when top actions commit.

Therefore, we decided that communication should take the form of paired send and reply. Paired send/reply can be provided by a synchronous procedure call mechanism or by an asynchronous mechanism, in which sending a message and receiving a reply are decoupled. We only require that for each request sent, a reply is expected later, and furthermore, that this pairing is part of the semantics of the communication mechanism.

It is worth noting that CSP-style communication,[12] in which the sending primitive guarantees that the message is delivered but not necessarily processed, does not remove the client's need for a reply. CSP-style communication can increase the likelihood that a successful execution of the sending primitive will result in a successful execution of the associated request, but the client must still receive a reply later to be sure the request was carried out.

Of course, since hardware and software are not highly reliable, there is a nonnegligible chance that the desired reply to a request will not be forthcoming. This implies that the client must have the ability to stop waiting for the reply. Even though requests are always paired with replies, the communication mechanism must permit the client to avoid carrying out the second part of the communication.

A difficult problem that arises here involves control over waiting and retransmission. A straightforward solution is based on the notion of a timeout: a duration of time after which the waiting process will either retransmit the request, or give up and report a failure. Although this mechanism might seem simple in principle, its performance is critically dependent upon two choices: how long processes wait, and the circumstances under which requests are retransmitted. These decisions must take into account the characteristics of the communication medium and the speeds of the communicating processors. We think that users do not have enough information to make these choices reasonably; only the system itself is in a position to do so. Therefore, a second major decision was that the communication mechanism should relieve the programmer of the burden of low-level timeouts by assuming responsibility for timeout and message retransmission.

Although users should not define timeouts based on assumptions about system performance, users do require the ability to terminate pending requests for a variety of reasons. It is important to observe that any mechanism based only on timeouts is inadequate, since elapsed time is just one of many possible reasons for terminating a request. For example, a client may send identical requests to multiple servers, accepting the first response and ignoring the others, or a client may terminate a pending request because a user at a terminal has hit the "break" key.

Another difficult problem is that there are many possible reasons why a reply has not been received. For example:

1) The server never received the request.
2) The server actually sent a reply, but the reply cannot be delivered because of a network partition.
3) The server is working on the request.
4) The server was working on the request when it crashed. In this case, before it crashed, it may have sent a subsidiary request to another server.

Whatever the reason, when a client stops waiting for a reply, it may be important to ensure that the aborted request will have no effect. For example, suppose that a file will be spooled for printing at a second spooler if the first one does not reply. If the file being spooled contains checks to be printed, it is crucial that it not be accepted for printing at both spoolers.

These considerations led to our third major conclusion about communication: if the client chooses not to wait for a reply, or if the system determines that a reply is unlikely to arrive, the activity associated with the request must be guaranteed to have had no effect. These semantics can be accomplished naturally by treating this activity as an atomic subaction of the action being executed by the client. The client can then abort the subaction when it stops waiting for a reply.*

Paired send/reply meshes well with subactions since each pair of messages delimits clearly the start and end of a subaction. This match between mechanism and desired behavior was one important factor in our choice of paired send/reply; the other was its perceived match to user needs as discussed earlier. As is always the case in programming languages, features in Argus are not independent of one another, nor are design decisions made linearly: paired send/reply matched to actions was accepted after a lengthy design process involving many iterations.

3.3 Other Communication Patterns

Paired send/reply is not the only communication pattern that arises in distributed programs. Other patterns can be implemented in terms of paired send/reply, but at a cost in efficiency and/or complexity in user code. We now go on to examine some other patterns in an attempt to estimate this cost. The patterns to be discussed are pipelining, forwarding, and conversations.

In pipelining, the client sends one request after another, and the server accepts and processes these requests one after another. Pipelining is desirable when either the client or the server (or both) does some processing that can be overlapped with the transmission of data. Although not compatible with synchronous paired send/reply, pipelining is compatible with asynchronous paired send/reply if the application requires that all the data sent be received. In applications like speech transmission, all data need not arrive at the server because missing data can be reconstructed if enough "neighboring" data arrive. This pattern is not compatible with our decision that the Argus system should try hard to deliver each request, so Argus may not provide the right support for such applications.

In forwarding, the reply to a request comes from a third party. For example, a client asks a server to do something. That server in turn sends a request to a second server. The reply from the second server is really a reply for the original client and is sent there directly, bypassing the first server and eliminating one message and one

---

* Note, however, that nested atomic actions are not used for local communication, such as procedure calls within a guardian.

message delay. With paired send/reply the response from the second server must go through the first. However, a closer examination reveals that the programmer of the first server usually does require some sort of reply, perhaps not the data returned by the reply but at least an indication that the request has been carried out successfully. Should such an extension prove to be worthwhile, it would be possible to extend paired send/reply to allow the second server to reply to the original client while informing the first server of the request's outcome.

A conversation occurs when two or more guardians communicate as "equals". In a (two-way) conversation, one party starts the discussion, the other replies, the first replies to that reply, and so on. As the conversation continues, each participant keeps track locally of the current state of the conversation. Paired send/reply has a hierarchical structure where the server acts as directed by the client. It is possible to implement conversations using paired send/reply, albeit unnaturally and perhaps with extra messages being exchanged.

In choosing paired send/reply we have chosen a mechanism that we believe fits the most common communication pattern in the application domain of Argus. We also believe that less common patterns occurring in the application domain can be programmed using paired send/reply with an acceptable cost in extra program complexity and execution overhead. However, validation of this latter claim must wait until we have more experience with applications.

3.4 Process Structure

To provide adequate performance, we believe that guardians must support some form of concurrency. In this section we justify this claim, discuss two approaches to providing concurrency within guardians, and describe our reasons for preferring one approach over the other.

Let us consider first the case of a client guardian. Here there is a spectrum of needs. The client may need to send a request while a previous request is outstanding; this pattern is needed, for example, in the case of pipelining. It is but a small step from this pattern to a pattern that also involves local processing. For example, upon receiving the reply to one request, the guardian may undertake a local computation to decide whether to continue waiting for the second response. Finally, a guardian might pursue purely local computations in parallel.

The case for concurrent processing is equally clear in the case of servers. Suppose a server is acting on a request and reaches a point where it cannot make further progress, either because needed data is locked or because it must wait for a reply to a request sent to another server. In either case the server should be able to turn its attention to a new activity rather than remaining idle. These requirements rule out the use of structures such as monitors,[13,4] which suffer from the "nested call" problem:[20,11] if the process having possession of the monitor makes a call that causes a delay, perhaps by calling another monitor where a delay is encountered, the higher level monitor cannot take part in any other activity. The nested call problem also exists for Ada tasks.[24]

There are two basic methods of providing concurrency. In the static model, there is a fixed number of processes per guardian. In the simplest case, a single process per guardian is provided, although more general schemes are possible in which the number of processes is determined when a guardian is created. The alternative is the dynamic model, in which processes are created dynamically as needed to handle concurrent demands. In the static model, it will usually be necessary to multiplex: a single process must switch its attention among concurrent activities. Multiplexing is most obviously needed when there is just one process, but cannot be eliminated in any static scheme. In the dynamic model, a process can be created for each concurrent activity, eliminating the need for explicit multiplexing.

The dynamic model has two important advantages over the static model. First, it is more general in that the program is structured in a way that can take advantage of any concurrency actually available in the hardware. Distinct processes could make use of different processors in a multiprocessor node, or could share the use of a single processor to avoid delays associated with paging. This generally arises because the abstraction presented by the dynamic model permits more freedom of implementation at the system level than does the static model.

Second, a well-structured dynamic process creation mechanism can lead to simpler and more understandable programs.* The dynamic model supports a methodology in which the programmer writes sequential programs each of which performs a separate activity. Writing programs in this way is straightforward - programmers are used to this style, and the resulting programs are easy to understand. Each such program can be proved to satisfy its specification using standard reasoning methods. By contrast, multiplexing is not similar to ordinary programming, and leads to complicated programs, especially when a high degree of multiplexing is needed.

Of course, if there is more than one process, the concurrent processes may need to coordinate with one other. Coordination among multiple processes can be accomplished

---

* It is critical that process creation and termination be well-structured; the undisciplined use of primitives such as fork can render programs difficult to understand.

through shared, synchronized data. Probably the main argument against multiple processes is that such coordination is needed; it is not needed in the single-process static model. Nevertheless, since all the sharing processes must be within the same guardian, reasoning about program correctness is local.

Another argument against the dynamic model is expense: processes must be created and destroyed, and the processor(s) must be multiplexed among the processes. However, all processes within a guardian would run in the same address space and same protection domain. The actual switching of the processor from one process to another is then like a cheap procedure call: the registers must be saved for one process and then loaded for another. If the cost of process creation and destruction turns out to be excessive, it might be reduced by measures such as reusing processes from a pool. All this should be inexpensive, and leaves only the scheduling itself. Whether this is expensive depends on whether a simple strategy is sufficient. We expect a simple strategy will be sufficient within something as constrained as a guardian, in which very few processes are likely to be ready to run at a time (many processes might be active but blocked waiting to receive a reply or use an atomic object).

4. STRUCTURE OF CLIENTS

We have already argued that client-server interactions take the form of paired send/reply. In this section we discuss process and communication mechanisms for clients; servers will be discussed in the next section. We begin by listing a set of design goals; most of these goals were discussed and justified in Section 3.

1) The paired send/reply must run as a subaction of the client action. This semantics allows the subaction to be aborted if communication fails; the client action survives the abort and can do some alternative computation.
2) The system, not the user, must make decisions about low-level timeouts and retransmission.
3) Clients require a general mechanism to terminate pending requests. The terminated request will be aborted, but the client should not have to wait until this abort has occurred (in other parts of the distributed system) before proceeding.*

---

* This requirement implies that activity on behalf of an aborted action may persist in other parts of a distributed system. We refer to such activities as "orphans". We have developed algorithms to eliminate orphans.[22]

4) Clients should be able to do concurrent tasks. A special case is the ability to do other work while remote requests are pending.

5) Clients having multiple pending requests may choose to handle responses in any convenient order (for example, in the order of arrival). Clients require the ability to receive any one of a number of pending responses, possibly of different types.

In the remainder of this section, we discuss two communication schemes for clients, and outline our reasons for preferring one over the other. Although each scheme satisfies our design goals, they differ in their respective degrees of safety and simplicity.

4.1 Reliable Send

The reliable send mechanism provides distinct send and receive primitives for originating requests and acquiring responses. A client process executing a send is blocked until the request message is constructed, but may then continue processing. The client acquires the response to a request by executing the receive primitive. After executing a send and before executing the receive, the client may undertake other activity, perhaps executing other sends and receives. An additional primitive is provided to terminate and abort pending requests.

Probably two forms of receive are needed, one that allows the client to wait for a response, and a second that allows the client to receive the response if it has arrived but to continue processing if it has not. In the latter form, an exception mechanism is a convenient way to distinguish the different cases,[19] i.e., the second receive terminates normally and returns the response if it has arrived, and terminates with the "not arrived" exception if it has not.

To satisfy the second design goal, the system must assume responsibility for delivery of requests. Each request is guaranteed to be delivered to the server with a certain (implementation-dependent) probability. If the system determines that the response is unlikely to arrive (because, for example, it is unable to communicate with the server's node), it aborts the subaction corresponding to the paired send/reply, and conveys this information to the client when the client executes the receive by signalling the "failure" exception. The receive primitive raises the failure exception only if no response is expected, and if attempts to retransmit the original request in the near future are highly unlikely to succeed. In the absence of these guarantees, it is difficult to imagine how programmers could decide whether to attempt retransmission, resume waiting, or give up. As it is, there is no point in immediate retransmission; thus there is no decision to make, and no point in having retransmission loops in programs.

It may appear that incorporating reliable send in a language is straightforward: just provide the primitives sketched above. However, when those primitives are examined in more detail, two problems arise. The first is a naming problem: how does the caller of the receive primitive indicate the request for which a response is desired? The second is a safety problem: is it possible to prevent the occurrence of common user errors, such as forgetting to receive responses, or attempting to receive the same response twice?

The following structured language mechanism solves both these problems. One might define a <u>request block</u> that executes a send on entry, executes a number of statements inside the block, and executes the associated receive on exit. An abnormal exit from the block would terminate and abort the request. Such a scheme provides safety and also a degree of simplicity by removing the need for receives to name the request for which a response is desired, thus increasing program clarity. Unfortunately, however, request blocks are not a satisfactory mechansim, since they do not nest properly. If request blocks are nested, receives must be executed in the reverse order of their corresponding sends, a restriction that is inconvenient and inefficient, and that violates our fifth design goal.

Since the request block cannot be used, we are left with fairly low-level and poorly-structured mechanisms. Requests must be named explicitly (presumably the send provides this name), and the responses received by name.[3,2] The receive primitive that waits for responses would have to take an argument list of request names, and a type-safe mechanism would have to be designed to handle each of the different kinds of responses. Safety could probably be achieved by some combination of static and dynamic checks, but rather than exploring the details of such a mechanism, we simply observe that safety concerns introduce considerable complication. Therefore, in the next section we discuss an alternative to reliable send that is both simple and safe.

4.2 Remote Procedure Call

The <u>remote procedure call</u> mechanism provides a single primitive for sending a request and receiving the associated response. A remote call resembles an ordinary procedure call: the invoking process is blocked until either a response message is received, or a failure exception is raised. As noted above, the remote call is executed as a nested atomic action, and a failure exception means that the request is unlikely to succeed if retried in the near future, and that no effects of the failed request will be observed (i.e., the subaction has been aborted). The remote call avoids the problems of the reliable send: it is safe, and does not require complicated linguistic mechanisms. Because it violates our fourth design goal of concurrency, an additional mechanism is required for creating concurrent processes.

We have chosen to use remote call combined with multiprocessing in Argus because we think the safety of remote call is very important, and because we prefer the dynamic model for the reasons discussed in Section 3. To provide for creation of new processes, we supply the coenter primitive. A coenter consists of a collection of arms, where each arm is block of code that may include remote calls. A process executing a coenter is blocked, and the arms are then executed by concurrent processes. When all the arms have completed, the original process is unblocked. The process structure can thus be visualized as a tree, where the only active processes are the leaves.

In designing the coenter we could have had the new processes all execute on behalf of the action performing the coenter. We chose instead to have each process execute as a distinct subaction to take advantage of the synchronization and recovery provided by the action system, and to avoid the need to synchronize processes operating on behalf of the same action. This decision implies that each arm may be aborted independently of the others, and that the net effect of executing the coenter is that the arms that commit appear to have been executed in some serial order. The decision also leads to a methodology for organizing the concurrency of a coenter: all interarm communication, and all communication between the arms and the parent action should be limited to atomic objects. In this case, the shared atomic objects will provide proper synchronization and recovery.

An additional mechanism is used to abort pending requests: if a statement in a coenter arm transfers control out of the enclosing coenter, it is assumed that the outcomes of any arms that have not yet completed are not of interest, and those arms are aborted. For example, a timed remote call can be constructed using a coenter with two arms: one arm performs the call and exits when it is complete, while the other arm exits when a chosen time interval has elapsed. The arm that finishes first decides the outcome of the coenter.

4.3 An Example

To illustrate how remote procedure call and concurrent processes might be used, let us examine part of the implementation of a fully replicated database. The client in this example is the front-end for the database, and the servers are the replicated copies of the data. The front-end is a guardian that mediates between database clients and the replicated copies; it accepts update and query requests from clients, and carries out the protocols necessary to execute the request: To read from the database, it suffices to read from any copy; to modify the database, it is necessary to modify every copy. Figure 1 illustrates two procedures that might be used by the front-end to query and update the database. This example can readily be extended to implement various voting strategies.[8]

```
query = proc(database:array[replica],key:key_type) returns (item_type) signals (unavailable)
    coenter % Create an arm for each replica.
        action foreach r: replica in array[replica]$elements(database)
            return(r.query(key)) % Return the result of the first successful query.
            except when failure (*): end % ignore failures.
        end % coenter
    signal unavailable
    end query

update = proc(database:array[replica],key:key_type,item:item_type) signals (unavailable)
    coenter % Create an arm for each replica.
        action foreach r:replica in array[replica]$elements(database)
            r.update(key,item)
            except when failure (*): % Signal to caller.
                        signal unavailable
                end % except
        end % coenter
    end update
```

Fig. 1:   Using Coenter to Manage a Fully Replicated Database

The query procedure sends concurrent requests to all replicas and returns as soon as it receives the first successful response. If all the handler calls fail, the query procedure signals an exception named unavailable to its caller. The update procedure sends concurrent requests to all replicas. The procedure signals unavailable to its caller as soon as the first handler call fails. If all the handler calls succeed, the update procedure returns normally to its caller.

Let us examine this program fragment in more detail. The names of the individual database guardians are kept in an array of replicas, here called database. Since the number of replicas may not be known at compile time, the foreach statement is used to define a coenter block having a variable number of arms. An iterator[16] is used to extract the elements of the array one at a time, and a new arm (subaction and process) is created (but not yet executed) for each element. In this example, if the array [replica]$elements iterator yields n replicas, n distinct arms will be created, each containing a call of one replica's update or query handler. The variable r to which the replica is assigned is local to each arm. Once the iterator has terminated and all the arms have been created, the arms are executed concurrently.

In the query procedure shown, each arm makes a handler call to a replica guardian. (The expression "r.query" denotes the query handler of the guardian named by the variable r). Once a single handler call succeeds there is no point in waiting for the others; thus as soon as one arm receives a response, the results are returned to the procedure's caller without further delay, committing the arm whose call was successful, aborting the arms whose calls are pending, and releasing any locks held by the aborted subactions. If a handler call fails, it raises an exception that is caught by the associated except clause, leaving the other, completed arms unaffected. It is clear in this instance that the failure of a remote call should not cause the entire query to abort; nested actions facilitate this kind of fault-tolerance.

The update procedure uses a complementary strategy. The update succeeds only if all component handler calls succeed. If a failure occurs, there is no point in waiting for the other arms to complete because the update attempt is doomed. In this event, the signal unavailable statement aborts all uncompleted arms, releasing their locks, and signals an exception to the procedure's caller. The caller can then abort the enclosing action, and the atomic action mechanism ensures that any modifications that may have been performed by the other arms will not be observed by subsequent actions.

Note that this scheme is subject to deadlock if two nearly simultaneous, conflicting update attempts lock the replicas in different orders. Nevertheless, the scheme might be attractive if updates were expected infrequently, and deadlocks thus expected even more infrequently. Deadlocks that do occur are broken when one of the clients times out. No timeout arm is provided in these procedures because we think that such decisions are made more naturally at a higher level, perhaps by a "shell" program that interacts directly with the user's terminal.

4.4 Remarks

Both remote call and reliable send allow the client guardian to do work in the interval between the request and the response. The difference lies in how that work is accomplished. In the reliable send scheme, the client explicitly multiplexes among distinct activities, while in the remote call scheme, the work is done by concurrent processes. The remote call scheme is more elegant and easier to program, since it resembles familiar procedure invocation. It is also much safer; the relationship between a process and its actions is implicit, thus avoiding a rich source of programmer errors. On the other hand, remote call requires that multiple processes be cheap.

A disadvantage of remote call is that it does not support pipelining. (Neither does the reliable send as described above, since we said nothing about order of delivery of messages; however, such a requirement could easily be added to reliable send). With remote call it is possible, for example, to partition a database, process and then transmit each partition in parallel. One can imagine that such a structure could be useful, but whether it could take the place of true pipelining is unclear.

5. STRUCTURE OF SERVERS

We now address the structuring of server guardians. The main issues here are how to schedule service requests (i.e., handler calls), and how processes carry out those requests. In addition, the server guardian may perform periodic background activities independently of any handler call. Associating a process or processes with such activities is also an issue.

As discussed in Section 3, there are two possible process structures for guardians: static and dynamic. In the most common form of the static model, a single process switches its attention among concurrent tasks. For example, when a server is ready to accept a new call, it might choose which call to accept, perhaps by some form of guarded command.[6] Mechanisms supporting the static model typically provide explicit control over the scheduling of incoming calls.

In the dynamic model, a guardian contains many processes, each of which performs a task. As discussed above, we chose a dynamic structure for client guardians. Since server guardians may also be client guardians, servers may contain multiple processes. Given this fact, can explicit scheduling be done by servers? Ultimately (after many attempts to provide explicit scheduling) we decided that explicit scheduling is usually not needed in the dynamic model. Furthermore, although user control over scheduling may be needed in some instances, we do not understand those needs sufficiently to provide a mechanism, even in the static model. These conclusions are justified later in this section.

Therefore, we decided on a full dynamic model for Argus, in which the scheduling of calls is done automatically by the Argus runtime system. This approach works as follows: Sometime after a handler call arrives, the system schedules the call and creates a process to run it.* The process is run as a subaction of the calling action. When the call has been serviced, a reply message is sent back to the caller, and the process and the associated subaction are terminated. The system is responsible for reliable delivery of the reply. The server is not explicitly informed of a failure to deliver the reply, because there is nothing useful it could do. If the reply is not delivered, the action will be aborted and the locks held at the server will be released automatically.

The decision to use an implicit approach for scheduling handler calls represented a tremendous step forward, for two reasons. First, it greatly simplified guardian definitions. Guardian definitions now look like Simula classes,[5] consisting primarily of definitions of "handlers", which resemble procedures. There is one handler for each type of request the guardian handles; the name of the handler is the name of the re-

---

* Processes are not created for procedure calls within a guardian.

quest. A handler may have arguments (the arguments that go with its request) and results (the results expected in a reply to its request). In addition, a guardian can have background code, which performs whatever periodic activity the guardian needs to carry out. The runtime system provides a process to run the background code. Of course, both handlers and the background code can use the coenter if desired to achieve greater concurrency. We have found guardians written in this way easy to program and easy to understand.

The second benefit of implicit scheduling is safety, in the sense of keeping actions distinct. As mentioned in Section 3, such safety was an important goal of the Argus design. The local data of different actions are disjoint, since the actions run in different scopes. It is never necessary for user code to make explicit use of action names or the caller's identity, reducing the complexity of the code and the chances for error. Instead, action information is associated implicitly with the process running the call. If processes share atomic data, synchronization is done automatically on the basis of the associated actions. When a process has finished, its action is terminated, and the reply is routed back to the caller automatically. Thus, there is no danger of a programmer becoming confused about which action a process is working on, since a process does not survive the termination of its single action. In contrast, such confusion is likely in multiplexing since a single process switches its attention among actions.

5.1 An Example

To illustrate how the dynamic process model facilitates server design, we will examine the implementation of a simple guardian. A more extensive example has been given elsewhere.[17]

An example of a guardian definition is shown in Figure 2. A guardian definition implements a special kind of abstract data type whose operations are handlers. The name of this type, and the names of the handlers, are listed in the guardian header. In addition, the type provides one or more creation operations, called creators, that can be invoked to create new guardians of the type; the names of the creators are also listed in the header.

The guardian in this example might be part of a banking system. It keeps track of a collection of accounts, perhaps the accounts managed by a particular branch office. Operations are shown to credit and debit accounts, to inspect account balances, and to open and close accounts. Atomic actions may be used to construct more complex operations that must preserve invariants: for example, by executing a credit and a debit as a single atomic action, it is possible to ensure that a funds transfer either succeeds or leaves the account balances unchanged.

```
bank = guardian is create handles debit, credit, query, open, close...

   % Define the following abbreviations:
   status_record = atomic_record[balance: int,...]
   table = atomic_table[account_id, status_record]

   % The table associates each account with a status record,
   % which contains the account's current balance and other information.
   % Initially, the table is empty.
   stable directory: table:= table$create()

create = creator () returns (bank)
        return (self)
        end create

credit = handler (account: account_id, amount: int) signals (not_found)
         account_status: status_record:= table$lookup(directory, account)
          resignal not_found
         account_status.balance:= account_status.balance + amount
         end credit

debit  = handler (account: account_id, amount: int) signals (not_found, insufficient_funds)
         account_status: status_record:= table$lookup(directory, account)
          resignal not_found
         if account_status.balance < amount then signal insufficient_funds end
         account_status.balance:= account_status.balance - amount
         end debit

query = handler (account: account_id) returns (int) signals (not_found)
         account_staus: status_record:= table$lookup(directory, account)
          resignal not_found
         return(account_status.balance)
         end query

open  = handler (account: account_id) signals (duplicate)
         table$insert(directory, account, status${balance: 0,...})
          resignal duplicate
         end open

close = handler (account: account_id) signals (not_found)
         table$remove(directory, account)
          resignal not_found
         end close

   ...

end bank
```

Fig. 2: An Example Guardian

The first internal part of a guardian is a list of abbreviations for types and constants. Next is a list of variable declarations, with optional initializations, defining the guardian state. Some of these variables can be declared as stable variables; the others are volatile variables. The stable state of a guardian consists of all objects reachable* from the stable variables; these objects, called stable objects, have their new versions written to stable storage by the system when top-level actions commit. After a crash, the system re-creates the guardian and restores its stable objects from stable storage.

The bank guardian's state consists of a single stable object, a directory that associates account identifiers with status records. The directory is an object of the atomic table abstract data type, which is parameterized by the types of the objects it contains (account ids and status records in the example). This type provides the following operations: create creates an empty table, insert inserts a new account/ status record pair in the table, remove removes an account/status record pair from the table, and lookup returns the status record associated with a particular account. Insert raises an exception if the account is already in the table, while remove and lookup raise an exception if the account is not in the table.

An account's status record is an object of type atomic record, which is parameterized by the names and types of its fields. The atomic record type provides operations to read and modify the record's fields. For the purposes of this example, we will be concerned with the field of the status record that contains the account's current balance. The following abbreviations are used in the program text: status record stands for the atomic record type, and table stands for the atomic table type. Both tables and status records are atomic objects, providing their own synchronization and recovery, and may be shared by concurrent actions.

Aside from processor speed, the only limitation on concurrency is contention for resources; a process executing on behalf of an action may have to wait for locks held by other actions. To analyze the concurrency provided by this guardian, it suffices to analyze possible lock conflicts. To understand this analysis, it is important to understand the distinction between process synchronization and action synchroniza-

---

* Argus, like CLU, has an object-oriented semantics. Variables name (or refer to) objects residing in a heap storage area. Objects themselves may refer to other objects, permitting recursive and cyclic data structures without the use of explicit pointers. The set of objects reachable from a variable consists of the object that variable refers to, any objects referred to by that object, and so on. In languages that are not object-oriented, the concept of reachability would still be needed to accommodate the use of explicit pointers.

tion. Process synchronization ensures that processes cannot view any concurrently shared object in an internally inconsistent state, while action synchronization ensures atomicity. Process synchronization requires only short-term exclusion; a semaphore or monitor could be used to give each process exclusive access to an object while operations are in progress. Action synchronization requires longer term exclusion; if one action modifies an object and a second action then attempts to observe the object's state, the second action must wait until the first action either commits or aborts. Because of the difference in the time scales, concurrency among actions is more important than concurrency among processes. Thus in the bank example, it is important that an action that cannot proceed because it cannot access the needed account not prevent other actions from running in the bank guardian. It is less important that processes within the bank guardian really run concurrently, although such concurrency might be useful if the guardian ran on a multiprocessor.

Two kinds of objects may be shared by actions using the bank guardian: the table and the status records of individual accounts. The degree of action concurrency provided by the guardian is thus determined by the kinds of synchronization provided by these types. Let us make the conservative assumption that each object has its own lock which may be held by an action in shared or exclusive mode, and which is released when the action commits or aborts. For the table type, the lookup operation acquires the lock in shared mode, and the insert and delete operations each acquire the lock in exclusive mode. Reading a field of a status record locks the record in shared mode, and modifying a field locks the record in exclusive mode. Since credit, debit and query requests affecting different accounts do not acquire conflicting locks, they may be processed concurrently. An action that modifies an account balance will conflict with other actions affecting that same account because one action will have to wait for another to release the lock protecting the account's status record. A request to open or close an account excludes all other requests, since it acquires an exclusive lock on the entire table. If these requests are expected to occur frequently enough that contention becomes a problem, the implementation of the table data type should be modified to provide a finer grain of locking, perhaps by placing an individual lock on each account/balance pair (an example of such an implementation has been given elsewhere[26]).

5.2 Remarks

We have argued that dynamic guardians are a good linguistic mechanism leading to simple and safe programs. Guardians may not, however, explicitly control the scheduling of incoming calls. In the static model, by constrast, the programmer is usually provided with some kind of scheduling mechanism. Using such a mechanism, two kinds of decisions are made. The first we might characterize as exclusion decisions: a decision is made not to schedule some call, perhaps because a resource it needs is not available at present. The second kind of decision is a <u>priority</u> decision: one call is accepted over another because it is judged to be more important.

In the static model, guarded commands[6] might be used to make explicit decisions about exclusion. Several difficulties emerge, however, if we consider how the guarded command might be used to implement the bank guardian described in Section 5.1. To ensure that distinct accounts can be updated concurrently, the guarded command's predicate must depend on the arguments of the call (e.g., the account id) as well as the caller's identity and the guardian state. These requirements imply that the Ada guarded command mechanism, which does not permit access to arguments, is not sufficiently powerful, although the more elaborate mechanisms proposed by Andrews[1] would probably suffice. Guarded commands also have efficiency problems. To ascertain the status of an account, the guarded command must look up the account's status record in the table each time the predicate is evaluated. When the call is accepted, the status record must be retrieved again.

In our model, exclusion decisions are made by the processes themselves, usually by waiting for a needed atomic object to become available. This approach avoids the efficiency problems associated with guarded commands because results of computations can be saved for future use, eliminating unnecessary predicate evaluations and other computations. Of course, it is true that the dynamic model will have blocked processes instead of queued messages, but blocked processes need not be expensive, since they just consist of activation records chained on queues.

The dynamic process model also avoids a problem that may arise in the static model: the problem of accepting the wrong request. In Ada, for example, if a task working on a request discovers that it is unable to make progress, it can accept a new request, but the new call must be completed before the old one can be continued, a requirement that we consider unacceptable. Note that despite the shortcomings of guarded commands, they are needed in Ada to compensate for the language's limited support for multiplexing.

Priority raises a different set of questions. One could imagine two reasons why application programmers might desire explicit control over process priority: to control bottlenecks, and to support activities that require particularly fast response. Although we find the notion that applications programmers will use priority to control bottlenecks somewhat suspect, it is clear that if a priority mechanism is to relieve rather than exacerbate bottlenecks, it must be fast and simple.

Language support for process priority raises a number of difficult issues. How does a process establish (or change) its own priority? What kind of security requirements are needed to ensure that processes do not consume a disproportionate share of resources? Priority also interacts with atomic actions. It may be necessary for high-priority actions to preempt locks held by other actions to ensure that the high-priority action will make progress. Such a guarantee is not possible otherwise, since two or more actions could deadlock repeatedly. We believe that some form of process priority may eventually be needed, but we do not understand the needs of applications well enough to provide a mechanism at present. There is no reason to believe that these problems can be addressed more easily using the static model.

6. CONCLUSION

This paper has been concerned with process and communication structure for distributed programs. Although we explained the approach taken by Argus in these two areas, the main point of the paper was not the particular solutions chosen, but the analysis of the issues that must be faced in these areas. We also discussed some alternative solutions, and explained the rationale for our choices.

In looking back on the design of Argus, it is clear that the basic decisions discussed in Section 3 constrained severely the set of options open to us. Given these options, it seems to us in retrospect that our decisions were fairly obvious. The decision process can be summarized as follows. We did not choose reliable send for the client's communication mechanism because it cannot be made simple and elegant. Furthermore, the concurrency it provides is not needed if there are multiple processes in a guardian. But we wanted multiple processes anyway because of the potential complexity of programs when a single process must be multiplexed.

Our decisions were motivated primarily by the perceived needs of Argus users: the programmers of distributed applications in the domain supported by Argus. We believe that distributed programs are inherently more complex than sequential programs. Consequently, we think that a high-level language intended for distributed programming must do more work for the user than would more traditional languages/systems. In addition, it is important to keep the mechanisms presented to application programmers as simple as possible, and as safe as possible, even at some cost in expressive power.

Because errors in synchronization and recovery are particularly difficult to detect and fix, considerable effort was expended to shape the language so that the atomic action structure would be largely implicit. This concern was crucial in our decision to provide a dynamic process structure in guardians: each action running within a guardian has its own process, and the lifetime of the process is the lifetime of the action. The implicit association of processes and actions relieves users of the need to keep track of action identifiers, and avoids the inherent complexity of synchronization and recovery in programs where a single process is multiplexed. Of course, our decision to provide multiple processes presupposes that processes can be implemented cheaply; we argued earlier that this is a realizable goal.

The desire for simplicity led us to omit mechanisms where we felt we lacked a clear understanding of the needs of applications. Rather than complicate the language with a large set of primitives, we decided to wait until we had acquired more experience and insight. For example, we provide no mechanism for user control of priority. In this respect, Argus users are in the same position as users of most operating systems, which also do not provide user control over process priority.

Like any such effort, the Argus design is a result of trading off expressive power for safety and simplicity. Our decisions were formed by considering the Argus application domain, and by working out examples. A final evaluation of the efficiency and expressive power of our design will be forthcoming when the Argus implementation, currently underway, is complete.

7. ACKNOWLEDGEMENTS

This research was supported in part by the Advanced Research Projects Agency of the Department of Defense, monitored by the Office of Naval Research under contract N00014-75-C-0661, and in part by the National Science Foundation under grant MCS 79-23769.

The authors gratefully acknowledge the contributions made by the members of the Argus design group, especially Paul R. Johnson, Robert Scheifler and William Weihl.

REFERENCES

1.   Andrews, G.R., "Synchronizing Resources," ACM Trans. Program. Lang. Syst., Vol. 3, No. 4, October 1981, pp. 405-430.

2.   Bartlett, J.F., "A 'NonStop' Operating System," Tandem Computers, Inc., Cupertino, California.

3.   Brinch Hansen, P., "The Nucleus of a Multiprogramming System," Commun. ACM, Vol. 13, No. 4, April 1970, pp. 238-265.

4.   Brinch Hansen, P., "Distributed Processes: A Concurrent Programming Concept," Commun. ACM, Vol. 21, No. 11, November 1978, pp. 934-941.

5.   Dahl, O.-J. et al., "The Simula 67 Common Base Language," Publication No. S-22, Norwegian Computing Center, Oslo, 1970.

6.   Dijkstra, E.W., "Guarded Commands, Nondeterminacy, and Formal Derivation of Programs," Commun. ACM, Vol. 18, No. 8, August 1975, pp. 453-457.

7. Eswaren, K.P., Gray, J.N., Lorie, R.A., and Traiger, I.L., "The Notion of Consistency and Predicate Locks in a Database System," Commun. ACM, Vol. 19, No. 11, November 1976, pp. 624-633.

8. Gifford, D., "Weighted Voting for Replicated Data," Proc. 7th ACM SIGOPS Symposium on Operating Systems Principles, December 1979, pp. 150-162.

9. Gray, J.N., "Notes on Data Base Operating Systems," in Lecture Notes in Computer Science 60, G. Goos and J. Hartmanis, eds., Springer-Verlag, Berlin, 1978, pp. 393-481.

10. Herlihy, M., and Liskov, B., "A Value Transmission Method for Abstract Data Types," ACM Trans. Program. Lang. Syst., Vol. 4, No. 4, October 1982, pp. 527-551.

11. Haddon, B.K., "Nested Monitor Calls," ACM Operating Systems Review, Vol. 11, No. 4, October 1977, pp. 18-23.

12. Hoare, C.A.R., "Communicating Sequential Processes," Commun. ACM, Vol. 21, No. 8, August 1978, pp. 666-667.

13. Hoare, C.A.R., "Monitors: An Operating System Structuring Concept," Commun. ACM, Vol. 17, No. 10, October 1974, pp. 549-557.

14. Lampson, B., "Atomic Transactions," Distributed Systems: Architecture and Implementation, in Lecture Notes in Computer Science 105, G. Goos and J. Hartmanis, eds., Springer-Verlag, Berlin, 1981, pp. 246-265.

15. Liskov, B., "Primitives for Distributed Computing," Proc. 7th ACM SIGOPS Symposium on Operating Systems Principles, December 1979, pp. 33-42

16. Liskov, B. et al., "CLU Reference Manual," Lecture Notes in Computer Science 114, G. Goos and J. Hartmanis, eds., Springer-Verlag, Berlin, 1981.

17. Liskov, B., and Scheifler, R., "Guardians and Actions: Linguistic Support for Robust, Distributed Programs," Proc. 9th Annual ACM Symposium on Principles of Programming Languages, January 1982, pp. 7-19, Revised version to appear in ACM Trans. Progr. Lang. Syst.

18. Liskov, B., Snyder, A., Atkinson, R.R., and Schaffert, J.C., "Abstraction Mechanisms in CLU," Commun. ACM, Vol. 20, No. 8, August 1977, pp. 564-576.

19. Liskov, B., and Snyder, A., "Exception Handling in CLU," IEEE Trans. Software Eng., Vol. SE-5, No. 6, November 1979, pp. 546-558.

20. Lister, A., "The Problem of Nested Monitor Calls," ACM Operating Systems Review, Vol. 11, No. 2, July 1977, pp. 5-7.

21. Moss, J.E.B., "Nested Transactions: An Approach to Reliable Distributed Computing," Ph.D. Thesis, Technical Report MIT/LCS/TR-260, MIT Laboratory for Computer Science, Cambridge, MA, 1981.

22. Nelson, B., "Remote Procedure Call," Ph.D. Thesis, Technical Report CMU-CS-81-119, Carnegie-Mellon University, Pittsburgh, PA, 1981.

23. Papadimitriou, C.H., "The Serializability of Concurrent Database Updates," J. ACM, Vol. 26, No. 4, October 1979, pp. 631-653.

24. Preliminary ADA Reference Manual, SIGPLAN Notices, Vol. 14, No. 6, June 1979.

25. Reed, D.P., "Naming and Synchronization in a Decentralized Computer System," Ph.·D. Thesis, <u>Technical Report MIT/LCS/TR-205</u>, MIT Laboratory for Computer Science, Cambridge, MA, 1978.

26. Weihl, W., and Liskov, B., "Specification and Implementation of Resilient, Atomic Data Types," Computation Structures Group Memo 223, MIT Laboratory for Computer Science, Cambridge, MA, 1982.

DISCUSSION

Chairman: H. Sandmayr (Brown Boveri, Baden, Switzerland)

J. Ludewig (Brown Boveri, Baden, Switzerland)
I am not sure where Argus should be put on the scale between a traditional language like Ada and an object-oriented language like Smalltalk. More generally, I would like to know your opinion about future development. Are we all moving towards object-oriented languages? Will we replace traditional languages by object-oriented languages within the next decade?

B. Liskov
I am not sure whether I can give you a satisfactory answer. If you think of an object-oriented specification, you can see that there is a very easy translation from such a specification to guardians in Argus. So in that sense, Argus is an object-oriented language. On the other hand, Argus does not have a subclass mechanism in it. And I am personally not convinced about the importance of a subclass mechanism. I have been trying to understand what exactly subclasses have to do with programming and where they are to fit into the programming process. Perhaps Dr. Matsumoto would have an answer for that. I think of Argus as a programming language, not as a specification technique.

Y. Matsumoto (Toshiba Corporation, Tokyo, Japan)
My opinion is that object-oriented languages do not cover the whole range of applications because in object-oriented thinking everything is divided into particles. In some of our applications, such as very large power-dispatching systems, one algorithm is used to optimize the overall process. In such an application we cannot divide the algorithm into distributed particles. We use object-oriented languages mainly for distributed programs.

D. Keller (Brown Boveri, Baden, Switzerland)
I want to make a short comment on object-oriented design and object-oriented languages. From my point of view, people often mix the terms object orientation and abstract data types together. Sometimes object-oriented design is said and what is actually meant is abstract data types. An object-oriented language, for me, is something like Smalltalk. I think people will move towards abstract data types, but not so strongly towards Smalltalk.

H. Aschmann (Brown Boveri, Baden, Switzerland)
I have a question concerning the replication of guardians. What are the mechanisms to coordinate handler calls, and especially calls that handlers of a replicated guardian may make? How is the consistency of stable data among replicas ensured?

B. Liskov

There is no special support provided for replication of guardians. The state of the replicas can be maintained consistently by keeping the replicated data in atomic objects. The synchronization provided by these objects will coordinate handler calls running in the guardians. If more coordination is needed, it must be programmed explicitly. This can be done, for example, by using additional atomic objects.

REQUIREMENTS ENGINEERING AND SOFTWARE DEVELOPMENT:
A STUDY TOWARD ANOTHER LIFE-CYCLE MODEL

Y. MATSUMOTO

Toshiba Corporation, Tokyo, Japan

ABSTRACT

The problem of how to increase both productivity and quality is of primary concern in software production. The waterfall life-cycle model is a strict sequence of phases where the sequential order is fixed. The paper describes a unified life-cycle model which is a derivative of the waterfall model. The unified model is a scheme for connecting the phases in the life cycle in a customized order and to bring about higher productivity. A distinctive feature of the unified model is the "semantic model" which is created at the top phase of the life cycle. The selected phases are arbitrarily interconnected to this phase in the order selected by the designer.

In the waterfall life-cycle model, consideration of implementation is delayed until after the requirements are completely specified. In the unified life-cycle model designers are allowed to devise a prototype model which satisfies user requirements, describe it and execute it during the requirement-specification phase. The semantic model is a representation of this model in a semantics which is compatible with that of the implementation. In other words, the semantics applied in describing the semantic model will be consistent with the semantics in which the programming of the software system for the target computer system is implemented. Applying one unique semantics throughout the life cycle is beneficial because the maintainability of the documents in the early phases could be improved to the degree in which the source text for those documents is maintained.

The semantic model is a set of objects interconnected with one another through the passing of messages. The paper describes a technique for defining objects from the requirements and their description by an object-oriented language called OKBL (object-oriented knowledge-based language). Each defined object may be transformed to a set of functions or it may be transformed to a set of program modules in the succeeding design phases. The final codes for the target computers may also be described in OKBL. Following a unique semantics throughout the life cycle leads to increased productivity for the following reasons:

- Describing specifications based on the same semantics throughout the whole life cycle increases traceability between phases and results in improvement in maintainability.

- Reusing existing objects simplifies the problems of reusability improvement.

- The time spent in the intermediate phases, where the transformations from requirements to programs are made, is shortened.

1. INTRODUCTION

Software design is a refinement process in which requirements specified in a problem domain are gradually transformed into programs. A software life cycle defines phases along a time dimension. As time passes, the design process progresses from phase to phase. The waterfall life-cycle model is a strict sequence of phases, in which no phase may be bypassed. This model has been widely applied in industry, but without complete success for the following reasons:

1) In order to study the implementability of the requirements being defined, designers usually direct their thoughts to "how" even before the requirements are completely finalized. In large-scale systems, it is usually unrealistic to delay "how" until the requirements are completely defined. Thinking "how" in the middle of the requirements phase has been the practice in order to create a prototype model and execute it for the purpose of increasing the implementability of the requirements to be defined and also the correctness of the semantic model or the prototype model.

2) While requirements are being analyzed, potential customers are willing to check that the requirements are being correctly followed and met by the software being designed. In addition, some customers are interested in prediciting the performance of the target system.

3) The development process often reuses a number of existing concepts and software modules. The reusing processes often connect a phase in the early

stage directly to a phase in the later stage of the life cycle; for example, if we select existing software which fits the requirements, we can jump into the detail design or programming phase.

In contrast to the waterfall model, the throwaway model allows phases to be defined in a customized, rather than a fixed sequential order. In the throwaway model, unnecessary phases are not executed and completed phases may be executed again if necessary.

The customization varies depending on the nature of the application, and each customization is based on long experience specific to the particular application. In our software factory,[18] a number of applications of different nature proceed in parallel and each one applies the most appropriate life-cycle model. Our desire of establishing an integrated concept to manage various hrowaway models in our factory brings us to derive a life-cycle model called the "unified life-cycle model".[17] In a unified life-cycle model:

- All phases are independently defined;

- Each phase has well-defined interfaces to other phases to allow interconnection of phases which have matching interfaces;

- Selected phases are arbitrarily interconnected by the designer, either in the sequential or customized order.

In the unified life-cycle model, the "semantic model" is created in the requirements phase in parallel with the creation of the requirements specification. The semantic model is a representation of the designer's scheme in a semantics that is compatible with the semantics applied to implementing the corresponding target software. The semantic model can be executed and, as a result, can be modified repeatedly until the behavior of the model satisfies the requirements. At the end of the requirements phase, the semantic model created should satisfy the requirements. The selection of the phase which succeeds the requirements phase depends upon the degree to which each object in the model is detailed. If, for example, an object in the semantic model represents an existing program module, this object may be transferred to the programming phase. If another object in the same semantic model describes a new concept, this object may be passed to the conceptual design phase.

The paper first provides a survey of various models applied to specifying requirements. Requirement specifications are considered as the descriptions of a semantic model. Secondly, our semantic model descriptions are introduced. Finally, semantic modelling in the language OKBL is presented in an example.

2. REQUIREMENTS SPECIFICATION

At the beginning of the life cycle, the target system is conceived as a conceptual object which is related to various conditions. The system may exchange information with the given environments and generate solutions for users under the given constraints and evaluation criteria. The items which users define are as follows:

1) The information which the system communicates with the outside environments

2) The solutions or results which will profit users when the target system is realized

3) The constraints and evaluation criteria by which design alternatives are evaluated

Requirements analysis is defined as a phase for analyzing the given problem, in which relationships between those items in (1), (2) and (3) are analyzed and defined. To do this, several traditional engineering technologies, such as those listed in Table 1, are applied. Results of the analysis are usually written in a text which may consist of a number of sentences in a natural language, drawings and tables. We call this text the "problem description" or "needs description". A problem description should contain details of the items which are described in (I), (2) and (3). In the real project, the designer studies the problem descriptions and then creates his conceptual model, describing it with templates. We call such a template "view". Several types of view have been proposed in a number of papers, the major ones in current use are as follows:

Contextual View: Sentences in a natural language in problem descriptions may contain subjective and objective words (e.g., nouns) which are considered to represent major components that will constitute the framework of the target software. We call these components elementary entities. We extract elementary entities from the given problem descriptions and define each relationship between elementary entities. The description is made in a syntactic form which is called contextual form.

The syntax of the contextual view applied in many previous works[23,3] has its base in the E-R (entities, relationships) model. Some systems do not allow the presentation to be described in precise semantics, whilst other systems[11,4] provide languages with which precise semantics can be presented in the way that we describe the semantics of the programming language using an attribute grammar.

TABLE 1

Techniques Applied in Requirements Analysis

Items	Technologies
Problem survey	Morphology, Delphi method, scenario, structural modelling, relevance matrix
Value system design	Value analysis, relevance matrix/tree
System synthesis	Relevance tree, problem-solving tree, structuring/clustering
System analysis	Modelling/simulation, relevance matrix
Optimization	Multipurposes optimization, dynamic programming
Decision making	Cognition, decision making, technology assessment
Scheduling	Critical-path method, network theory, queue

Functional View: The given problem description may contain a number of verbs each of which represents a fragment of the functions associated with a certain elementary entity in the system. We analyze all the functions and the associated entities. The functions may then be clustered into a finite number of processes with reference to each associated entity. A class of syntax to describe processes and relationships between processes is called "functional view". The semantics applied to the functional view is that all processes are in concurrence and streams of data passed between processes are also in concurrence. A number of systems[20] are based on this view.

<u>Data View</u>: In data-dominant systems, such as information-processing systems for office business, a target system is generally modelled as a combination of sets of data structures and functions to transform each set of data structure into another set of data structure. In these systems, we could define requirements by describing the structures and semantics of input data, output data and historical data to be stored, and then define the functions which transform the defined data. Several systems[12,4] have applied this view.

<u>Dynamic View</u>: In control-dominant systems, such as process-control systems, the timings to activate or terminate major functions of the target systems are related to the real-time clock or defined events which may occur in the outside world. In these systems, requirements have been modelled using state transition,[21,16] stimulus/response or parallel computations.[1]

Using either one of the above views or a combination of them, designers define a first-hand scheme for target implementation, called the prototype.[2] In general, the semantics applied to the description of a prototype is not precisely compatible with the semantics applied to implementing the target software for which the prototype was designed.

The author's contribution is a proposal to introduce a view based on an object-oriented model during the prototyping, under the assumption that the model is finally implementable as the target software in our real-time operation environment, without changing its semantics. This specific class of the prototype we call the semantic model. The semantics applied to describe the object-oriented semantic model is precise and consistent between the model and the corresponding target-software system. The application of a unique semantics through all phases of a project life cycle may clarify traces between descriptions made at different phases of the life cycle. It should be remembered that it is current practice that the programs already in operation are generally maintained using only the source text of those programs. Very few of the documents made in the early phases can participate in the maintenance. This is a result of the lack of semantic consistency between early-phase documents and the target codes. The reasons for introducing the object-oriented model as the semantic model are listed as follows:

- <u>Good Affinity with the Conceptual Model</u>: The semantics of the object-oriented model has better affinity with the conceptual model held by the designers than the other models.

- <u>Easy Implementability</u>: Implementing the object-oriented model under the real-time software environments currently provided by industrial process-computer systems is feasible without difficulty.

3. OBJECT-ORIENTED MODEL

The semantic model has been defined as a model described with a semantics that is compatible with the semantics of the target software, and semantic modelling is a process for creating a semantic model. A significant number of contributions has been made concerning the model and the modelling process for the phases of requirements specification and design. Many of the existing techniques have been classified into the following categories:

Requirements model:	Described in the preceeding section
Design model:	Bubble charts,[22] CSP,[9] automata-based model, Petri net-based model, parallel-process model structured charts,[22] structured diagrams,[10] Warnier-Orr charts,[19]
Program model:	Abstract model based on abstract data type,[13,6] concurrent model[8,5]

Someone responsible for a large project and familiar with a model such as CSP, a state transition-based model, a Petri net-based model, a concurrent processes-based model or an abstract data-type model, would find that the semantics of these models is too precise to describe the specifications in the early phases, while the semantics of the other models is different from the semantics applied when implementing target software.

Our semantic model is a description based on a definition of the semantics for those descriptions in terms of an object-oriented parallel-computation model.[7,24] In this model the computations are performed by concurrent passing of messages among procedural entities called objects. Each object has its own state and defines a set of operations which changes the state when the object becomes active. The object is in one or other of the binary modes, active or inactive. When an object is activated by a passed message, it invokes one of its operations defined by the message and may change its state or send the next message defined as a result of its computations.

In our computation model, computations are performed by concurrent passing of messages among procedural modules called objects. Each object models a conceptual unit which integrates data and related functions. The relationships between data and functions which will be incorporated into the same object are analyzed using views defined in the preceeding phase. Each object which contains data and associated functions has its own processing power and local memory. Each object has its own description which determines what messages it can accept and what computations it performs. When a described message arrives, the object may either perform a computation, send another message to the object including itself, create a new object, or

change its memory according to the description. After it completes the described
functions, it becomes inactive until it receives the next message.

Though the generation of messages in a system of objects may take place concurrently,
we assume message arrivals at an object to be linearly ordered: No two messages can
arrive at the same object simultaneously, and a single message queue sorted in the
arrival order is assumed for each object. When a message arrives at an inactive object
and no message remains in the message queue, then the arrived message is accepted
by the object. If the object is active, or any preceeding message is in the queue, the
arrived message is placed at the end of the queue. A message may include the name
of the object to which the return message is to be sent by the recipient. That is, an
object may send a message to an unknown object and know from the return message
which object received the message.

Various semantics can be conceived in defining the passing of messages. In our
semantic model we apply three types of semantics all of which are based on the common assumption called the arrival-ordering-preservation assumption. This type of the
message-passing semantics has been devised by Yonezawa.[25] Since we became involved
in the process-control business about twenty-five years ago, we have applied this
semantics in synchronizing control between asynchronous real-time tasks. Our experience has led us to favor Yonezawa's arrival-ordering-preservation assumption.

When two messages are sent to an object T by the same object O, the time ordering
of the two message transmissions (according to O's clock) must be preserved in the
time ordering of the two message arrivals (according to T's clock).

    1) Unacknowledged-type message passing denoted by [T<=M]
       Suppose object O is activated and sends a message M to object T. In un-
       acknowledged-type message passing, O does not wait for the answer that T
       receives M but continues computation after the generation of M, if it still
       has script for execution. This type of message passing increases concur-
       rency between objects. This message-passing type can be further divided
       into two subtypes which may correspond to different implementations. In the
       first subtype, O does not continue its computation until the arrival of the
       message is confirmed. In the second subtype it is not confirmed, which may
       allow higher concurrency than the first, but may reduce robustness against
       unexpected errors.

    2) Acknowledged-type message passing denoted by [T<==M]
       O sends a message M to T and waits not only for the response that T
       receives M, but also for the return message sent by T. If T does not re-
       turn a message, O waits until the timing that the current activation of T

caused by M terminates. This type of message passing is similar to a traditional "procedure call", but differs in that T may continue its activation
after it sends a message back to O. This message-passing type provides a
convenient means for synchronizing objects concurrently activated in a
parallel construct. This will be discussed in a later section.

3) Mailback-wanted type message passing denoted by [X:[T<=M]]
   After O sends a message M, O proceeds with its computation without waiting
   for acknowledgement by T. When T receives the message M, T performs its
   computation and puts the results into a mailbox X. O comes back, and looks
   into the mailbox and finds the result. If no result is found, O makes repeated visits to the mailbox until it gets the result. X is a specified memory
   area internal to O.

In addition to the message-passing semantics of Yonezawa, we implement others which
are derived from the scheme presented by CCITT Recommendation X.409(1983), some
of which are as follows: Suppose that O sends a message M to multiple objects simultaneously. If the objects which receive the message are many and unidentified, the
type of message passing is called "broadcast passing". If the receiving objects are
identified and multiple, the type of message passing is "multicast passing". Unac-
knowledged-broadcast passing of the message M is denoted by [[ ]<=M]. Unacknowl-
edged-multicast passing of the message M to the objects A, B and C is denoted by
[[A,B,C]<=M]. If object O waits until it receives acknowledgements by all objects of
the message M sent in multicast passing, the passing is denoted by [[A,B,C]<==M].
If object O waits until it receives acknowledgement by either one of the objects that
are guarded for selection[5] in multicast and broadcast passing, the denotations for
passings are:

$$[|A,B,C|<==M] \text{ for multicast passing and } [| \quad |<==M] \text{ for broadcast passing}$$

A message consists of two parts, an RR-part and a C-part as in ABCL, and the syntax of the message description is almost same as that of ABCL.[15] The RR-part describes the request by the message initiator. If the message is a return message, it
describes the reply by the responder. The C-part describes the name of the object to
which the next message created by the receiver is to be sent. In our notational
convention, a message is expressed by a pair of which the first and the second parts
are separated by "=>". The first part is the RR-part, and the second part is the
C-part and is of the form:

$$[<RR-part> => <C-part>]$$

If the C-part is not needed for specification, it may be left blank, or the message
form may be:

$$[\text{<RR-part>}]$$

The C-part is not necessarily specified in acknowledged-type message passings where
the object to be responded to is determined.

4. SEMANTIC CONSISTENCY BETWEEN SEMANTIC MODEL
AND TARGET SOFTWARE

The application software which is implemented under the real-time environment on the
target computer system consists of a multiple number of tasks (or processes) of which
the states are controlled by the real-time operating system. In a system of average
size, the number of tasks may be three to five hundred. A task can request the
operating system to activate the next task it needs. The timing of the activation can
be designated by the request primitives. Under the operating system, tasks may
exchange event signals with which the tasks participating in this exchange of signal
can synchronize each other. The semantics applied to the signal passings among tasks
have been defined in various ways since we became involved in the process-control
business. A typical semantics has been defined in ADA.

The consistency between the semantics of a model created in the requirements phase
and the semantics applied to constructing the target software which corresponds to
this model is defined in terms of the following transformation rules:

- A task in a real-time multiprogramming environment implements an object defined in the semantic model.

- A task may be created, activated, suspended, terminated or destroyed in
  accordance with each behavior of the corresponding object, including object
  creation, message generation, message receipt, script invocation or garbage
  (discarded objects) collection.

- A message passing is transformed into a sequence of primitives that are processed by the operating system in such way that the semantics of the message
  passings described in the preceeding section is completely realized by the
  task-control functions.

- The semantics represented by each name should be compatible with that of the
  same name in the corresponding task descriptions.

- The relation between each superclass and its subclasses is implemented by a set of tasking primitives which coordinates superseding and subsidiary tasks.

5. OKBL

In order to describe a semantic model, or the behavior of the objects in a system, in a precise and concrete manner, we developed a language called OKBL (object-oriented knowledge-based language). OKBL, which has its base in ABCL (An object-Based Core Language),[15] is an experimental programming language to describe prototypes as well as the target software which is implemented in the environments of target computer systems for process control.

5.1 Object Definition

Each object description beginning with the object name has a fixed set of message patterns which it can accept. The description of the computation which is performed in the acceptance of each message pattern is called "script". The object may have its own local memory in which its "state" is stored. To write a definition of an object in OKBL, we use notation of the following form:

```
[object:<object-name>
  (state: <representation of memory>)
  (receive[<pattern>] <script>)
  . . . . .
  (receive[<pattern>] <script>)]
```

In the description, (state:...) declares representation of local memory and initialization. Each script corresponding with each message pattern is a set of executable codes which computes, refers variables, and manipulates list structures. The following is a simple example of the object description. It is an object which models the behavior of an automatic ticket vending machine.

```
[object: aTicketVendor
  (state: [sum := 0])
  (receive: [throwin: X yen:]
    [sum := (+ sum X)]
    (display$sum sum))
  (receive: ([push: *y yen:] (>= sum *y))
    (sell$ticket *y)
    (return$change (- sum *y))
    [sum := 0])
```

```
  (receive: [reset:]
    (return$change sum)
    [sum := 0])
]
```

The state in this object is "sum" and 0 is bound to it at the start. The object can accept three types of the messages shown. In each denotation of the message pattern, the symbol with ":" at its end is the constant data. The symbol which begins with an upper case character or "*" is the variable. The value transmitted will be bound to each symbol in the pattern denotation. The symbols containing "$" appearing in the scripts describe functions which are activated. The object to represent a class may be described in a simple form presented in the following:

```
[object Class_of_Something
  (receive:   [<message pattern of request for creating class
            >]
            <object description>
  )
]
```

The following example describes a class of Ship.

```
[object: CreateShip
  (state:   [*default-x-position* := 0.0]
            [*default-y-position* := 0.0]
            [*default-x-velocity* := 3.0]
            [*default-y-velocity* := 4.0]
            [*default-mass*       := 1.0]
  (receive: [make-instance: ! Rest]
[object:
  (state:
   [x-position   := (if (memq 'x-position: Rest) then
                    (cadr (memq 'x-position: Rest))
                    else *default-x-position*)]
   [y-position   := (if (memq 'y-position: Rest) then
                    (cadr (memq 'y-position: Rest))
                    else *default-y-postion*)]
   [x-velocity   := (if (memq 'x-velocity: Rest) then
                    (cadr (memq 'x-velocity: Rest))
                    else *default-x-velocity*)]
   [y-velocity   := (if (memq 'y-velocity: Rest) then
                    (cadr (memq 'y-velocity: Rest))
```

```
                     else *default-y-velocity*)]
  [mass          := (if (memq 'mass: Rest) then
                     (cadr (memq 'mass: Rest))
                     else *default-mass*)])
(receive: [speed:]
   (sqrt (plus  (times x-velocity x-velocity)
                (times y-velocity y-velocity)))
)
(receive:  [direction:]
   (atan y-velocity x-velocity)
)
(receive:  [position:]
   (list x-position y-position)
)
   (receive:  [mass:]
      mass
   )
   (receive: [move: X hours:]
    [x-position  := (plus x-position (times x-velocity X))]
    [y-position  := (plus y-position (times y-velocity X))]
    )]
)]
```

For example, to create an instance called MyShip, it is necessary to describe the following lines:

```
  [MyShip = [CreateShip <== [make-instance :
             x-position: -3.0 y-position: -4.0 mass: 3.7]]]
```

The symbols specifying the positions and the mass are bound to "Rest". If any position or mass is not specified in the instantiation, the default values defined in the state descriptions are applied.

5.2 Superclass

Superclass is explicitly described in OKBL as follows:

```
[object: Create_Bird
 (receive: [new:]
   [object: aBird
   (receive:  [can_fly?:] "YES")
   (receive:  [kind?:] "BIRD")])]
```

```
[object: Penguin
 (super: [Create_Bird <== [new:]])
 (receive: [can_fly?:] "NO")]
```

There are two objects shown, of which the first creates superclass with respect to
the second. That is, the superclass Penguin is created when Penguin sends message
[new:] to Create_Bird. When Penguin receives message [can-fly?:], it returns "NO".
If it receives message [kind?:], Penguin transfers this message to the superclass and
returns "BIRD".

5.3 Implementation

Each object defined in OKBL is implemented by a task embedded in the real-time
multiprogramming operating system prepared for industrial process-control systems.
Tasks can be activated in concurrence and the passing of messages is implemented by
a sequence of primitives which have been long used in process-control systems. The
memory-management system controls memory protection for each task.

6. SEMANTIC MODEL CREATION

The phase for creating a semantic model is comprised in the phase for requirements
definition. In this phase, requirements are analyzed and semantic modelling made in
parallel with this process.

Semantic modelling is a process for creating a semantic model in terms of the syntax
and semantics provided by OKBL. In our semantic model, all presentations previously
made in the contextual, functional, data and dynamic views are unified. A semantic
model once built may be executed in order to check the match with the requirements
and implementability of the requirements; this may be called "prototyping", but prototyping in the conventional sense does not necessarily mean building a model with a
semantics which is unique throughout the life cycle.

In using OKBL, each "script" within specifications at early phases may be described
in a more abstract context, then it may be refined and transformed into concrete descriptions in the subsequent phases.

6.1 A Study by Example

The proposed method for semantic modelling is best illustrated with a real example: in this case, a simplified model of a guidance system for electric power network operators. As shown in Figure 1, the simplified electric power system consists of the following types of components:

Power source : P1, P2 and P3, each representing a group of power-generating resources.

Substation   : S1, S2 and S3, each representing a station where transmission lines with different voltages are interconnected through transformers.

Distribution
station   : D1, D2 and D3, each station distributing electricity to large consumers or other distribution stations of lower voltage.

Load   : U1, U2, U3 and U4, each representing a group of consumers.

Downline   : L1, L2, L4, L5, L7, L8, L9, L10, L11, L12, L13, L14, each representing a transmission line connecting components of different types.

Paraline   : L3, L6, each representing a transmission line connecting two substations.

Note: A switch, consisting of a circuit breaker and the disconnecting switches associated with it, is connected to each end of a downline or paraline.

Suppose that any one component fails. It is disconnected from the network automatically by the actions of the protective relays followed by additional manual operations by the human operators. Afterward, the network is reconfigured by operating the necessary switches (to set up detours around the failed components) so that the power supplied to customers will continue after the failure. The reconfiguration is made manually by human operators. The problem given in this example is to design software to generate guidance messages for the operators of the control center to operate switches for reconfiguring the network. The message should contain a list of switches which must be operated and a procedural indication to be followed by the operators in order to recover the network. It is presumed that the failed components were completely identified in the preceding stage and the information transmitted to the system.

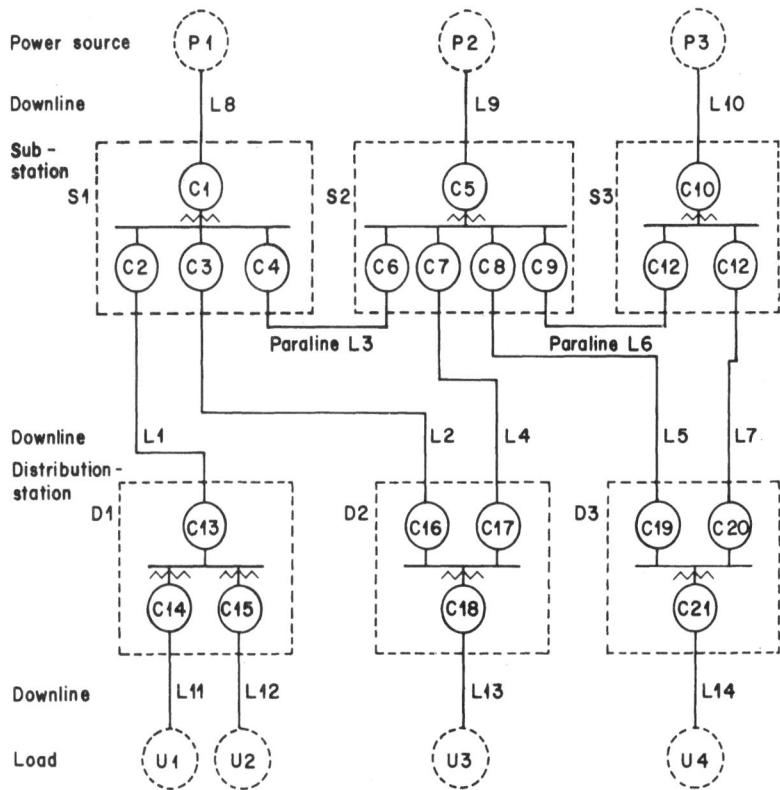

Fig. 1: The Example System

There are several constraints which should be considered in the real problem, a few of which are aimed at simplifying the problem. The constraints applied are as follows:

1) Each "Load" has its own priority index. The share amount of the power permitted for each load must be decided with reference to each priority index. If the total power received is insufficient to supply the total demand, or if the power flows are restricted by the maximum capacities of intermediate components, the power consumption of the load with the lowest priority has to be suppressed or cut off.

2) The power flow in paraline is bidirectional, whilst the power flow in downline starts from the end near to the power source. Every line has a maximum capacity which may not be exceeded. In the actual problem, we run the load-flow calculation program to obtain flow values at each point of the network. For simplicity, we adopt the rule that a line with larger capacity can carry a larger load flow.

3) We assume that only one network component failure is given to the system
for each computation.

The requirements should be analyzed from as many aspects as possible. Here, in this
example, we apply contextual view, functional view and data view. For simplification,
we solve the problem in the sequential mode without using dynamic view.

Contextual View: The elementary entities defined may be the network components:
substations S1, S2, S3 ..., loads U1, U2, U3 ..., which are shown in the network
configuration diagram. Switches are not independent entities but are included in the
appropriate substation or distribution station. The elementary entities are classified
according to the component types in the problem description.

Functional View: Functions are classified with reference to each of the elementary
entities. A cluster of functions may belong to an elementary entity. Each function in
the cluster is defined corresponding to an input data which reaches the elementary
entity. A function may generate output data or it may change stored data.

Data View: A presentation of each network state, the attributes associated with each
component, input data and output data is a data structure in the data view. A data
structure may be related to other data structures through the entities to which both
data structures belong. Otherwise, a data structure may be related to the second data
structure through the functions by which the former is transformed to the latter.

After analysis of the various views, they are unified and a single model created. Semantic modelling is a process for unifying views and creating a model with a precise
semantics which will be consistent throughout the life cycle. In semantic modelling, a
cluster of the relationships between data and functions is analyzed and this cluster
may constitute an object. The object assignments for this example obtained as a result
of semantic modelling are shown in Figure 2.

We have as the object a Supervisor which overrides the system. We have six classes
each of which creates the corresponding instances listed. An instance inherits the
attributes of the corresponding class. The Supervisor knows all data concerning the
network components and creates instances when the system is initialized. Then the
Supervisor receives from the user the name of the component to be disconnected. The
Supervisor sends the first message to inform the instance which represents this
component that the component must be disconnected. The instance receives the message and activates its own scripts and, if necessary, sends the next messages to the
objects which are connected to it. When all operations have been completed, the
Supervisor receives the last message that acknowledges completion and then terminates
the system.

object:	
Supervisor	

class:	instance:
Create_Substation	Substation_S1 Substation_S2 Substation_S3
Create_Distributionstation	Distri_D1 Distri_D2 Distri_D3
Create_Downline	Line_L1, Line_L2, Line_L4, Line_L5, Line_L7, Line_L8, Line_L9, Line_L10, Line_L11, Line_L12, Line_L13, Line_L14
Create_Paraline	Para_L3, Para_L6
Create_Powersource	Power_P1, Power_P2, Power_P3
Create_Load	Load_U1, Load_U2, Load_U3, Load_U4

Fig. 2: Object/Class/Instance of the Example

A class is illustrated by an example description. The following is the description of Create-Paraline which is a class that creates instances of L3 and L6. Unless otherwise specified by the syntax of OKBL, the denotations in the scripts are based on Franz Lisp.

```
[object: Create_Paraline        --The class of Paraline
 (receive: [create: *name]      --*name is the name of instance
                                 --to be created. This message
                                 --is sent by Supervisor
  [object: aParaline            --The scripts starting from
                                 --this line are instantiated
                                 --par *name.
   (state:   [NODEline:=(*array 'NODE t 2 2)]
             [NAME:=*name]) --These two lines declare the
    --memory NODEline which is an array sized 2 by 2.
    --In the array the following data are stored.
   (receive: [init: node: N-list]
     (store (funcall NODEline 0 0)(nth 0 (car N-list)))
```

```
(store (funcall NODEline 0 1)(nth 1 (car N-list)))
(store (funcall NODEline 1 0)(nth 0 (cadr N-list)))
(store (funcall NODEline 1 1)(nth 1 (cadr N-list))))
--When this message is received, the values bound in
--N-list shown in Figure 3 are passed by supervisor.
--NODEline is filled with the values in N-list.
(receive: [DISCONNECT:Mode Sender:Sname]
--This is the message to inform that the network com-
--ponent of which the name is bound in Mode has to be
--disconnected. The name of the message sender is bo-
--und in Sname. The scripts followed describe proce-
--dural steps which are executed for disconnecting the
--indicated component.
(var:TARGET)
(caseq Mode
   (YOU                --In case that the component to
                       --be disconnected is YOU or the
                       --component represented by you,..
   (do ((i 0 (1+ i)))
       ((eq i 2))
       (if (((funcall NODEline i 1) 0)
         then [Mode := 'SOURCE]
              [(funcall NODEline i 0) <==
               [DISCONNECT:Mode Sender:NAME]]
            (store (funcall NODEline i 1) 0))))
   (SOURCE             --In case that the component that
                       --exists in the source side to
                       --the component represented by
                       --this instance is to be discon-
                       --nected,......
   (if (eq (funcall NODEline 0 0) Sname)
    then [TARGET := (funcall NODEline 1 0)])
   (if (eq (funcall NODEline 1 0) Sname)
    then [TARGET := (funcall NODEline 0 0)])
   [Mode := 'SOURCE]
   [TARGET <== [DISCONNECT:Mode Sender:NAME]]
   (store (funcall NODEline 0 1) 0)
   (store (funcall NODEline 1 1) 0))
   (SINK               --In case that the component that
                       --exists in the sink side to the
                       --component represented by this
                       --instance is to be disconnected,
```

```
    (if (eq (funcall NODEline 0 0) Sname)
     then [TARGET := (funcall NODEline 1 0)])
    (if (eq (funcall NODEline 1 0) Sname)
     then [TARGET := (funcall NODEline 0 0)])
    [Mode := 'SINK]
    [TARGET <== [DISCONNECT:Mode Sender:NAME]])))
  (receive: [      --other message continues.
    .......
    .......
```

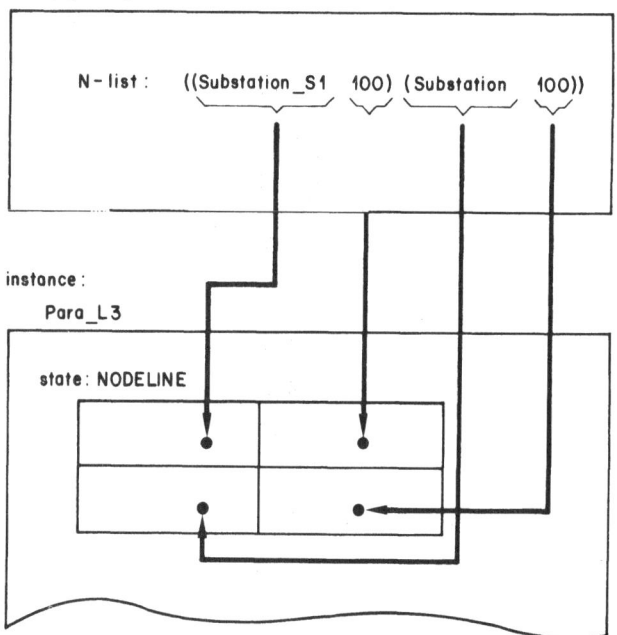

Fig. 3: Instantiation of the State in Para_L3

7. CONCLUSION

A unified life-cycle model which has its base in the semantic model is shown in Figure 4. The figure shows how the semantic model is incorporated into the life-cycle
model. It presents a semantic model created in the phase of the requirements analysis,
and also shows that the adjacent phases are selected depending upon the designer's
selection. The selection will be made depending upon the degree of depth to which
the semantic model is detailed.

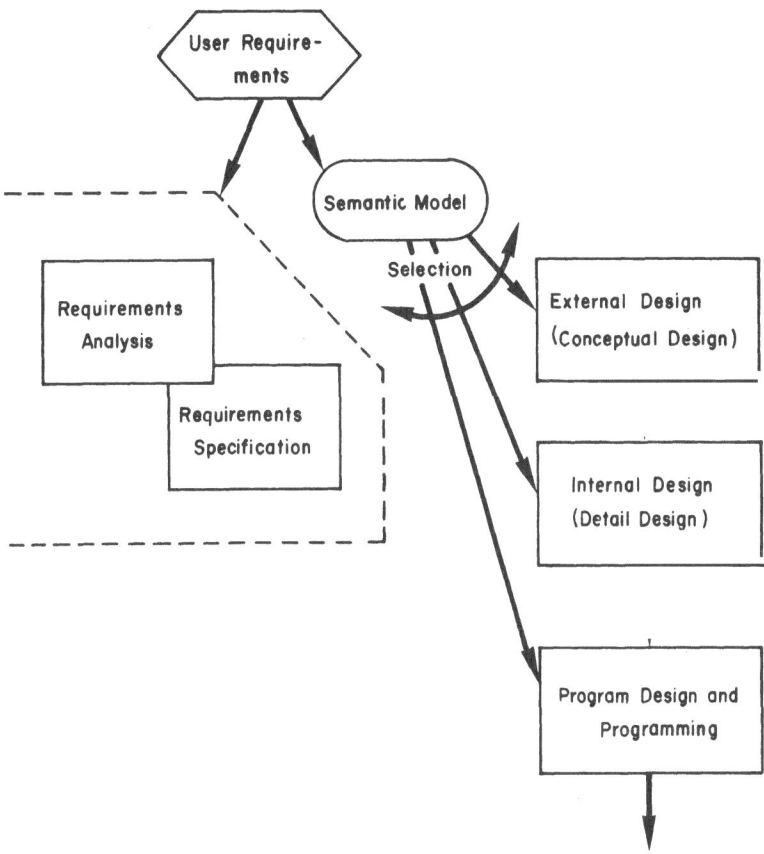

Fig. 4: Unified Life-Cycle Model

Introduction of the semantic model results in the following advantages:

1) In current average life cycles, specifications are written in a semantics different from that of the implementation. For this reason specifications are rarely revised when program codes are modified. The semantic model is described in a semantics unique throughout the whole life cycle. This serves to increase the possibility that descriptions of the semantic models are maintained as well as the program codes.

2) By execution of the semantic model, a prototyping is attained. In other words, either correctness of the semantic model or the match between user requirements and the model is improved by execution.

3) Because the semantics of the model is the same as that of implementation,

execution of the semantic model improves implementability of the requirements defined.

4) An object is an abstraction which unifies procedures and associated data
   structure, and therefore provides good reusability.

ABCL[15] is an experimental language which has been running for about a year. OKBL
is an extension of ABCL in a real-time environment and test usage of the serial
version of OKBL started in August 1985.

8. ACKNOWLEDGEMENTS

The author expresses his thanks to Prof. A. Yonezawa who acted as a consultant
during the development of OKBL, and to Mr. T. Maeda who assisted the author in
implementing the environment for OKBL.

REFERENCES

I.   Alford, M.W., "A Requirements Engineering Methodology for Real-Time Process-
     ing Requirements," IEEE Trans. Software Eng., Vol. SE-3, No. 1, 1977, pp.
     180-193.

2.   Balzer, R.M., Goldman, N.M., and Wile, D.S., "Operational Specifications as the
     Basis for Rapid Prototyping," ACM Sigsoft Software Engineering Notes, Vol. 7,
     No. 5, 1982, pp. 3-16.

3.   Biewald, J., Goehner, P., Lauber, R., and Schelling, H., "EPOS-A Specification
     and Design Technique for Computer Controlled Real-Time Automation Systems,"
     Proc. 4th Int. Conf. on Software Engineering, 1979, pp. 245-250.

4.   CODASYL Development Committee, "An Information Algebra Phase I Report,"
     Commun. ACM, Vol. 5, No. 4, 1962, pp. 190-204.

5.   Dijkstra, E.W., "Cooperating Sequential Processes," in Programming Languages,
     F. Genuys, ed., Academic Press, New York, 1968.

6.   Guttag, J.V., Horowitz, E., and Musser, D.R., "Abstract Data Types and
     Software Validation," Commun. ACM, Vol. 21, No. 12, 1978, pp. 1048-1064.

7.   Hewitt, C.E., and Baker, H., "Actors and Continuous Functionals," in Formal
     Description of Programming Concepts, North-Holland, New York, 1978, pp.
     367-387.

8.   Hoare, C.A.R., "Towards a Theory of Parallel Programming," in Operating Sys-
     tems Techniques, C.A.R. Hoare and R.H. Perrott, eds., Academic Press, New
     York, 1973.

9.   Hoare, C.A.R., "Communicating Sequential Processes," Commun. ACM, Vol. 21,
     No. 8, 1978, pp. 666-677.

10. Jackson, M.A., Principle of Program Design, Academic Press, New York, 1975.

11. Katayama, T., "HEP: a Hierarchical and Functional Programming Based on Attribute Grammar," Proc. 5th Int. Conf. on Software Engineering, 1981, pp. 343-353.

12. Liskov, B.H., and Berzins, V., "An Appraisal of Program Specifications," Computations Structure Group Memo 141, Laboratory for Computer Science, MIT, 1976.

13. Liskov, B.H., Snyder, A., Atkinson, R., and Schaffert, J.C., "Abstraction Mechanisms in CLU," Commun. ACM, Vol. 20, No.8, 1977, pp. 564-576.

14. Ludewig, J., "Zur Erstellung der Spezifikation von Prozessrechner-Software," Doctoral Dissertation, Technical University Munich, 1981.

15. Matsuda, H., and Yonezawa, A., "ABCL User's Manual," Internal Memo, Dept. of Information Science, Tokyo Institute of Technology, 1984.

16. Matsumoto, Y., "A Method of Software Requirements Definitions in Process Control," Proc. IEEE COMPSAC'77, 1977, pp. 128-132.

17. Matsumoto, Y., "Some Experiences in Promoting Reusable Software: Presentation in Higher Abstract Levels," IEEE Trans. Software Eng., Vol. SE-10, No. 5, 1984, pp. 502-513.

18. Matsumoto, Y. et al., "SWB System: A Software Factory," in Software Engineering Environments, H. Hunke, ed., North-Holland, 1981, pp. 305-318.

19. Orr, K.T., Structured System Development, Yourdon Press, New York, 1977.

20. Ross, D., "Structured Analysis (SA): A Language for Communicating Ideas," IEEE Trans. Software Eng., Vol. SE-3, No. 1, 1977, pp. 16-34.

21. Salter, K., "A Methodology for Decomposing System Requirements into Data Processing Requirements," Proc. 2nd Int. Conf. on Software Engineering, 1976, pp. 91-101.

22. Stevens, W.P., Myers, G.J., and Constantine, L.L., "Structured Design," IBM Systems Journal, Vol. 13, No. 2, 1974, pp. 115-139.

23. Teichroew, D., and Hearshey, E.A. III, "PSL/PSA: A Computer-Aided Technique for Structured Documentation and Analysis of Information Processing Systems," IEEE Trans. Software Eng., Vol. SE-3, No. 1, 1977, pp. 16-33.

24. Yonezawa, A., "Specification and Verification Techniques for Parallel Programs Based on Message Passing Semantics," Ph.D. Thesis, MIT, 1977.

25. Yonezawa, A., Matsuda, H., and Shibayama, E., "An Object-Oriented Approach for Concurrent Programming," Research Report C-63, Dept. of Information Science, Tokyo Institute of Technology, November 1984.

26. Yonezawa, A., and Matsumoto, Y., "Object-Oriented Concurrent Programming and Industrial Software Production," in Formal Methods and Software Development, Vol. 2, H. Ehrig, et al., eds., Springer-Verlag, Berlin, 1985, pp. 395-409.

DISCUSSION
Chairman: H. Sandmayr (Brown Boveri, Baden, Switzerland)

J. Ludewig (Brown Boveri, Baden, Switzerland)

I have a question on the syntax and representation of your specifications. Is the syntax you showed meant to be the syntax actually to be used, or do you expect to develop an additional representation for communicating these specifications?

Y. Matsumoto

The description shown is very easy to use for the programmer, but difficult for the system engineer. So in the very early stages of the design process, we also use a natural language for describing operations to be embedded in each "script" line.

M. Glinz (Brown Boveri, Baden, Switzerland)

You emphasized the importance of reusing software. How do you integrate the reuse of software into your software-development method?

Y. Matsumoto

Our method for the reuse of software is not theoretical or mathematical; it is very practical. We have a reusable software center in our factory, which collects all the reusable code. The module specifications are even modified there in order to be more comprehensible. The software designer can search for the module he requires using the terminal. He can input some keywords, and the descriptions of modules that correspond to those keywords will appear on the screen. The designer can then choose a module and ask the reusable software center to deliver all documents concerning the module. Afterwards the designer modifies the reused module such that it fits in the new application.

T. Sakaguchi (Mitsubishi Electric Corporation, Amagasaki, Japan)

I would like to make a comment on the reuse of software. We are developing a translator that converts a FORTRAN program to an equivalent C program. We consider such a translator as a means for the reuse of software. Our initial experience has shown that the speed of the program in the target language is slower than in the source language. Do you have some experience with a similar translator, in particular concerning performance?

Y. Matsumoto

In one instance, we transformed a Lisp program to an equivalent FORTRAN program, in order to transport a knowlegde-based program developed on a host computer to a target computer for real-time process-control use. Improvement in execution speed was the intention in such cases. Some parts of the FORTRAN program are very efficient, as the questioner suggests.

SOFTWARE QUALITY ASSURANCE

K. FRUEHAUF, J. LUDEWIG and H. SANDMAYR

Brown Boveri, Baden, Switzerland

ABSTRACT

Quality assurance has been in existence for a long time in many fields of engineering, but in software engineering the idea is still fairly new. In this paper we describe the current state of Software Quality Assurance (SQA), and propose a classification of techniques. The paper consists of three parts: Firstly, the fundamental terms are defined, and the state of the art identified. This includes a survey of standards and proven techniques, as published in journals and textbooks. Secondly, practical experiences with SQA techniques used at Brown Boveri are presented, and their successes and failures discussed. It is difficult to obtain reliable data for the benefits of SQA, therefore only some general conclusions can be reached regarding its impact on quality and productivity. Thirdly, SQA consists of many different components which are difficult to order according to their relative importance. With the aim of establishing common understanding and terminology we have defined several levels of SQA which may serve as a general framework.

1. INTRODUCTION

"... consistent with the precepts that quality is conformance to requirements and prevention of defects, it is the responsibility of quality to act as the independent instrument of management in auditing all aspects of software development and maintenance through the review of plans, specifications, designs, test documentation, configuration control, and programming standards. Quality must also assume its traditional responsibilities in vendor surveillance of procured software, qualification and acceptance of all software, certification of tools used for software testing, defect analysis, and quality improvement analysis."[7]

Quality assurance is not a popular topic. Few people enjoy inspecting the work of others, and many of their "victims" persist in their role of artists who should not be subject to control procedures. Yet there is a growing need for software of guaranteed high quality and, hence, for SQA.

All large companies and organizations have recognized the need for SQA, and many experiences have been published already. Despite the large number of papers we know little about the true results. Many authors discuss their successes from various points of view, but either omit or sanitize failures, others hide their experience behind confidentiality (perhaps for good reasons). Critical views of SQA in practice are usually published only after an employee leaves his company when, e.g., a disastrous SQA-history may be reported.[4]

2. SOFTWARE QUALITY ASSURANCE - THE STATE OF THE ART

2.1 Terms and Definitions

2.1.1 Software Quality

The following definitions are based upon IEEE recommendations:[15]

Software consists of computer programs, procedures, rules, documentation and data pertaining to the operation and maintenance of a computer system. (This definition is a synthesis of two slightly different definitions with the words "and maintenance" added.)

Quality is the totality of features and characteristics of a product or service that bears on its ability to satisfy given needs.

The term "quality" has some intuitive meaning which is unrelated to any requirement; e.g., we may say that a product is good for a particular purpose, or just good. In the latter case, we have in mind some basic level of quality.

Software Quality is the totality of features and characteristics of a software product with regard to its ability to satisfy given needs, e.g., conform to specifications (the first out of four different IEEE definitions). A number of such features and characteristics have been identified.[2] Since some incorporate others, they may be arranged in a hierarchy. Figure 1 shows all but the lowest levels of that hierarchy.

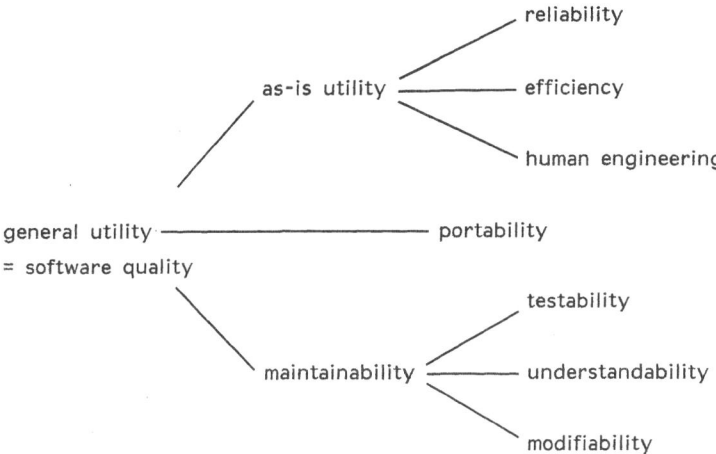

Fig. 1: Software Quality Characteristics Tree (partial)

Following the definition given above (i.e., assuming that software is entirely defined by its requirements), one may simply say that software quality is the degree to which the software meets its requirements. However, there are two problems with this apparently more precise definition:

- Since quality is based on the requirements, the requirements must exist, not only in the customer's head, but in some written form. In practice, there are requirement specifications, but these are far from complete. In particular, they do not address any qualities but functional performance and timing behavior, and not even these are fully covered. Hence, we must assume "default requirements". Note that the default requirements which are generally accepted in a company determine the reputation of its products.

- To date, we do not have metrics which are well known, generally accepted, and sufficiently meaningful for measuring qualities. Therefore, we cannot express software quality in a simple figure which is determined in an unbiased manner. As a consequence the degree to which the requirements are met remains undefined.

2.1.2 The Range of Software Quality Assurance

Every activity aimed at improving software quality may be considered as part of SQA. This broad definition, however, is not of great value because it turns out to be very close to the definition of software engineering. Since we consider SQA to be a component of software engineering (probably its most important component) we prefer to restrict the term SQA to activities which do not <u>directly</u> contribute to the final product, but provide standards, checklists, acceptance procedures, etc. specifically introduced for raising and controlling software quality. <u>Building</u> the system is <u>not</u> an SQA-activity.

Application of advanced tools for software design, for instance, is not part of SQA (though it is certainly beneficial to software quality). Even testing is not, because it aims at finding deficiencies, i.e., it is part of the building process. However, the way in which certain tools are used, and how a particular test procedure is performed, may well be regarded as part of SQA.

2.2 Quality Assurance of Hardware and Software

In the traditional engineering disciplines quality assurance has been practiced for many years. Its goal is to ensure that products have a high level of quality (<u>quality improvement</u>) which can be defined (<u>quality control</u>).

Neither of the two is sufficient in the absence of the other: while quality control does not enhance the quality of the product (except by the threat to programmers, resulting from better visibility), quality improvement in general cannot avoid poor performance by individuals (at least in a large organization) and therefore the quality of a few products may still be far below average. These substandard goods must be identified and singled out. In practice it is often impossible to perform both methods of quality assurance. Third party software components can only be subject to quality control (but not quality improvement). Furthermore, it is not always possible to demonstrate the qualities of a software component with an acceptable degree of effort.

Though we cannot always validate the quality of our product, we are certainly able to check the quality of the procedure by which it is produced. Hence, quality improvement of the product is achieved mainly by quality control of the production.

The nature of software differs in several ways from other products, such as electronic equipment:[7]

- Hardware is built from standard components of known quality, while software is often built from scratch;

- Mass production of hardware is subject to divergence; quality control must deal with every example of the product. Copies of software are precise replicas of the original;

- Software is (usually) far more complex than hardware; therefore, there is no way of exhaustive testing;

- The design of hardware must acknowledge physical limits (e.g., material strength and the like). Such limits are well known and carefully observed. Since there are virtually no physical limits for software design, our mental abilities are the most restrictive limits. These are usually overestimated.

Current trends will reduce the differences between the characteristics of hardware and software:

- Integrated circuits are just as complex as software is (i.e., VLSI-people are facing most of the problems of SQA);

- There are now approved software components, at least at the level of compilers and run-time systems (for the DoD-language Ada);

- QA of hardware development faces the very same problems as does SQA.

Timing behavior of software (efficiency) can be observed with relative ease; therefore, quality control may take care of this aspect. However, functional performance can only be observed by exhaustive testing, which is not feasible. Hence software (just like a VLSI-chip) must be functionally correct on the very first attempt. Most other qualities, such as readability, are observable, but extremely difficult to enhance and must also be built in.

2.3 Standards

"Standards are like morals; they are rules adopted by some segment of society, within some context of social behaviour, to regulate activity - everything else being equal. But, they are not the same in all contexts, nor should they be regarded as hard and fast within any given context. We erect them so that people will usually behave in a predictable way. In a system development context, however, another valid purpose for a standard is to subject a proposed deviation from the standard to public scrutiny in order to decide whether the deviation is justifiable within the total system context."

G.H. Mealy (1969)

Standards for SQA have been developed[14,15] in order to provide well-defined terms and techniques. Many large organizations apply such standards, usually extended by special regulations. Using standards has several advantages:

- Formulated by experts, they represent the state of the art.
- Immediate availability.
- Discussions regarding the "right" rules and guidelines can be largely avoided.
- They can be incorporated in contracts.

A comparison of standards for SQA has recently been published:[22]

MIL-STD-52779A   and
MIL-STD-1679A   from U.S. DoD
FAA-STD-018   from U.S. Federal Aviation Administration
A.N.S. 4.3.2.   from ANSI
RTCA-DO-178   from Radio Technical Commission for Aeronautics
IEEE-STD 730   from IEEE (IEEE, 1981)
NATO-AQAP-13   from NATO
GAM-T-17   from French Ministry of Defense

Workers in the IBM-Federal Systems Division[3] point out the important role of tight standards (i.e., company standards) in introducing and maintaining a high level of software technology; their experiences indicate that high standards lead to a significant payoff.

2.4 Metrics

"To measure is to know."

J.C. Maxwell

Every engineer strives for exact, reliable data. To obtain such data requires a metric consisting of:

- A rule (what should be measured or counted)
- An algorithm (how the results are combined)
- A unit name (what the results are called)
- An interpretation (what they actually mean)

The area of a rectangular piece of land, for instance, may be measured by

- Counting the number of steps in each direction
- Multiplying the two numbers
- Attaching the unit "square yard" to the result
- Explaining the outcome as a steady measure of the quantity of seed required and fruit grown in this area

This procedure is not applicable to measuring the area of computer chips, because they will all turn out to be zero.

One approach[11] to the capture of software has been through a fairly small set of measurable properties (number of operands, number of operators, etc.) and derived measures. Other authors[9,20,21] provide large sets of metrics.

Metrics in general, and for software quality in particular, need to be

- Measurable   (with as little effort as possible)
- Sufficiently exact (i.e., reproduceable)
- Expressive   (i.e., leading to a spectrum of results)
- Meaningful   (i.e., results correspond to other useful measures)
- Widely accepted   (i.e., understandable and useful for many people)

To date, no software metrics meets all the requirements listed above. Simple ones, such as the number of lines of code, are easily measured but their value is questionable. Should a program of 2n lines cost twice as much as one of n lines? Or less, because the programmer did not seek a more concise solution?

Other terms are closer to what we are interested in, but there is no agreement on how they can be measured or computed. For instance, metrics of complexity are highly desirable, but neither the "classical" one by McCabe[19] nor any other has been very successful to date.[12] Reference may also be made to a recent report[13] which contains a survey of 50 software quality measures. Walters[23] from General Electric emphasizes the role of metrics and tools for measuring software quality. However, most of the simple metrics currently being used (such as the quantitative measures) are not related to program quality.

All simple metrics can be misused; when they become more important, people will adjust their programs to them (e.g., somebody whose salary depends on the number of lines of code produced will eventually replace all subprogram calls by the code of the subprograms). This problem will lead to a refinement of metrics.

"Bang! bang! Maxwell's Silver Hammer came down upon her head ..."
J. Lennon and P. McCartney (1969)

2.5 The Software Quality Assurance Department

Let us imagine a company whose products consist either partially or entirely of application software. Therefore there is a SQA-department whose tasks include the following:

1) Providing standards and guidelines for SQA

2) Supporting development and engineering groups in taking measures for SQA

3) Supervising all SQA-activities

4) Collecting and reporting data on software quality

5) Elaborating proposals for improvement (e.g., regarding training or organizational matters)

The SQA-department is highly independent of its clients, i.e., developers cannot put pressure on the SQA-staff. In companies such as BBC, it is part of the general quality assurance department or a secondary organization with special reporting lines, but staffed in such a way that both quality-assurance know-how and software-engineering knowledge are present.

2.5.1 Standards and Guidelines for Software Quality Assurance

The existence of a (sufficiently precise) SQA-plan is vital for the success of any SQA-activity. Though such plans often contain some rules that are specific to a project, there is a large, invariant kernel, that may be constantly used. KET-7[18] is such a BBC-internal kernel (or framework) based on an international standard.[14]

The SQA-plan overlaps with rules for project management and documentation, or even contains such rules explicitly. This is necessary because responsibilities must be clear, and because management and documentation issues are most important for SQA.

2.5.2 Support for Software Quality Assurance

The SQA of a particular project is organized and performed by that project not by the SQA-department. The project will, however, call for support when:

- Starting a new project (i.e., setting up a SQA-plan)
- Training in SQA is needed
- Encountering serious problems about quality or quality assurance

2.5.3 Control of Software Quality Assurance Activities

The effectiveness of SQA should be regularly evaluated by the SQA-department. This is achieved by audits. An audit is a check on whether rules (in particular those stated in the SQA-plan) are actually observed. Without audits SQA will sooner or later degrade to mere lip service.

2.5.4 Data Collection and Distribution on Software Quality

Collection of data on software quality is necessary for a number of reasons. Such data will help to:

- Recognize poor quality components at an early stage
- Improve estimates of software reliability and cost
- Convince top management of the virtues of the SQA-department

In order to be useful, data must be collected in a uniform way, to ensure the consistency of statistics based on figures from different departments. Therefore a precise guideline is necessary (e.g., KET[17]).

2.5.5 Elaboration of Proposals for Improvements

Since the SQA-department can observe the quality figures of many projects, it has a unique insight into problems. Though it cannot solve such problems, it can report them, and recommend improvements. These may range from additional training, acquisition of new tools, or changes in the project organization. The independent position mentioned above should permit frank criticism by the SQA-staff.

2.6 Software Quality Assurance and the Management

Experience shows that management is the most important factor in SQA. Many authors stress this point. Crawford and Hiering[5] from Bell Laboratories describe a strategy of early error detection through the use of formal inspections. Collection of data, vital for monitoring the process of software development, requires (besides much training) a commitment not only from staff, but also from management. Baker and Fisher[1] from the U.S. Army conclude that the project manager himself must be finally responsible for SQA.

Even the best airplane cannot fly in a vacuum; it needs the air under its wings, most urgently for takeoff and climbing, to a lesser degree when it has reached its cruising altitude. For SQA, this invisible support is a strong management. (Note that hot air reduces the climbing rate!)

At all times, and under all circumstances, the management must confirm the priority of (sensible) SQA, just as the government must endorse the role of (wise) judges and police forces; otherwise, they lose control. The worst situation is a corrupt government which makes its own laws appear ridiculous, and similarly, a management whose orders are inconsistent with their own SQA rules.

3. PRACTICAL SOFTWARE QUALITY ASSURANCE

The previous section contains a general survey of SQA. A number of tasks are assigned to an independent SQA. This section contains the results of such SQA activities in BBC. We begin with an explanation of our approach to the assignment of
responsibilities concerning product quality. This is followed by the presentation of
four examples of our SQA activities. The first case is the use of an international
standard for Software Quality Assurance Plans (cf. Sections 2.3 and 2.5.1) and the
second illustrates the use of metrics (cf. Sections 2.4 and 2.5.4). The third case
represents the management support function of our SQA department and the last case
indicates the SQA duty of making proposals for improvements (cf. Section 2.5.5).

3.1 The Approach

The entire responsibility for product quality lies with the development team. The role
of independent Software Quality Assurance (SQA) is to support the development team
in formulating and achieving the quality objectives. The main task of SQA, however,
is to assess the degree to which the quality objectives are met. We assume that this
is a distinguishing characteristic of our approach. Thus the following sections do not
include any notes on reviews or tests, on approval or rejection by SQA. We consider
these tasks to be inherent parts of the development effort.

3.2 Case A: Software Quality Assurance Plan

The Software Quality Assurance Plan (SQAP) is the statement of quality objectives in
terms of procedures applied, documents provided, standards followed, and tools used
in developing software. Additionally, the responsibilities for carrying out the various
quality assurance related activities are stated.

The following characteristics of our environment have an impact on the SQAP:

- The software is embedded, i.e., it is part of a larger delivery comprising
  computer and telemetry equipment.

- The software product, a system of the BECOS family,[10] is sold ten to a hundred times. The product must, however, be adapted to the needs of the actual
  customer. Thus it is neither a one-off "customer-tailored" product nor a mass-
  produced (by copying) "off-the-shelf" product.[8]

- The department producing the software is part of a larger company.

From the latter it follows that the SQAP must fit into the overall corporate Quality

Assurance System, which traditionally covers only manufacturing. The fact that we have to deal with embedded systems makes this requirement more important. The tradeoff between "standard product" and "customer-tailored product" necessitates the distinction between product development and product customization. In either case a project team is established, but the project objectives and consequently the content of the SQAP will differ substantially.

Our solution to the problem is as follows:

- The corporate Quality Assurance System is extended by the simple rule: "Every project comprising software development must prepare a SQAP." As more and more SQAPs emerge, common sections can be identified and this simple rule can be replaced.

- The SQAP is issued at the beginning of software product development. The responsibility for issuing and maintaining the SQAP is with the independent Software Quality Assurance but the actual content is formulated in close cooperation with the project team which is deeply involved in setting its own quality objectives. For product development projects we have applied the IEEE Std 730-1981 for Software Quality Assurance Plans.[14]

Notes on our use of the IEEE Std 730-1981:

- We strictly followed the required table of contents and used exactly the same headings. In the actual content however we used the corresponding terms already established in our environment. This should facilitate checking for compliance with the standard and also enhance the understanding of the SQAP by the development team.

- We resolved the often difficult distinction between tasks and responsibilities by defining the responsibilities as measurable objectives for the project function, e.g., the Product Development Team Leader has the responsibility of achieving the goal:

  "The ratio of activity duration in calendar days to the effort in person days is less than two."

- The standard requires a number of redundancies.[22] We simply made the choice of where to put the relevant information and referred to it from other sections.

- At release time, as stated in the plan, we conduct an audit in order to assess the SQAP effectiveness. The audit leads to the revision of the SQAP and to corrective actions for the development project.

- Our plan currently comprises around 30 pages and 60 measurable (and audited) objectives. Other users of the IEEE standard are invited to provide us with figures which would enable us to make comparisons.

Conclusions regarding the IEEE Std 730-1981:

- The term "plan" in the title is misleading. The standard does not require (and thus an SQAP will not contain) estimates of effort or schedules which are usually associated with the term "plan".

- The standard should be revised in order to eliminate redundancies.

- The standard is applicable to software development projects without obligations to customers (see below). We would not recommend its use for the latter.

Providing "customer-tailored products" involves more than the IEEE Std 730-1981 requests. Communication between customer and supplier and the responsibilities assigned to both must be regulated. Together with the regulations for the interface with product development - a product release will be the basis for the customization - this is the main additional requirement. The CSA Preliminary Q396.1-1982 Standard for Software Quality Assurance Program Part I,[6] seems to be more appropriate for these types of projects. We are currently preparing the first SQAP for a customer-tailored product development based on this standard. We expect to be able to report our experiences in a year's time.

3.3 Case B: Continuous Investigation of the Product

A software product undergoes continuous evolution. It reaches degrees of maturity - releases - at which it can be utilized by end users. For each release we determine a (currently small) set of metrics chosen with the following main objectives in mind:

- It should be possible and reasonable to relate them to the effort required in order to improve planning.
- It should be easy to measure them using simple tools.
- They should provide an indication of the discipline in the development team.
- It should be easy to derive additional information helpful to the developers.

The two following examples illustrate the type of tasks we perform.

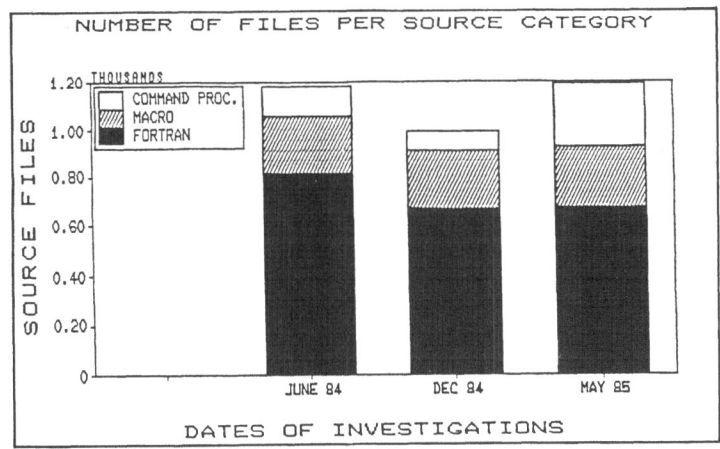

Fig. 2: Number of Source Files per Programming Language

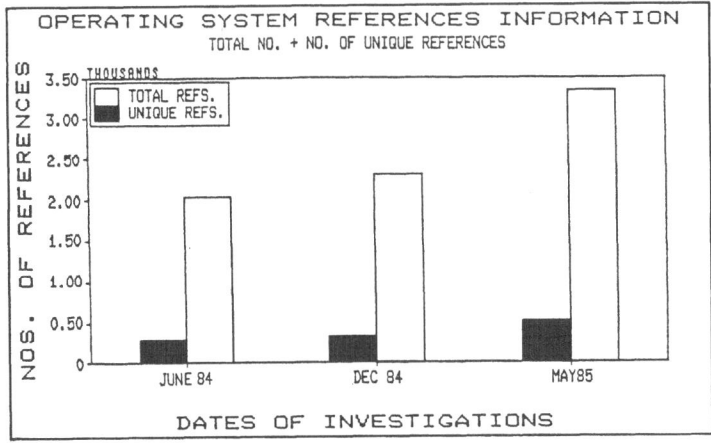

Fig. 3: Number of Referenced Operating System Facilities

The obvious observation from Figure 2 is the extraordinary increase in the number of command procedures for the last release. Investigation revealed that all the command procedures were stored in three different places. It is obvious that we avoided some maintenance effort, but the actual finding is that the configuration management does not function to the required level of satisfaction.

The increase in referenced operating system facilities for the latest release (see Figure 3) has a simple explanation: A new driver was added to the system. In this case the explanation is satisfactory and whilst we gained little it was beneficial to the developers. They received from SQA a cross-reference showing in which files the different operating system facilities were referenced. This is a typical by-product of the counting exercise. For the next release a transfer to a new major release of the operating system was planned: The cross-reference was a valuable input for the estimation of the required effort as well as assisting in the actual completion of the task.

The above findings are certainly not such that they would prevent project disaster. A cautious enlargement of the set of metrics will - by increasing the effort for applying them - lead to more cost-effective findings.

3.4 Case C: Independent Effort Estimation

A factory acceptance test results in an acceptance or rejection of the customer-tailored product by the customer. Frequently it will result in a deficiency list enumerating the nonconformities with the (explicitly stated or implicit) requirements. The project team delivers its estimates for the clean-up work. In strategic projects an independent estimate will be provided by the SQA.

In one particular case, the deficiency list contained 160 items. The project team estimated the effort for correction at 200 person days. SQA's independent estimate of 960 person days was regarded as grossly excessive. It was a rough estimate based on a classification of such deficiencies as cosmetic, normal, and risky, and using a constant effort estimate for each class of deficiencies. The classification was biased by the (lack of) experience by SQA.

The effort directly accounted for correction was 98 person days, which the project team considered to be inaccurate by a factor of two, thereby confirming their original estimate. A final investigation of the work account and of the product suggested that the actual effort for correction was about 900 person days, i.e., very close to the SQA estimate.

It must be noted that this was the first time we applied this approach. The conclusion that SQA is better in estimating effort would be wrong. The lesson to be learnt is that both SQA and the project team will improve the quality of their estimates if historical data is available, and if they have to make estimates regularly. The use of different approaches and the increase in accuracy due to regular exercise will certainly be of benefit to all the parties involved.

3.5 Case D: Staff Education

SQA checks the achievement of quality objectives and therefore provides immediate insight into the weaknesses of the product and of the development process. It is an SQA task to request the use of appropriate methods and tools, as well as to provide input for staff education in order to raise the quality level. Although staff education is not an SQA activity, we have included it in this paper because we firmly believe it is one of the most effective ways of achieving higher quality. Our educational efforts are threefold:

- In a regular series of lectures with discussions we aim to "broaden the mind". The topics are not necessarily related to daily work, but deal with matters of our market segment or with software engineering. It is a rule, for instance, that persons attending conferences, workshops, and so forth, summarize their impressions in a presentation.

- The purpose of a tutorial is to teach a very narrow topic. A method or a tool is taught and applied to the current work of the participants. The aim is "to learn the craft", to master the topic using current tasks in order to put it into immediate practice.

- Broader areas are taught in courses. This "care for roots" is provided on demand. A new operating system or a new programming language are typical topics for courses. The aim is to obtain basic knowledge of a broader area. However, an extensive on-the-job training is necessary to master the whole topic.

4. A GAMET OF SOFTWARE QUALITY ASSURANCE

4.1 The Scenario

The explanations and examples given so far have shown that SQA is concerned with

- Standards for the development process, their application, and their adaption when new insights are gained from experience;

- The quality of <u>verification and validation</u>, i.e., with the adequacy of the methods and tools used to assure consistency and comprehensiveness among requirements and the results of the different development phases;

- The quality of <u>test and evaluation</u>, i.e., the quality of the end product and its proof, the test plan and its adequacy and comprehensiveness;

- The quality of <u>configuration control</u>, i.e., the adequate management and control of changes to the software;

- <u>Nonfunctional attributes</u> of the software product such as growth constraints, maintainability and portability requirements, adequacy of documentation, etc.

SQA serves a dual purpose. It is a sensor for the performance of an organization and as such delivers a necessary input to line and project management in order to achieve the final target: a reasonable level of quality. At the same time, the presence of a sensor increases the awareness of the different quality aspects by those involved in the development process.

4.2 The Levels

The management point of view is crucial when deriving the different levels of SQA implementation. The question to answer is: "How much can I afford <u>not</u> to do ?". The "how much" is mainly influenced by the size and financial risk of a project.

The <u>lowest level</u> of SQA implementation centers around standards for the development process:

- Selection and tailoring of existing standards to the needs of the particular environment

- Supporting the implementation of the selected standard by rules, guidelines, and organizational aspects

- Auditing the effectiveness of the standards

Depending upon the size of the organization this level of SQA can be achieved by one person, maybe even as part-time job.

The quality aspects of verification and validation, test and evaluation, and configuration control are the main additions on the <u>middle level</u> of SQA implementation and start again with audits.

The top level SQA implementation considers not only nonfunctional attributes but also provides data on the performance of an organization or a project team. The state of the art of measuring performance in the software development process requires careful use and interpretation of these values. In general, their absolute value is meaningless, i.e., they cannot be used to compare organizational units. However, the change of a certain measurement over the project development cycle or over a sequence of projects can provide useful indications, e.g., for the improvement of test quality or the performance of an organization.

5. CONCLUSION

Starting from an overview of SQA-literature, we tried to isolate what we consider to be essential SQA activities. By this selection we also attempted to differentiate between SQA and software-development activities. We presented three cases for the application of SQA practices in our environment and described our staff education concept which was strongly influenced by the SQA. Finally, we tried to define levels of SQA implementation in an organization. We intentionally omitted the allocation of our organization to one of the levels. The reader is encouraged to do so based on the cases presented in this paper. We would appreciate learning on which level we are. The benefit of SQA depends very much on its utilization by the management. A sensor on its own cannot achieve anything; in order to be effective the connection to an actuator is a must. We do not claim the discovery of a unique approach to SQA. We are well aware that more could be done - but what we do, we prefer to do correctly, and to improve slowly but surely.

ACKNOWLEDGEMENTS

P. Read has taken the trouble to polish our English (of an earlier version). P. Martindale provided the Figures 2 and 3.

REFERENCES

1. Baker, E.R., and Fisher, M.J., "Software Quality Program Management," IEEE Intern. Conf. on Communications, Boston, Mass., 1983, pp. 741-744.

2. Boehm, B.W., Brown, J.R., and Lipow M., "Quantitative Evaluation of Software Quality," 2nd Intern. Conf. on Softw. Eng., IEEE, 1976, pp. 592-605.

3. Britcher, R.N., Moore, A.R., and Segal, M.A., "Technology Transfer: The Key to Software Engineering Standards," in IEEE (1983 b), 1983, pp. 33-36.

4.   Cook, C.M., "Lessons Learned from Implementing a Software Quality Assurance Section," 3rd Software Engineering Standards Application Workshop, (SESAW-III), San Francisco, California, October 1984, pp. 60-67.

5.   Crawford, S.G., and Hiering, V.S., "Software Quality Control and Assurance," IEEE Intern. Conf. on Communications, Boston, Mass., 1983, pp. 713-717.

6.   CSA, "Standard for Software Quality Assurance Program," part I, CSA Preliminary Standard Q396.1, 1982.

7.   Dunn, R., and Ullman, R., Quality Assurance for Computer Software, McGraw-Hill Book Company, New York, 1982.

8.   Frühauf, K., and Sandmayr, H., "Quality of the Software Development Process," in IFAC Safecomp 1983, Cambridge University Press, pp. 145-152.

9.   Gilb, T., Software Metrics, Winthrop Publishers, Inc., Cambridge, Mass., 1977.

10. Goudie, D.B., Davis, M., and Spatz, A., "The BECONTROL Family of Supervisory Network Control Systems," Brown Boveri Review, Vol. 71, September/ October 1984, pp. 406-415.

11. Halstead, M.H., Elements of Software Science, Elsevier North Holland Inc., New York, 1977.

12. Harrison, W., Magel, K., Kluczny, R., and DeKock, A., "Applying Complexity Metrics to Program Maintenance," IEEE Computer, Vol. 15, No. 9, 1982, pp. 65-79.

13. Höcker, H., Itzfeldt, W.D., Schmidt, M., and Timm, M., "Comparative Descriptions of Software Quality Measures," GMD-Studien Nr. 81, Gesellschaft für Mathematik und Datenverarbeitung mbH, Postfach 1240, D-5205 St. Augustin 1, 1984.

14. IEEE, "Standard for Software Quality Assurance Plans," IEEE Std 730-1981 (Revision of ANSI/IEEE Std 730), 1981.

15. IEEE, "Standard Glossary of Software Engineering Terminology," IEEE Std 729-1983.

16. IEEE, "Proc. of the 2nd Workshop on Software Engineering Standards," (SESAW-II), San Francisco, California, May 1983.

17. KET-7, "Richtlinie zur Erfassung von Fehlern in Software-Produkten," KET-7 QS-Information No. 7, BBC, unpublished, 1984.

18. KET-7, "Rahmenrichtlinie für die Qualitätssicherung von DV-Software," KET-7 QS-Information No. 9, BBC, unpublished, 1985.

19. McCabe, T., "A Complexity Measure," IEEE Trans. Software Eng., Vol. SE-2, 1976, pp. 308-320.

20. McCall, J.A., and Matsumoto, M., "Metrics Enhancement," Rome Air Development Center, TR-80-109, 1980.

21. McCall, J.A., Richards, P.K., and Walters, G.F., "Factors in Software Quality," NTIS AD-A049-014, -015, -055, 1977.

22. Meekel, J., and Troy, R., "Comparative Study of Standards for Software Quality Assurance Plan," 3rd Software Engineering Standards Application Workshop, (SESAW-III), San Francisco, California, October 1984, pp. 60-67.

23. Walters, G.F., "Developing Software Product Standards Based upon Metrics,"
     IEEE Intern. Conf. on Communications, Boston, Mass., 1983, pp. 727-731.

DISCUSSION

Chairman: O. Klammer (Brown Boveri, Copenhagen, Denmark)

M. Morganti (ITALTEL, Castelletto di Settimo Milanese, Italy)

I have a comment about the difference between hardware and software quality assurance. It seems to me that you are comparing two different phases of the life cycle: you are comparing software design with hardware production which is in a sense surprising because software has production, where you can put the wrong modules together or make a bad release kit by missing documentation. If you compare the two design phases, they are much more equal. In fact, you could find out that hardware quality assurance is as poor as software quality assurance. And you would be surprised to find that metrics that apply to one apply to the other too. So, in fact, I do not see the two problems as different.

J. Ludewig

I completely agree. But I think that in many areas of hardware the influence of production on the quality is very strong, which is usually not the case in software. So that may be the difference.

W. Giloi (Technical University of Berlin, Berlin (West), Germany)

Just a brief comment. In the 1970's there was a famous bestselling novel written by Pirsig and entitled Zen and the Art of Motorcycle Maintenance. It concerns a philosopher who went out of his mind trying to find out what the meaning of the notion of quality is. So let us be careful.

D. Allison (Stanford University, Stanford, CA, USA)

I would like to return to the various tasks of a software quality-assurance department you listed. What I want to point out is that there appears to be no actual testing and validation being done by the software quality-assurance department. Likewise the standard you cite does not include any requirement for anything other than the planning of such testing.

J. Ludewig

We think that testing and validation is part of the production process. The quality-assurance department should just make sure that testing and validation actually happen and should audit them, but they are not executed by the software quality-assurance department.

K. Frühauf

It is important that tests are documented, and that the quality assurance department has access to those data and can evaluate the results of tests, derive statistical data, and make suggestions for improvements.

T. Lalive d'Epinay (Brown Boveri, Baden, Switzerland)

You mentioned that a software quality-assurance team ought to be highly independent. On which level would you establish such a quality-assurance team? Should it be a group, a section, a department, or a division?

K. Frühauf

Quality assurance is a form of consulting. I think it is not very important on which level it is in the organization. We can understand software quality assurance as a piano. By collecting data, we tune the piano, but the management must play on that piano. The sound depends mainly on the management.

FUNCTIONAL SPECIFICATION OF PROCESS-CONTROL SOFTWARE

F. KAUFMANN, D. SCHILLINGER and U. SCHULT
Brown Boveri, Baden, Switzerland

ABSTRACT

Functional programming languages are based on the mathematical concept of recursive functions. Such languages define the desired result of a computation as a static input-output mapping. Thus, programs expressed in a functional language can be considered as an executable specification. In contrast to procedural languages, the sequence of computation is not explicitly specified, which thereby facilitates the construction of programs.

At the BBC Research Center, the concepts of functional languages are applied to the development of a graphical programming environment which is used for the construction of process-control software. This programming environment is intended to be used by control engineers who may have little knowledge of computer science. Programs are constructed as data-flow graphs which correspond to the function charts traditionally used for the description of process-control mechanisms. The computer-supported graphic system is used for both programming and documentation. The graphic source is translated automatically into executable code.

1. INTRODUCTION

A process-control system is a computer-based system (including the software) used to control some type of plant. Roughly speaking, the process-control system can be said to provide control functions and an interface for human interaction. This interface is a means to indicate instantaneous values and to receive desired values. In this way an operator may observe and manage the plant.

Figure 1 may be viewed as a traditional control loop, i.e., the methods of classical automation control can be applied to the control system. However, the major difficulty in carrying out a process-control project lies in coping with the huge complexity of the application, rather than stability problems or other aspects of automation control.

Process-control software has to be tailored to each particular plant. Consequently, to carry out a process-control project economically, a large effort has to be expended on methods to allow efficient development of the software. An obvious method for this purpose consists of dividing the software into the project-independent and the project-specific parts. The idea is, of course, to reuse some project-independent parts in subsequent projects. The customer should be encouraged to contribute to the project-specific software, whereas the manufacturer should focus on problems which are common to all projects and on the infrastructure to support the customer's needs. Accordingly, the following activities must be supported:

- Programming of control functions
- Picture generation for control displays
- Configuration of dialogues
- Configuration of data-management systems

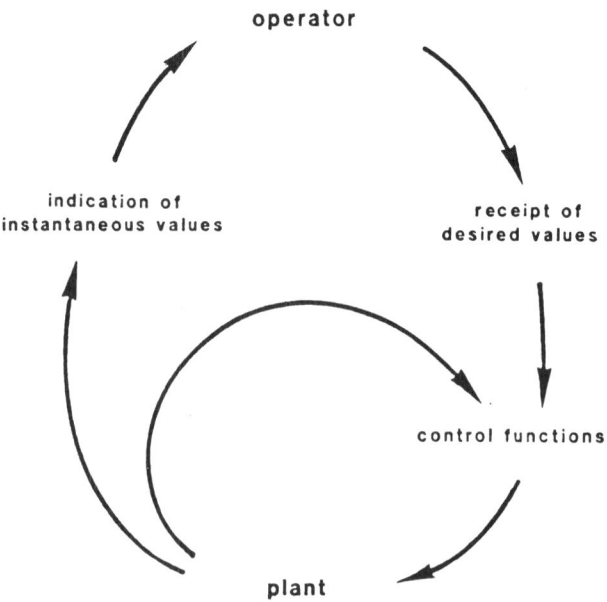

Fig. 1: Tasks Performed by a Process-Control System

Specification by means of a problem-oriented description can assist in carrying out all the tasks listed above. Starting from the specification, some parts of the process-control software can be automatically generated or configured. For these, the conventional coding can thus be avoided. This approach requires a set of computer-based tools, each of them supporting one specific activity. The existence of such tools allows the software to be configured by an engineer who need not be familiar with computer science. Instead, such an engineer is supposed to have a substantial knowledge of the particular plant. As a consequence, graphics, forms or tables are used as a problem-oriented description for coping with the above tasks, rather than a conventional programming language.

This paper focusses on the computer-aided programming of the control functions for which a programming method is proposed. Programs are specified by means of an application-oriented language based on the concepts of functional languages. The programmer specifies only the logic of the program (what) and is not burdened with operational details (how). A programming tool supports the construction of programs. The program is represented graphically (function chart) and can be translated automatically into a conventional (sequential) machine program.

2. THE PROCESS-CONTROL ENVIRONMENT

The conditions found in the domain of process control differ drastically from those of a typical data-processing environment. In a data-processing system, the effect of a program becomes visible on completion of a "run". The produced results are represented as values. In contrast, a process-control system is "forever" in control of the plant and the data is handled as streams of values.

In a process-control environment the aspect of real time must be considered. Current information is sampled at certain points in time. Fixed-time slices are available for processing the current data. In most cases, the computation is carried out cyclically, which causes a constant load of the processing units even in emergency situations (worst case).

A process-control system is preferably based on standard microprocessor hardware. In consequence, the process-control hardware profits from advances in microelectronics. Moreover, a process-control system based on standard components can easily be configured according to a wide range of applications.

In any phase of a process-control project the software is subject to modifications or extensions. Therefore, not only the development but also the commissioning, debugging and maintenance of the software has to be supported by appropriate tools. The

development and debugging phases are off-plant activities, whereas the other phases
are carried out on the plant site. The computer-based tools provided on the plant site
are supposed to run on a small, movable machine: the use of personal computer
hardware has to be considered.

A complex process-control system may include several hundreds of processors, each
running its own program. These programs are stored in different memories and permanent storage. Installing a new software version requires certain programs to be
replaced. The consistency of the software documentation and the programs actually
installed has to be guaranteed. The concept of a self-documenting system is a solution
to this important problem. This approach assumes the documentation to be reconstructed out of the programs actually installed. The current documentation can be requested and displayed by means of a programming system which may be connected to the
process-control system.

3. PROGRAMMING METHOD

The proposed programming method differs in many respects from a conventional approach. The program is constructed graphically with no other representation accessible to the user. Commissioning, debugging, and maintanance of software are based
on the graphical representation. The graphics serves not only as the source-program
representation but also as a final document used for describing control functions.
Thus, the graphics may be called an executable documentation.

We have to consider both the documentation and the language aspects. The documentation aspect affects the appearance (syntactical form) of the graphics. The graphical
representation must be in line with the function chart which has been used for decades in automation control. The function chart defines the behavior of a controller in
terms of transfer functions. Several proposals have been published[3] for an international standard on this kind of documentation.

The graphics also provides the same properties as a high-level programming language.
In particular, the program does not reflect the target machine on which the code will
actually be executed. Therefore, the user is not allowed to optimize the program by
taking advantage of features offered by a specific target machine.

The proposed language is based on the concepts of functional languages. The programmer defines the data dependency between predefined building blocks (functions)
provided by a library. Thereby the concept of stepwise refinement is applied. The
functional approach allows a high degreee of independence from the target system.
The functional language assumes a model target system which offers any amount of

computing power and degree of parallelism desired. Moreover, the target system guarantees the consistency of the data at any time. However, a real target system is far from that ideal. Consequently, a program must be tailored according to the real target system in order to run efficiently. Since this optimization affects the operational behavior of the program without changing its logic (Figure 2), the programmer is not concerned with the optimization.

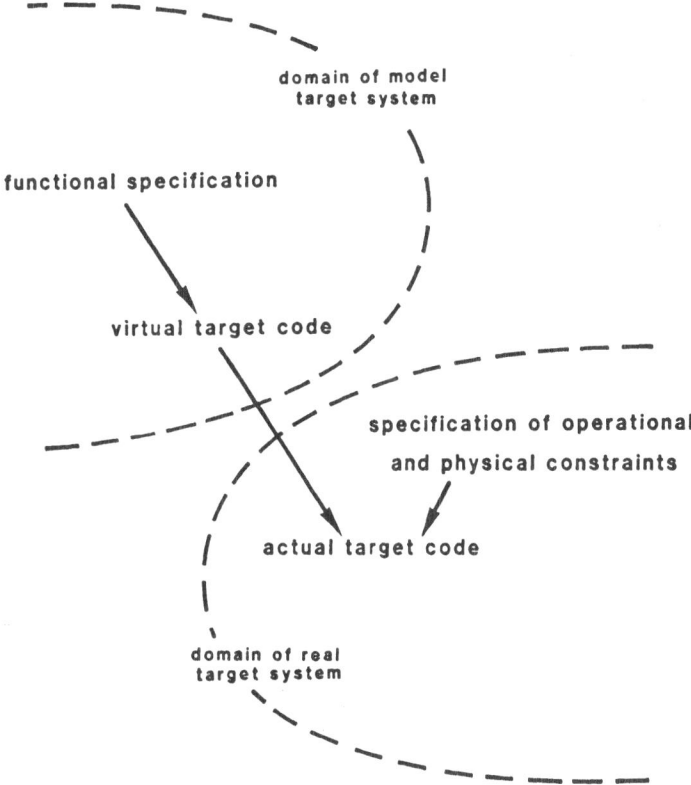

Fig. 2: The Domain of the Model and Real Target System

3.1 Concepts of Functional Languages

Research in the area of programming has shown that programs may be viewed as mathematical <u>functions</u>.[5] Functions in the mathematical sense map some input set onto some output set. Functions between sets of functions are called <u>functionals</u>. Mathematical functions are free from side effects, i.e., the execution of a function has no implied secondary effects on its environment. By considering programs as functions, it becomes possible to apply mathematical methods to them. The semantics can be specified in a concise form, which is difficult for conventional instruction sequences.

With this aspect in mind, conventional (procedural) programs are functions that map the set of program states onto itself. However, even primitive language constructs (e.g., a jump instruction) lead to very complex functionals and sets of functions.

The concept of viewing programs as mathematical functions has led to an alternative to the strictly operational programming model that conventional programming languages support. With respect to semantics, we have to distinguish between two points of view:

- The operational, and
- The functional (mathematical)

The operational semantics gives an enumeration of the executing steps (<u>how</u> a program is executed). The functional semantics determines the externally visible (input/output) behavior of the program (<u>what</u> it does).

Functional languages support the mathematical point of view and, thus, neglect operational details. Programming in a functional language consists of creating new functions out of old ones by applying some functionals (<u>combining forms</u>) to some functions.

Within the present paper two combining forms are used. For any functions f, g and any object x they are defined (in FP-notation[1]) as follows:

- composition: $f°g:x = f:(g:x)$
- construction: $[f,g]:x = <f:x,g:x>$

Hence, the <u>composition</u> represents the nesting of functions, whereas by means of the <u>construction</u> several functions are collected to form a list of functions.

3.2 The Use of a Functional Language in the Process-Control Environment

The response of a process-control system is usually based on the current input as well as on past inputs. Thus, the system has to remember its previous behavior, which means that some state information has to be retained. A process-control system has a behavior similar to that of a finite-state machine. The state information (local data) is currently updated taking into account both the current input and the previous state information, whereby the output is determined from the state information.

The true mathematical function, as introduced in the previous section, does not consider such state information. Nevertheless, a different point of view allows the concept of functional languages to be applied. In the process-control environment the focus is not on current inputs and outputs but rather on a sequence of outputs related to a sequence of inputs by means of a transfer function. This transfer function corresponds to a mathematical functional which maps functions (in time) to functions (in time) and the state information can therefore be discarded.

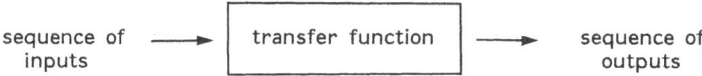

sequence of inputs   ⟶   transfer function   ⟶   sequence of outputs

The term function denotes a functional and the term signal a sequence of values as described above.

4. FUNCTIONAL CONTROL LANGUAGE

4.1 Language Summary

A program in terms of the proposed functional control language is a collection of separately handled parts called segments. Each segment is represented as a graphic entity. There are two kinds of segments: The program segment which invokes a function and the library segment which contains the definition of a function. A function defined by a library segment is accessible to all other segments of the program. A segment is constructed as a combination of functions. The function is the only active element provided by the language. Any manipulation of data results from the application of a function (without side effects). The language assumes data of different type, but in any case the data is viewed as a signal, that is, a stream of values.

4.2 Signals and Types

The data handled by the functional control language are streams of values (signals). Each signal is updated as the result of a certain function and accordingly has a single source. A signal has no explicit declaration. The scope of a signal is implicitly specified by means of a prefix heading the signal name. The scope is either a segment, the program, or even the entire process-control software. A signal may assume a current value out of the domain determined by the corresponding type. A signal and its source are always of the same type. Since the type is bound to the source, a signal does not need a type definition. The functional control language provides a set of predeclared functions (sources) and accordingly a set of predeclared types. There are basic types which include scalar values, and structured types in the form of records. The record type consists of several components of possibly different types.

4.3 Functions

A function maps one or more input signals to one or more output signals. The function exclusively influences its environment by use of its output signals. Therefore, it is said to be free of side effects. There are two kinds of functions: the <u>simple functions</u> which form the atomic elements of the language, and the <u>compound functions</u> which can be decomposed into a combination of more primitive functions. Further, a function may retain some state information. The occurrence of state information can result from cyclic nesting of functions, which is a special case of recursion. In any case some additional information is required in order to set up the initial state. Each function is identified by a unique name and has a well-defined interface which does not depend on the implementation of the function. In particular, the interface specifies the formal arguments including their types. The formal arguments are substituted by actual arguments when the function is invoked (Figure 3). Some functions (e.g., arithmetic functions) do not expect a fixed number of arguments. Different instantiations may assume a different number of arguments. The function is said to have variable arity.

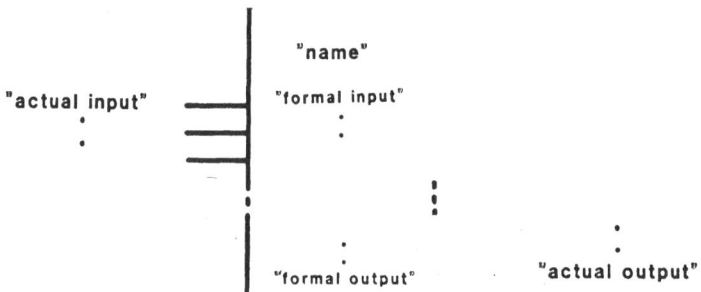

Fig. 3: Graphical Representation of a Function

4.4 Program Construction

The functional control language provides two combining forms: the composition and the construction. The graphical representation of a combination of functions is commonly known as a <u>function chart</u>. A function chart represents the decomposed form of a compound function and can also be viewed as the definition of a compound function.

A hierarchically structured program can be obtained by means of compound functions (Figure 4). On any level of abstraction the compound functions can be decomposed

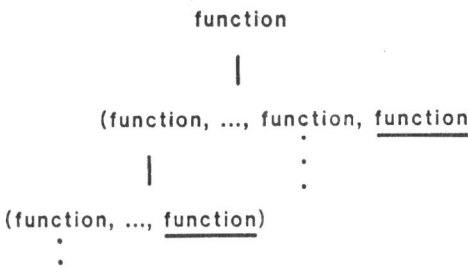

Fig. 4: Hierarchical Program Structure

(refined) to a combination of lower level functions. The program constitutes the top of the hierarchy. The simple functions (underlined) do not allow further decomposition and therefore form the base of the hierarchy.

The separately handled parts of a program are called segments. There are two kinds of segments: the program segment and the library segment. The program segment is the executable entity. A program segment implements a function applied to actual data at program runtime. Usually the program segment implements a compound function and thus can be represented as a function chart. A program may include several program segments which communicate by means of shared data. Each data item is produced by a single program segment.

The library segment is a definition entity. It defines a function as a subprogram. This function is accessible to all other segments of the program by use of its unique name. When the library segment defines a compound function, it is represented as a function chart (Figure 5). Of course, there are formal names (formal arguments) referring to the signals to be accessed.

The library segment may also define a simple function, which does not use the functional control language for implementation. The implementation has to meet some conventions to prevent side effects. This fact must be respected if a conventional, procedural language (e.g., PASCAL) is used.

In particular, this section concentrates on simple functions which behave corresponding to a finite-state machine and can preferably be described by a state/transition chart.[4] The state/transition chart is viewed as an application-oriented program which is represented in graphical form. It consists of states and transitions. Each transition

compound function

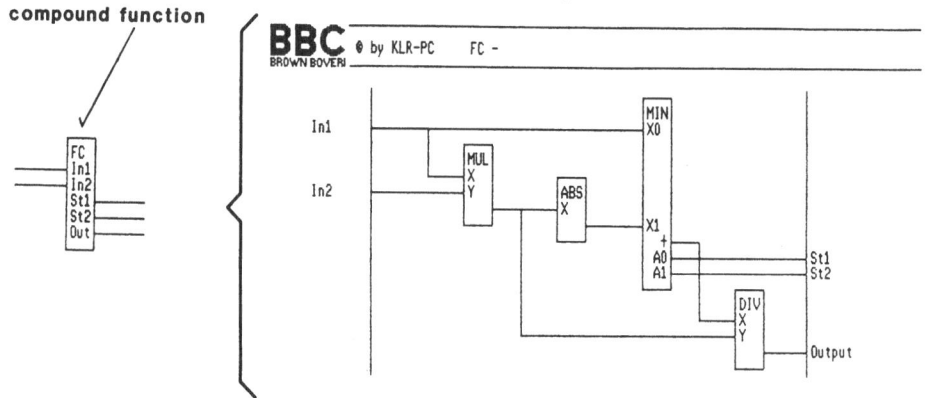

Fig. 5: Compound Function Defined as a Function Chart

simple function

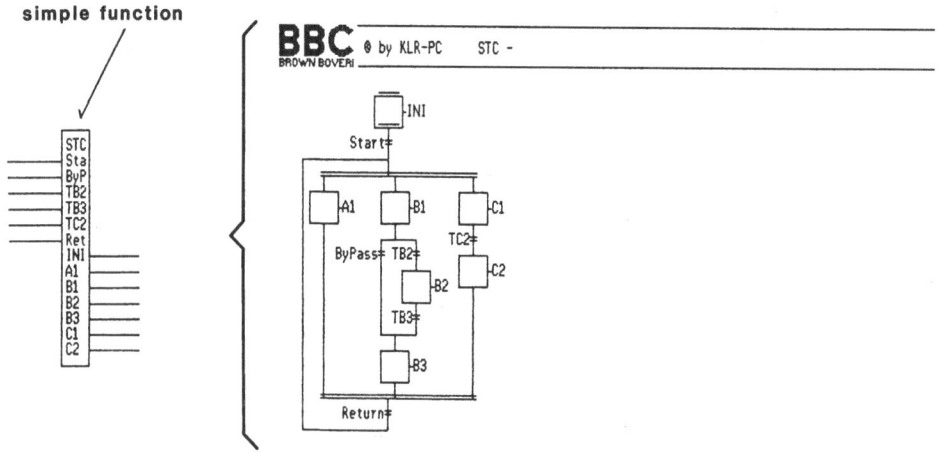

Fig. 6: Simple Function Defined as a State/Transition Chart

is associated with a so-called firing condition. A true condition causes a token to be moved from the current state to the successor state marking it as active. The state/transition chart is derived from the Petri net.[4] In contrast to the Petri net it enforces block structure.

A library segment represented as a state/transition chart defines a simple function

(Figure 6). The firing conditions form the inputs of this function and the outputs indicate whether or not a state is active.

5. THE PROGRAMMING TOOL

A prototype of a programming system which supports the proposed graphical language has been implemented. The prototype executes on a commercial personal computer. It comprises two syntax-controlled graphics editors, a cross compiler and a cross back-compiler which support a target system based on the Intel 8086 processor family. The prototype is being used as a kernel for the development of a marketable product. In this section we outline some key aspects of the implemented programming system.

5.1 Approach

We distinguish between two fundamentally different methods for generating a graphical representation on a computer system:

- Composed (CAD approach) or
- Synthesized (low-cost approach)

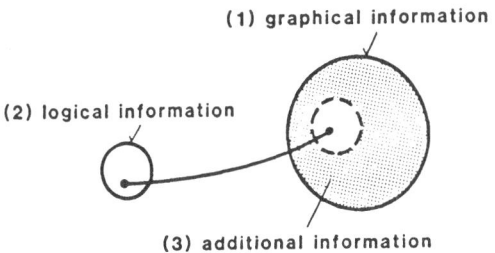

(1) graphical information

(2) logical information

(3) additional information

Fig. 7: Informational Diagram of a Graphics

As shown in Figure 7, the essence of each graphical representation is some logical information (2). The purpose of the graphics is to facilitate reception of the logical information by the reader. Obviously, the graphics contains some additional information (3) which contributes less (or even nothing) to the logical information and hence is secondary. For composed graphics, the graphical information (1), including both the logical and the additional information, is user defined. This is in accordance with the traditional notion of "drawing". The computer may extract the logical information from the graphical one. Graphics with different appearances may contain the same logical information.

For synthesized graphics, the user defines the logical information. The graphical information is generated (synthesized) by the computer out of the logical one by means of a fixed algorithm. Hence, some logical information is depicted in a unique way which cannot be influenced.

A large system (CAD system) to handle a large amount of information and to realize an expensive dialog with the user is needed to support graphics composition. In addition, the extraction of the logical information may be very costly.

A graphics synthesis can be implemented on a small low-cost system. The amount of information to be handled is small and the user dialog is simple. Of course, this approach restricts the user's freedom, however, this restriction is utilized to enforce optimized graphics (e.g., with regard to crossing lines) which are compatible with international standards.

For present purposes, we shall limit consideration to synthesized graphics. Some connected objects (functions or states and transitions) build the logical information. The graphics layout is of minor importance and hence may be synthesized by a graphics generator.

5.2. The Graphics Editor

The graphics editor uses an internal representation (realized with a linked list) of the logical information (i.e., the user program). After each editing operation, the graphics on the screen is updated by the graphics generator (Figure 8). In this way the user gains the impression of directly manipulating the graphics. This approach provides a simple, clear editor dialog. The emphasis lies on specifying the logical information (the program) and not on defining any insignificant information. Hence, efficient programming is assisted. After an editing session, the internal representation of the user program is transformed into the machine program. The machine program is the only representation which is saved on the disk for further use.

For reediting a program, the machine program is retransformed into the internal representation by means of the backcompiler. The two mappings performed by the compiler and by the backcompiler are exact inverses: the regenerated internal representation is always identical with the original. Consequently, the regenerated graphics is also identical with the original.

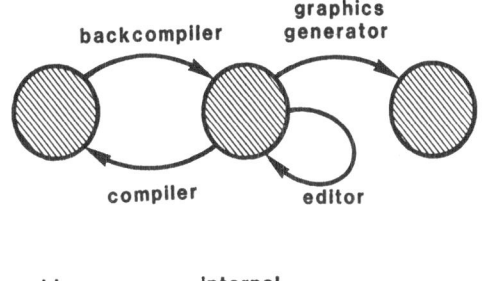

backcompiler   **graphics generator**

compiler   **editor**

machine program   **internal representation**   **graphics**

Fig. 8: The Internal Program Representation and Its Transformations

5.3 The Compiler: Code Generation for the Target Machine

The compiler transforms the logical program information (the internal representation) into the machine program. For this purpose, the compiler adds some operational information and may be said to provide the operational semantics of the user program.

In accordance with the semantics of functional languages, the partial order within the logical program representation allows parallel execution of independent subexpressions (subtrees). However, because the program has to be executable on a single processor, it must be executed sequentially. Thus, the compiler has to generate a total ordering (which respects the given partial order) among the logical objects.

To allow treatment of infinite data streams, the code of all segments is arranged in an overall loop. The loop is preceded by an initialization part which defines the initial values of all states within the program. The initialization part is executed once for the start up of the program before entering the loop. The initialization part is generated implicitly by the compiler and so is not directly accessible to the user.

A one-to-one translation of the internal representation into the machine program leads to threaded code,[2] which consists of a sequence of procedure calls. This threaded code can partly be replaced by machine code for optimization purposes. The rest of this section gives additional details of the threaded code generated for function charts and state charts, respectively.

Function Chart: Each object (function) within the internal representation yields a procedure call (Figure 9). The bodies of the procedures are located in the libraries described in Section 4.4. The calls are arranged in a sequence compatible with the total order outlined above.

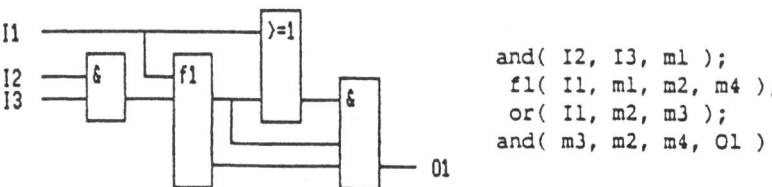

and(I2, I3, m1);
f1(I1, m1, m2, m4);
or(I1, m2, m3);
and(m3, m2, m4, O1)

Fig. 9: Translation of a Function Chart into Threaded Code

State Chart: Each transition generates a call for a procedure named "Transition".
This procedure may be viewed as a transition with a variable number of preceding
and succeeding states. Thus, its parameter list has three entries: A Boolean signal
(condition) to control the firing, a list of input states and a list of output states.
The semantics of the procedure "Transition" is described by the following implementation:

```
procedure Transition( condition,
                      list of input states,
                      list of output states );
begin
    if     'all input states set' and 'condition fulfilled'
    then   'reset all input states';
           'set all output states'
    else   'do nothing'
    endif
end Transition
```

All the other logical objects (states, branchings, etc.) generate some implicit parameter references but no explicit code. Figure 10 shows the translation into threaded
code of the four syntactical blocks (sequence, alternative branches, parallel branches, and loop).

A fundamental difference between the function chart and the state chart with regard
to the threaded code generated is noteworthy. The function chart is translated on the
function level: each function generates a call. Thus, the resulting code again has the
character of a functional language (e.g., no side effects). This is not true for the
state chart. Here, the logical objects are not functions and, hence, the threaded code
generated has side effects.

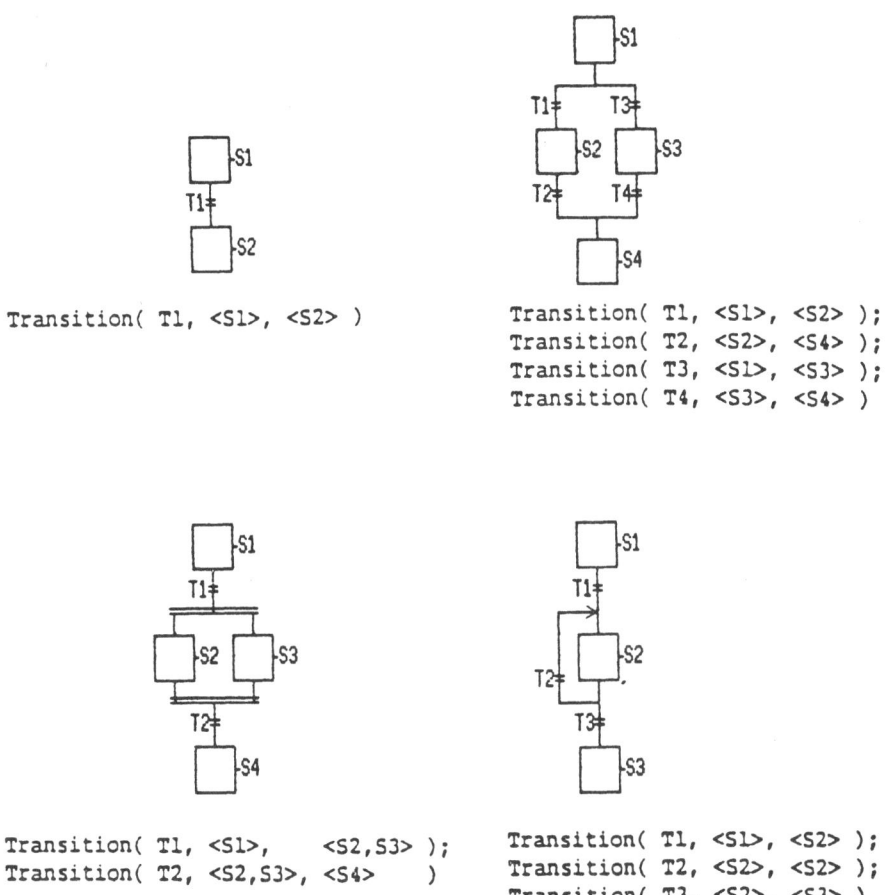

Transition(T1, <S1>, <S2>)

Transition(T1, <S1>, <S2>);
Transition(T2, <S2>, <S4>);
Transition(T3, <S1>, <S3>);
Transition(T4, <S3>, <S4>)

Transition(T1, <S1>,   <S2,S3>);
Transition(T2, <S2,S3>, <S4>   )

Transition(T1, <S1>, <S2>);
Transition(T2, <S2>, <S2>);
Transition(T3, <S2>, <S3>)

Fig. 10: Translation of the State Chart into Threaded Code

6. CONCLUSION

The programming method introduced in this paper has been applied to the specification of control functions. Several projects have been carried out to prove the method
under real conditions. The experience has shown that the proposed language supports
a programming style which is well suited to cope with software of high complexity.
This programming style allows the complex problem to be decomposed into several
subproblems which are less complex and can be solved independently. In this sense,
this paper may be regarded as a contribution to the topic of software engineering.

REFERENCES

1. Backus, J., "Can Programming Be Liberated from the Von Neumann Style? A Functional Style and Its Algebra of Programs," Commun. ACM, Vol. 21, No. 8, 1978, pp. 613-641.

2. Bell, J.R., "Threaded Code," Commun. ACM, Vol. 16, No. 6, 1973, pp. 370-372.

3. IEC Sub-Committee 65A, "Draft-Programmable Controllers," Central Office of the IEC, 3, Rue de Varembe, Geneva, Switzerland.

4. Peterson, J.L., Petri Net Theory and the Modeling of Systems, Prentice-Hall, Englewood Cliffs, N.J., 1981.

5. Scott, D., and Strachey, C., "Toward a Mathematical Semantics for Computer Languages," in Proc. Symposium on Computers and Automata, J. Fox, ed., Polytechnic Institute of Brooklyn Press, New York, 1971.

DISCUSSION
Chairman: O. Klammer (Brown Boveri, Copenhagen, Denmark)

V. Haase (Technical University of Graz, Graz, Austria)
How much impact does the method you presented on practical software engineering have in your company?

D. Schillinger
The tool I showed has been implemented at the Brown Boveri Research Center. It has been transferred to a division, where it is being further developed into a selling product.

K. Ohmaki (Federal Institute of Technology, Zurich, Switzerland)
Do you intend to support the validation of graphically constructed programs? Do you provide automatic tools that check whether the program meets some given input/output relation?

D. Schillinger
Not currently. However, it is possible to create different levels of abstractions by means of the functional decomposition presented. This decomposition technique facilitates the manual inspection of the program.

LOGIC PROGRAMMING

J. KRIZ and H. SUGAYA
Brown Boveri, Baden, Switzerland

ABSTRACT

Logic programming is based on formal mathematical concepts of first order predicate logic. A logic program is a static description of the problem in the form of normalized logic statements (Horn clauses). The execution of a logic program relies on automated theorem proving. In contrast to conventional procedural languages, in which the sequence of computation is specified explicitly by control structures, in logic programs information is provided only about "what" is to be solved, not "how" it is to be solved. Hence, logic programs represent executable specifications and are well suited for fast prototyping.

The concepts of logic programming have not yet been fully realized, because of the large requirements on memory and execution time of conventional computers. The programming language Prolog is based on logic programming, but employs various additional logical features in order to be a practical programming language. Recent advances in compilation and hardware design are continuously improving the feasibility of logic-based programming languages for industrial applications.

At the BBC Research Center, the system Modula--Prolog is being used for fast prototyping and knowledge engineering. Modula--Prolog permits the arbitrary combination of Modula-2 and Prolog programs and hence utilizes the advantages of both procedural and logic programming. Modula--Prolog has been applied to the development of knowledge-based expert systems for the configuration and diagnosis of technical systems.

1. INTRODUCTION

Mathematical logic has always been strongly interrelated with computer science. Since the logic lends itself to exact description of objects and reasoning about them, it has been applied for problem specification, knowledge representation, data description, program proving, etc. At the same time, logic itself, especially the mechanization of theorem proving, has been the subject of investigation in computer science.

In 1965, Robinson[19] published a unification algorithm for theorem proving based on the resolution principle. The algorithm decides mechanically the validity of an arbitrary logic formula in the normalized clause form by a nondeterministic refutation proof. Since then, various improvements and modifications of this algorithm have been developed and implemented. At the beginning of the 1970s, Kowalski[11] introduced the concept of procedural interpretation of logical formulas and stated explicitly the idea of logic as a programming language. At the same time, Colmerauer[5] and Roussel[21] designed and implemented the programming language Prolog that is based on a subset of the first-order predicate logic and employs an efficient but restrictive form of the Robinson algorithm.

Logic programming refers to specification and solution of problems by means of mathematical logic. It differs significantly from conventional procedural (imperative) programming in which a program is a procedure specifying the sequence of execution steps to be carried out by the computer. Logic programming features the declarative (descriptive) programming style: programs are specified as static input/output relationships or predicates without stating the execution sequence explicitly. Since the computation is not overspecified by prescribing a fixed sequence of execution steps, it can be carried out in an arbitrary order that is consistent with the operational semantics. Declarative languages are thus suitable for parallel execution.

Declarative languages offer various potential advantages over procedural languages and represent an innovative, promising approach to software engineering. They are based on precise and well-understood mathematical concepts of recursive functions and first-order predicate logic. Thus, declarative languages support the construction of concise and precise programs, program verification, testing and debugging. Declarative languages are also appropriate for fast prototyping: a logic program represents a formal and executable specification of the problem.

Some problems associated with declarative languages prevent their use for general purpose programming. Pure declarative programs can be executed only inefficiently on conventional von Neumann computers. They require large amounts of memory and execution time. The usual, but unsatisfactory, approach to this efficiency problem is to introduce various nondeclarative features, such as global variables, assignments, and control mechanisms. Current research addresses the problem from two directions. First, program transformation and compiling systems are developed for automatically mapping declarative languages into machine languages. Second, new computer architectures, such as data-flow machines, parallel computers and inference machines, are designed to match the operational semantics of declarative languages.

The best-known representative of logic programming languages is Prolog. Although Prolog is an abbreviation of PROgramming in LOGic, it is not the ultimate realization of the concept of logic programming. First, it is only a particular implementation of a subset of the first-order predicate logic called Horn clauses. Second, it uses a sequential variant of the resolution algorithm, which is efficient but generally may not necessarily find a solution even though it exists. Third, it is an extension of Horn clauses by extralogical features for controlling the execution and achieving global side effects. Nevertheless, Prolog has gained widespread acceptance: it has been used as a specification language, for the implementation of expert systems, for education, for industrial process control, in the field of conventional and deductive databases, for natural language interfaces, for man-machine dialog, and so on. Prolog has been chosen as the core language for Fifth Generation Computer Systems.[10]

At the Brown Boveri Research Center, logic programming is being investigated and used for Artificial-Intelligence (AI) applications. The Modula--Prolog[17] system has been developed in cooperation with the Federal Institute of Technology, Zurich (ETHZ). This system comprises an interpreter for the "Edinburgh" Prolog in Modula-2.[27] Moreover, it provides a capability for combining programs written in Modula-2 and Prolog. Thus, the user may write his own predefined predicates in Modula-2, for example, for numerical computation or input/output handling. The Modula--Prolog system has been used for various applications: for implementation of knowledge-based systems and expert-system shells for configuration and diagnosis of technical systems, for knowledge representation in frame-based and rule-based structures, and as an intelligent front end for relational databases and grid files. The interactive programming and fast prototyping features of Prolog have considerably eased the development of the software systems.

The present paper is organized as follows. Sections 2 and 3 are conceived as a brief tutorial on logic programming and Prolog. In Section 2, the concept of logic programming with Horn clauses is introduced. Syntax and both declarative and procedural semantics of logic programs are explained. Computations are elucidated by AND/OR derivation trees that suggest parallel execution. Robinson's unification algorithm is described in detail. The relations of logic programming both to mathematical logic and to functional programming are discussed. In Section 3, the programming language Prolog is sketched and its relation to pure logic programming with Horn clauses is clarified.

Sections 4 through 6 include research contributions of the Brown Boveri Research Center to the development of Prolog-based tools and knowledge-based systems. Section 4 is dedicated to Modula--Prolog, an integrated Prolog environment based on Modula-2. Section 5 gives a brief overview on the implementation of knowledge-based systems in Prolog. Techniques for knowledge representation in rules and frames are explained by an example of a realized system. In Section 6, a knowledge-based diagnostic system for analog circuits, which is currently being developed at the Research Center, is presented.

2. HORN CLAUSE LOGIC PROGRAMMING

2.1 Syntax and Semantics

In this section we review logic programming with Horn clauses which is the theoretical base for Prolog. Let us start with an example of a logic program in Prolog syntax:

way(a, b).   (F1)
way(b, c).   (F2)
route(X, Y) :- way(X, Y).   (R1)
route(X, Y) :- way(X, Z), route(Z, Y).   (R2)

Fig. 1: Example of a Logic Program

This program states that there is a way from a to b and from b to c. Furthermore, there is a route from X to Y if there is either a way from X to Y, or a way from X to an intermediate point Z and a route from Z to Y.

A logic program consists of a set of program clauses that are either facts F (assertions) or rules R (implications). A fact expresses an atomic relationship between individuals such as constants (a, b, ...), or variables (X, Y, ...). Generally a fact is an atomic formula and has the form

$$p(t_1, t_2, \ldots t_n)$$

where p is the name of the relation (predicate name) and t_i are terms. A term is either a constant, a variable, or a structure $f(t_1, t_2, \ldots t_n)$ where f is a functor (function name) and t_i are terms. Since an atomic formula and a term have the same structure, they are often not distinguished in logic programming.

A rule has the form

$$A :- B_1, B_2, \ldots B_n,$$

where A and all B_i are atomic formulas. The logical meaning of the rule is:

$$(B_1 \text{ and } B_2 \text{ and } \ldots B_n) \text{ implies } A.$$

A is called the head of the rule and all B_i comprise the body.

A logic program is invoked by conjecturing that a goal clause G,

$$G = G_1 \text{ and } G_2 \text{ and } \ldots G_n,$$

where G_i are atomic formulas, is a logical consequence of the program clauses F and R. When supplied with the question (query or goal)

$$?- G_1, G_2, \ldots G_n,$$

the logic system tries to prove

$$F \text{ and } R \text{ imply } G,$$

and if it succeeds it reports the binding made to the variables during the proof. Examples of goals and corresponding answers are given in Figure 2.

Goal (question)	Answer
?- way(a, b).	YES
?- route(X, b).	X = a
?- route(b, a).	NO
?- route(X, c).	X = b; X = a
?- way(b, X), way(X, b).	NO

Fig. 2: Example of Queries

All variables which appear are implicitly universally quantified in the program clauses F and R, and existentially quantified in the goal clause G, e.g.,

R1 means: for all X, Y: way(X, Y) implies route(X, Y),
?- way(a, Y). means: there exists Y, such that way(a, Y) holds.

2.2 Resolution and Unification

In the procedural interpretation of logic, the clause

$$A :- B_1, B_2, \ldots B_n$$

means "to solve A solve B_1, B_2, ... B_n" and is considered to be a procedure with the name A and the body consisting of the subprocedures B_1, B_2, ... B_n. A goal G (query)

$$?- A_1, A_2, \ldots A_m$$

is solved by replacing the procedure calls A_i (subgoals) by the respective procedure bodies (matching program clauses) and trying to derive the empty goal. One computation step in this derivation is specified by the following algorithm:

1) Choose an arbitrary subgoal A_k to be solved first

2) Find a program clause, say A, that matches A_k after an appropriate variable substitution s (unification). If no such clause exists, then the subgoal A_k fails. Report failure. Otherwise continue with 3).

3) The problem is reduced to the solution of the new goal G'

$$?- A_1', \ldots A_{k-1}', A_{k+1}', \ldots A_m',$$

if the clause A is a fact, or to

$$?- A_1', \ldots A_{k-1}', B_1', \ldots B_n', A_{k+1}', \ldots A_m',$$

if the clause A is a rule. The new clauses A_i', B_j' are derived from the old clauses A_i, B_j by applying the unifying substitution s. This step is called resolution. If the new goal G' is empty, report success.

Two terms A and B can be unified if the appearing variables can be substituted by terms such that the resulting terms A' and B' are identical. This is called variable binding or variable instantiation. When several unifying substitutions exist, of particular interest is the "simplest" substitution that is unique and this is called the most general unifier.

The derivation (of the original goal G) consists in successively applying steps 1) to 3). Three outcomes are possible:

  a) A failure is reported: a failed derivation.
  b) A success is reported: a successful derivation.
  c) The process does not terminate: infinite derivation.

The original goal succeeds if there exists a successful derivation. The bindings to the variables of G appear as output of the computation. Since more than one successful derivation may exist, the computation is nondeterministic. If the computation were performed on an ideal parallel nondeterministic machine, only those choices in 1) and 2) would be made that lead to a successful derivation (if it exists). Finding a successful derivation by a sequential machine requires searching in the derivation tree. If there is no successful derivation, the goal has no solution and the system answers NO. If the goal G possesses a solution, then there is a successful derivation by the resolution algorithm. However, the algorithm may not detect that G is not solvable and the derivation is infinite (which cannot be effectively decided).

2.2.1 Unification of Two Terms: Robinson's Unification Algorithm[20]

The recursive procedure $unify(t_1, t_2)$ either computes the most general unifier of the terms t_1 and t_2 if it exists or terminates with "failure". The unifier is represented by a set of variable bindings "bind".

Let bind be initialized to the empty set { }
PROCEDURE $unify(t_1, t_2)$:
IF one of the terms is a variable, say X
THEN let t be the other term;
IF t ≠ X THEN
IF X occurs in t THEN failure
ELSE bind:= composition of bind and {X:= t};
ELSE let $t_1 = f(x_1, x_2, \ldots x_n)$, $t_2 = g(y_1, y_2, \ldots y_m)$,
n ≥ 0, m ≥ 0 (including constants);
IF f ≠ g or n ≠ m THEN failure
ELSE FOR k:= 1 TO n DO
let x'_k, y'_k be x_k, y_k after the substitution bind;
$unify(x'_k, y'_k)$;

We now demonstrate the unification algorithm on the example in Figure 1 and the query ?- route(X, c). In the derivation tree (Figure 3) we abbreviate "route" by "r" and "way" by "w".

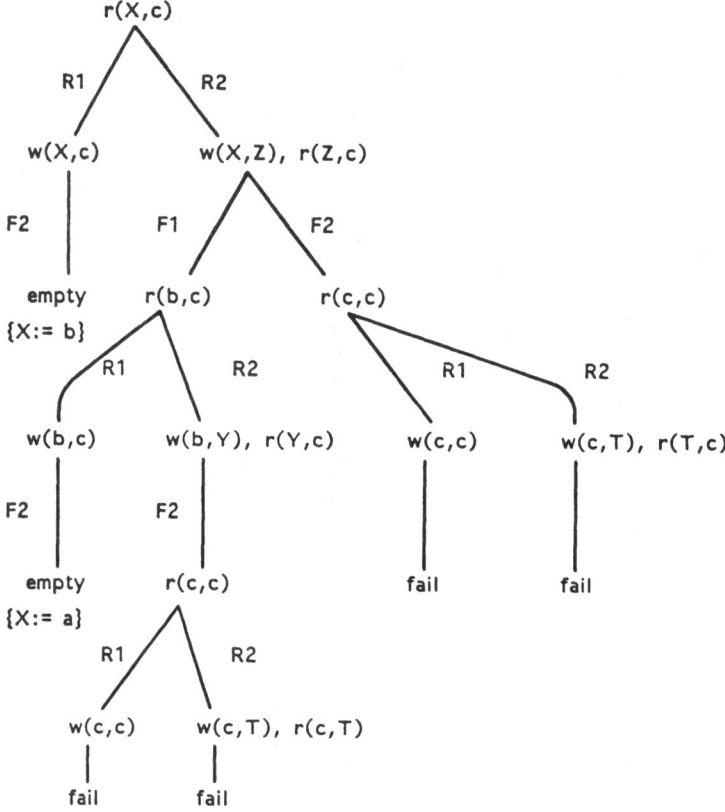

Fig. 3: A Derivation Tree

A derivation tree can also be seen as an AND/OR tree (Figure 4). The nodes are goals: unit goals (predicates) and composite goals (conjunction of predicates). There are AND branches from a composite goal to its subgoals. There are OR branches from a goal G to (the tails of) those clauses whose heads match G. More precisely, there is an OR branch from a unit goal G to the composite goal A_1, A_2, ... A_n if A:- A_1, A_2, ... A_n is a program clause and G and A are unifiable. There is also an OR branch from a unit goal G to the leaf node "empty" if G is unifiable with a program assertion.

A goal is satisfied if either one of its OR-subgoals or all of its AND-subgoals are satisfied. The resolution method corresponds to a top down traversal of the AND/OR derivation tree (backward reasoning). Another logical inference method is the natural deduction in which the tree is traversed bottom up (forward reasoning).

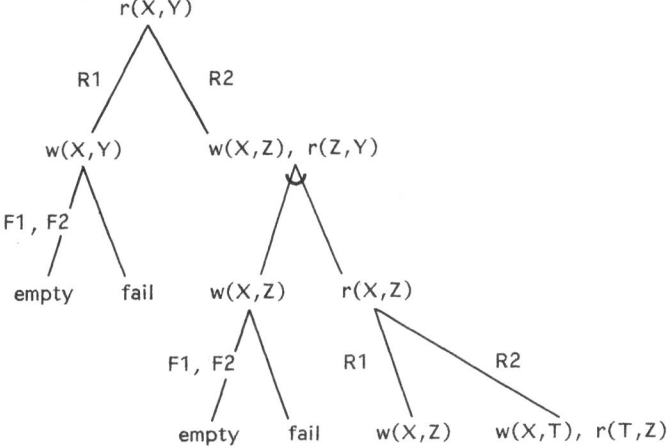

Fig. 4: AND/OR Derivation Tree ($\wedge$ AND Branches, < OR Branches)

The AND/OR tree suggests a parallel execution of logic programs.[13] There are obviously two main kinds of parallelism:

- AND parallelism: the subgoals of a goal are proved simultaneously. Unfortunately, the subgoals cannot be satisfied independently because they can share variables. Various attempts have already been made to exploit AND parallelism.[23]
- OR parallelism: the program clauses whose head match the current goal are pursued in parallel.

2.3 Relation to Mathematical Logic

Concepts of logic programming are strongly related to mathematical logic. A clause in the predicate logic denotes the formula

$$A_1 \text{ or } A_2 \text{ or } \ldots A_k \text{ or } not(B_1) \text{ or } not(B_2) \ldots \text{ or } not(B_n), \qquad (1)$$

where A_i and B_j are atomic formulas, and all variables are implicitly universally quantified by the quantifier ALL. This formula is logically equivalent to

$$(B_1 \text{ and } B_2 \ldots \text{ and } B_n) \text{ implies } (A_1 \text{ or } A_2 \ldots \text{ or } A_k). \qquad (2)$$

Clauses (2) can be classified as follows:

> Horn clauses, $k \leq 1$
> > - Program clauses (headed clauses), $k = 1$
> >   • Facts, $n = 0$
> >   • Rules, $n > 0$
> > - Goal clauses (headless clauses), $k = 0$
> > - Empty clauses (contradictions, "false"), $n = 0$, $k = 0$.

Note that the (goal) clause

$$\text{ALL } (x_1 \ldots x_j) \ (\text{ not}(B_1) \text{ or not}(B_2) \ldots \text{ or not}(B_n))$$

is logically equivalent to

$$\text{ALL } (x_1 \ldots x_j) \text{ not } (B_1 \text{ and } B_2 \ldots \text{ and } B_n),$$

as well as to

$$\text{not EXIST } (x_1 \ldots x_j) \ (B_1 \text{ and } B_2 \ldots \text{ and } B_n).$$

It can be shown that each well-formed formula of the predicate logic can be transformed into a conjunction of clauses C_i:

$$CF = C_1 \text{ and } C_2 \ldots \text{ and } C_m,$$

where each C_i is of the form (1). The main step in this transformation is the elimination of the existential quantifier EXIST by introducing the Skolem functions (Skolemization). For example, the formula

$$\text{ALL } x \text{ EXIST } y \ (f(y))$$

is transformed into

$$\text{ALL } x \ (f(s(x)),$$

where $s(x)$ is a Skolem function and replaces y. CF is called the clause form of the logic formula and has the important property that it is satisfiable (consistent) if and only if the original formula is satisfiable. Although each formula can be transformed into the clause form, it cannot generally be expressed by a conjunction of Horn clauses and consequently it cannot be described by a logic program. However, if a problem can be solved by Horn clauses, it can also be solved by a logic program in which only one clause is headless (goal).[4] In the following, we describe the resolution algorithm for this case.

From the logic point of view, the resolution algorithm is a refutation proof using the resolution as the only inference rule. If the program consists of the clauses P_1, P_2, ...P_n and the goal is G, then the system tries to prove the validity of

$$(P_1 \text{ and } P_2 \ldots \text{ and } P_n) \text{ implies } G \tag{3}$$

by assuming nonvalidity and arriving at a contradiction (refutation). The variables in P_i are universally quantified, those in G are existentially quantified. The negation of (3) is the clause form

$$P_1 \text{ and } \ldots \text{ and } P_n \text{ and } not(G) \tag{4}$$

with all variables universally quantified. Let G be

$$G = A_1 \text{ and } A_2 \ldots \text{ and } A_m,$$
$$\text{i.e., } not(G) = not(A_1) \text{ or } not(A_2) \ldots \text{ or } not(A_m)$$

and one of the program clauses P_i be

$$A :- B_1, B_2, \ldots B_n$$
$$(A., \text{ respectively})$$

such that one subgoal A_k matches A after a unification. Then (4) is inconsistent if and only if

$$P_1 \text{ and } \ldots \text{ and } P_n \text{ and } not(G) \text{ and } not(G') \tag{5}$$

is inconsistent, where

$$G' = A_1' \ldots \text{and } A_{k-1}' \text{ and } B_1' \ldots \text{and } B_n' \text{ and } A_{k+1}' \ldots \text{and } A_m'$$
$$(G' = A_1' \ldots \text{and } A_{k-1}' \text{ and } A_{k+1}' \ldots \text{and } A_m', \text{ respectively})$$

and the prime denotes the unifying substitution. The step from (4) to (5) is the resolution. Special cases of the resolution are the following inference rules:

From $not(A)$ and $(B \text{ implies } A)$ conclude $not(B)$ (modus tollens)

From $not(A)$ and A conclude "empty"   (contradiction)

Robinson's original resolution algorithm proceeds as follows:

1) At each step: let G be either the original or one of the derived goals. Resolve $not(G)$ with an arbitrary program clause whose head is unifiable

with G. Add the resolvent clause to the set of derived goal clauses.

2) If the empty goal clause can be derived, then (4) is inconsistent and hence (3) is valid.

2.4 Relation to Functional Programming

Logic programming is, roughly speaking, a generalization of functional programming to relational programming, because predicates correspond to relations which are general-ization of functions. Functional programming is based on the concept of mathematical function that is defined as a mapping f from one set (domain D_f) to another set (range R_f). Exactly one element $y = f(x)$ from R_f is attached to each element x of D_f. A functional language is specified by a set of basic functions and a set of con-struction forms (functionals) that allow complex functions to be built up out of basic functions. A fundamental construction form is the recursion. Sometimes a distinction is made between functional and applicative programming. In applicative programming, functions may additionally be defined using variables, i.e., expressions like f(X) are used, where X stands for any element of D_f. We apply the term functional program-ming to this type as well.

Since a functional program is a static description of a mathematical function, it pos-sesses a clear denotational semantics. In order to be executed, an operational seman-tics has to be defined as well, which essentially describes how values are substituted for arguments of functions (e.g., call-by-value, call-by-name) and how recursion is resolved by rewriting terms. The execution of a functional program is always deter-ministic and proceeds in one direction. The result of the execution is unambiguous and does not depend on the order in which independent functions are evaluated. No search in the solution space is required.

A function f(X), such that f(X) = Y, may also be written as a relation pf(X, Y) denoting the set of pairs (x,y) such that x is in X, and y = f(x). It is of theoretical interest that each computable function can be represented by Horn clauses.[13] Note that in functional programming we deal with the object function f(X) such that pro-viding a particular input x, the output y = f(x) is computed and returned. On the other hand, the output object Y appears explicitly in logic programming in contrast to functional programming. We demonstrate this on the example of appending two lists, X1 and X2. A functional program is

```
app(X1, X2) =  IF empty(X1) THEN X2
                ELSE cons(head(X1), app(tail(X1), X2))).
```

The functions empty, cons, head, tail have the obvious meanings and correspond to the LISP functions null, cons, car, cdr, respectively. For example, a call of app([1, 2], [3]) returns the list [1,2,3].

In a logic program, the result Y has to appear explicitly, because relations between parameters are specified:

```
append(X1, X2, X2) :- empty(X1).
append(X1, X2, Y)  :- head(X1, H), tail(X1, T),
                      append(T, X2, Z), cons(H, Z, Y).
```

The meaning is:

1. Clause: if list X1 is empty then append of X1 and X2 is X2 (third
           argument of append).

2. Clause: if H is head of X1 and T is tail of X1,
           and append of T and X2 is Z, and cons of H and Z is Y,
           then append of X1 and X2 is Y.

The goal ?- append([1,2], [3]), Y) is satisfied by binding Y = [1,2,3] which is the answer. Moreover, the logic program is a true relation. There is no difference between input and output parameters and all variables X1, X2, Y in append are treated equally. The following queries are also possible:

```
?- append(X, [2], [3,2]).        X = [3].
?- append([1], X, [2]).          NO
?- append(X1, X2, [1,2]).        X1 = [], X2 = [1,2] ;
                                 X1 = [1], X2 = [2] ;
                                 X1 = [1,2], X2 = [].
```

This feature of logic programs is called invertibility. In Section 3.3 a Prolog program of append is given.

3. PROLOG

3.1 Inference Method

The programming language Prolog[5,4] is based on logic programming with Horn clauses. In order to be a practical programming language, various restrictions and extensions with respect to Horn-clause programming were introduced. In particular, Prolog em-

ploys a special resolution algorithm for theorem proving and provides various extralogical extensions.

Prolog uses a restricted form of the "linear input resolution". The algorithm of Section 2 is implemented as follows:

   1) Always choose the first subgoal A_1 to be solved first (left-to-right strat-
      egy)

   2) Search program clauses that match A_1 in the order they are written in the
      program (top down)

   3) Same as point 3) of the algorithm in Section 2.2

Searching in the derivation tree is done according to the depth-first strategy. If a
derivation fails, some variable bindings are undone by backtracking. The resolution
in Prolog is very efficient, but has the drawback of noncompleteness in the sense
that it may get stuck by pursuing an infinite derivation even though a successful
derivation exists. For example, if the rule (R2) of the program in Figure 1 were
changed to the logically equivalent form

    route(X, Y) :- route(X, Z), way(Z, Y)

the Prolog system would loop infinitely on the question ?- route(X, c). Similarly, if
the clauses (5) and (6) in the definition of the predicate "not" (Section 3.2) were
interchanged, the semantics would change totally. Thus the order of clauses in a
Prolog program, as well as the order of subgoals in a clause, is significant in contrast to pure logic programming.

3.2 Built-in Features

The extension of Horn clauses is achieved by predefined operators and "built-in"
predicates that have the following main purposes:

   - Influence the control of the theorem prover: cut-operator "!", "true", "fail"
   - Enhance the expressive power by
      • Introducing negation by failure: not-operator "not"
      • Introducing variables for predicates and clauses: "=..", "call"
   - Perform input/output: "read", "write"
   - Manipulate the logic program: "assert", "retract"
   - Manipulate Prolog structures: "clause", "functor"
   - Define operators: "op"
   - Debug programs: "trace", "spy"

- Evaluate arithmetic expressions: "is"
- Simplify the notation for regular structures: list notation "[...]".

We now describe some of these features in more detail:
"!": cut prunes the derivation tree.

Example:

A :- B, ...	(1)
A :- C, ...	(2)
B :- D, !, E	(3)
B :- F, ...	(4)
?- A, ...	

Interpreted as a goal, cut always succeeds. If E fails as a subgoal, then the back-tracking is performed neither in D nor in B (4), such that B in (1) fails and the backtracking continues with (2).

"true":   always succeeds once (fails on backtracking)

"fail":   always fails

"call":   call(C) succeeds if and only if the clause represented by the variable
           C succeeds. In first-order logic it is not possible to have variables for
           formulas.

"not":   not(P) succeeds exactly if P fails. It is equivalent to

not(P) :- call(P),!, fail.	(5)
not(P).	(6)

It is demonstrative to explain this program. Let P succeed. Then not(P)
fails because of "fail" in (5), and (6) is not considered for backtracking
because of cut. Let P fail. Then, not(P) fails at the first attempt of
applying (5), but succeeds due to (6). We note that the not-operator is not
equivalent to the logical negation and that no negative facts can be asserted
in a Prolog program.

"[...]":   List notation. For example, [a, b, c] denotes the list of the elements a, b,
           c. a is the head and [b, c] is the tail (list). [X|Y] denotes a list with the
           head X and tail Y. Due to the relational nature of Prolog, the notation [X|
           Y] permits the construction of a list given its head and tail (comparable to
           cons in LISP) as well as the splitting of a list into its head and tail (compa-
           rable to car and cdr in LISP).

"=..":   This operator is called "univ" and is used for transforming terms into lists
           and vice versa. X=..L means that L is the list, the head of which is the
           functor of X and the tail of which is the arguments of X. Hence, $f(t_1,t_2)$
           =.. [f, t_1, t_2].

"read":   read(X) always succeeds and binds X to the Prolog structure on the input
           file.

"write":   write(X) always succeeds and writes the instantiated variable on the output
           file.

"assert": assert(C) adds the clause represented by the variable C to the logic pro-
           gram. The program modifies itself by this side effect.

"retract": retract(C) deletes the clause represented by the variable C from the logic
           program.

"op":   ?- op(P, T, F) defines the operator with the name F, the precedence P and
           the type T. T specifies the position and associativity of F, and has one of
           the following symbolic forms:
                xfx, xfy, yfx, yfy: infix operator
                fx, fy: prefix operator
                xf, yf: postfix operator
           x and y stand for arguments and define the parsing structure (a prece-
           dence grammar). Assuming the precedence given by brackets is already
           resolved, x (or y) means that the operators in the argument must have
           lower (or lower or equal) precedence than F.

"is":   Evaluation. The goal "X is 1+2" is satisfied by binding X to 3.

3.3 Programming Techniques

Logic programming differs considerably from programming in conventional languages.
The programmer can concentrate on the logic of the problem and need not be concerned with implementation details. The style is declarative: it is stated what is to be
solved in a static way, not how it should be solved on a computer. There are no
program variables that can carry values as in conventional programming. Thus, there
is also no assignment statement and no side effects. Logic programs are formal static
descriptions of problems. They can be seen as executable specifications and are hence
well suited to fast prototyping. The order of clauses and goals is irrelevant and the
execution of the theorem prover can be performed in parallel.

Prolog departs from this ideal of logic programming in several points:

- The order of clauses and goals is relevant. Due to the depth-first search, Prolog is more restrictive but also more efficient on conventional hardware.
- Searching in the derivation tree can be influenced by the cut. This further enhances the efficiency, but the clarity of the program deteriorates.
- The expressive power is enhanced by the introduction of negation as failure. The not-operator, however, is not equivalent to the logical not.
- It is possible to maintain a global state by means of program clauses that can be manipulated during the program execution. This simplifies some programming problems, but introduces side effects and destroys the idea of the goal as a logical consequence of the program.

We demonstrate the programming style with some typical examples.

member(X, Y) is true if X is a member of the list Y:

```
member( X, [X|_]).
member( X, [_|Y]) :- member( X, Y).
```

The symbol _ denotes a "don't care" variable. The first fact says that X is a member of a list whose head is X. The second clause means: if X is member of Y, then X is member of Y preceeded by an arbitrary element.

append(X, Y, Z) is true if the list Z is the concatenation of the lists X and Y:

```
append( [], Y, Y).
append( [Xh Xt], Y, [Xh Z]) :- append( Xt, Y, Z).
```

sum(N, S) is true if S is the sum 1+2+...+N:

```
sum( 1, 1) :- !.
sum( N, S) :- N1 is N-1, sum( N1, S1), S is S1+N.
```

The cut is introduced to enforce the unique solution of sum(N,S) and to prevent backtracking.

counter(N) is true if the goal count has been called N times:

```
counter(0).
count( N) :- retract( counter(N1) ), N is N1+1, assert( counter(N) ).
```

This is a Prolog implementation of the assignment and update statement "N:=
N+1". In the Prolog program the value of N is stored as the global program
assertion counter(N).

w(P,S): write all elements X of the set S which satisfy P(X):

```
w(P,S):- member(X,S), G=..[P|X], call(G), write(X), fail.
w(_,_).
```

Example: pos(X) :- X > 0.
          ?- w(pos, [1, -2, 3]).
          Answer: 1, 3.

This example illustrates the use of backtracking to simulate an iteration loop. In
the first clause, "member" generates an element X of the set S. Then, the
predicate "P(X)" is constructed using univ-operator and called using "call". If
the call succeeds then X is written. In any case, the goal fails due to "fail" and
backtracking resumes at "member" which generates the next element of S, etc.
When all elements of S are tried, "member" fails and w(P,S) succeeds because of
the second clause, also called "catch all" clause.

4. PROLOG SYSTEMS

4.1 Self-contained Prolog-Programming Environment

Programming languages can be used efficiently only if a "programming environment" is
provided that facilitates the programmer's work in developing, translating, executing
and maintaining programs. Typical tools are editors, compilers, linkers, debuggers,
and version maintenance. In conventional systems, these tools are not a part of the
programming language, but are usually provided around the operating system. This
has various undesirable consequences. Since the environment has to support various
languages, it is very rudimentary, cannot be optimized for a particular language, is
machine-dependent, and cannot be modified by means of the programming language
itself. That is why a great attention has been paid to the development of language-
based programming environments.

An example of a language that is often embedded in an environment of its own is
LISP. A LISP environment usually consists of an interpreter, compiler, editor, and
debugger, and allows easy switching between the various activities. Furthermore,
many high-level functions are provided written in LISP. Another example of a lan-
guage-based environment is APSE, ADA Programming Support Environment, which is
being developed in ADA based on KAPSE (K for "Kernel").

The Prolog language already includes many built-in features that make it a rudimentary programming environment. We shall mention some of them with reference to the
"Edinburgh Prolog":[4]

- Editing a Prolog program

> consult(F): Read in clauses from file F and add them to the previous clauses.
>
> consult(user): Read in clauses interactively from the user's terminal and add
> them to the previous clauses.
>
> reconsult(F): Read in clauses from file F. If a clause with the head P is read,
> all previous clauses with the head P are superseded.
>
> assert(C), retract(C): See definitions in Section 3.2.

- Listing

> listing(C): All clauses with the he d C are listed.
>
> listing: All program clauses are listed.

- Debugging

> trace: An exhaustive tracing of the called predicates is carried out.
>
> spy P: Tracing of the calls of the predicate (term) P.

Further helpful predicates may easily be defined either in Prolog or by modifying the
interpreter.

4.2 Modula--Prolog

A promising approach to an integrated Prolog environment is being investigated at the
Brown Boveri Research Center. The Modula--Prolog system has been developed in
cooperation with the Institute for Informatics, ETHZ.[17] It contains an interpreter for
the "Edinburgh Prolog"[4] written in Modula-2 and allows the user to combine arbitrarily Modula-2 and Prolog programs. The resolution algorithm is based on the techniques of structure sharing[3] as adapted by Sugaya.[24] Modula--Prolog (MP) is realized
as a set of modules, from which the user may configure his own Prolog system. The
machine-dependent features are confined to a few designated modules, such that the
system is highly portable to all computers that support Modula-2. To date, the system
has been installed on Lilith, VAX/VMS, PC-AT/MS-DOS, and Macintosh computers.

The structure of the system is depicted in Figure 5.

Terminal Input/Output

The Terminal Input/Output is embedded in a Modula-2 program which handles the
dialog between a Prolog user at the terminal, and the Prolog functions such as the

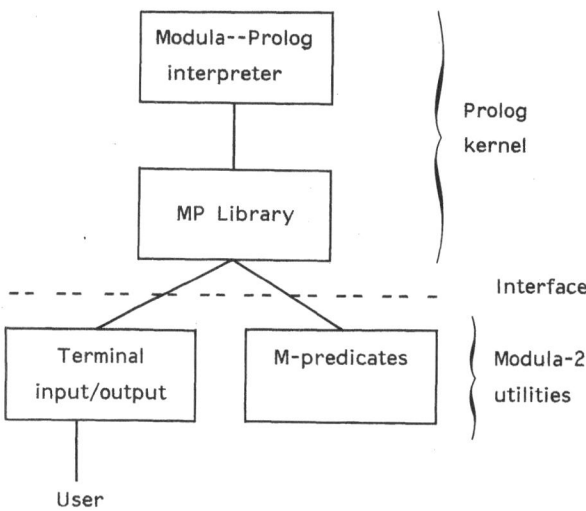

Fig. 5: Modula--Prolog System

"Parser" and "Prover", provided in the MP Library. Thus, a Prolog system can be tailored to different user requirements. For example, Prolog input or output may be redirected, and advanced dialog techniques using multiple windows can be implemented without changes at the level of Modula--Prolog. An implementation of the interface, which behaves in the same way as usual Prolog interpreters, is provided as a default configuration. The Terminal Input/Output program automatically consults a start-up file that may contain frequently used predicates and user-defined operators.

The Prolog system accepts queries from the user, answers YES or NO and outputs variable bindings, if any. The execution of a query can be interrupted to abort, continue, or change the tracing mode.

MP Library and Prolog Interpreter

The MP Library contains all objects of the Prolog interpreter and of the interface between Prolog and Modula-2. Only the definitions of the objects are exported to the user, the implementation is hidden. The main functionality of the MP Library is as follows:

- Prolog input and output
- Calling Modula-2 from Prolog and vice versa:

- Prolog parser and prover with the unification algorithm
- Defining new built-in predicates and implementing them as Modula-2 modules (M-predicates)
- Assembling and disassembling of Prolog terms
- Utility procedures
- Memory management of local data

M-predicates

The user may implement Prolog-predicates in Modula-2 using the features provided by the MP Library. These M-predicates are called from Prolog as built-in predicates. M-predicates are especially useful for programming dedicated input/output routines or for speeding up frequently used predicates. As an example, experiments have shown that the very frequently used predicate "member(X,L)" runs up to four times faster when implemented in Modula-2 as M-predicate.[25]

Expandable Modula--Prolog Environment

We are further developing the Modula--Prolog system and adding new features. Expansion of the system is relatively straightforward due to the modular design and the interface to Modula-2. Current developments concern in particular the integration of the following features:

- Grid File

  A grid file[18] is a data structure for efficient storage, access and manipulation of a large set of n-tuples with a predictable worst-case behavior. At the ETHZ, Prolog has already been used as an intelligent front end to the grid file.[1,16] Grid files can also be used for efficient storage and retrieval of Prolog facts.

- Modules

  The Prolog program is a flat list of clauses. The comprehensibility, maintainability and efficiency can be increased by introducing a modular structure. In the current version modules are implemented as files and can be imported using the "consult" predicate. Further improvements, e.g., modularization based on the class concept,[26] are also being studied.

- Man-Machine Interaction

  The interaction of the user with the Modula--Prolog system will be further improved by incorporating general dialog techniques, such as windowing and menus as well as dedicated features specific to Prolog.

5. KNOWLEDGE-BASED SYSTEMS IN PROLOG

In this section we discuss the application of logic programming to knowledge representation and reasoning. Knowledge-based systems consist of two major software
components: a knowledge base and an inference program called "shell" (Figure 6).
The knowledge base contains the specific domain knowledge of a problem. The inference program solves problems by deducing new knowledge using the stored knowledge
and inference rules. It is problem independent in the sense that it relies only on the
structure of knowledge representation but not on the specific content of the knowledge base.

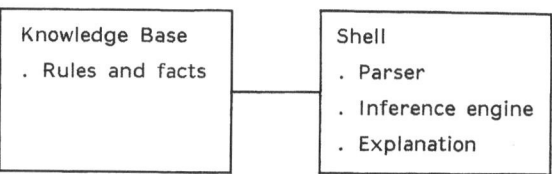

Fig. 6: Structure of a Knowledge-Based System

Knowledge is represented by if-then rules in rule-based systems and by frames with
slots in frame-based systems. Knowledge is preferably acquired in some user-friendly
representation language and then translated into an internal representation by the
knowledge-based system.

The shell contains procedures for parsing the knowledge representation and constructing the knowledge base. The program also contains inference procedures for solving
problems by reasoning from the premises to the conclusion (forward) or from the
conclusion to the premises (backward). The reasoning history, together with the
knowledge base, can be used to provide an explanation for the behavior of the program. This helps clarify the reasoning steps since the program guides the dialog with
the user to query some facts about the problem.

The concepts of logic programming, e.g., symbolic computation, pattern matching and
theorem proving, are well suited to the development of knowledge-based systems.
Prolog has already been successfully employed for the realization of expert systems.[8,9] Knowledge can be represented in if-then rules or frames other than the
facts and rules represented directly in Prolog predicates. The inference method can
be either a forward or a backward reasoning, or a combination of both.[15] The explanation facility can be used for debugging rules interactively in addition to the simple
tracing of resolution steps.

In this section, we briefly describe rule-based and frame-based systems and show how Prolog facilitates the construction of such systems. The concepts and techniques described here are based on a Brown Boveri system, realized in Prolog, for the determination of optimal routes in a grain-silo plant.

5.1 Rule-Based Systems

In a rule-based system, the knowledge base is made up of productions of the form

> rule N: if Premise then Conclusion.

Premise and conclusion are either a simple proposition or a combination of simple propositions using "and" and "or". As an example, we represent the knowledge used for routing in an industrial plant.

In a grain-silo plant, grain is stored, cleaned, weighed, or chemically treated. Transport routes, consisting of conveyor belts, elevators, slide gates, and diverter gates, are searched when grain is to be transported from the intake to a silo. The sample network below shows that at the diverter gate 510, grain must be sent to conveyor belt 201 if it is to be cleaned, and is otherwise sent to conveyor belt 202.

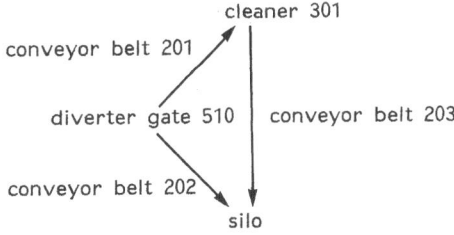

Fig. 7: Transport Network

The logic of the structure and the routing of the network can be represented in production rules as

> rule 1: if   current_component is diverter_gate_510
>   and transport_service is cleaning
>   then route is conveyor_belt_201.

```
rule 2:   if    current_component is diverter_gate_510
                 and transport_service is silo
           then route is conveyor_belt_202.


rule 3:   if    current_component is conveyor_belt_201
           then route is cleaner_301.


rule 4:   if    current_component is cleaner_301
           then route is conveyor_belt_203.
```

Rule 1 states that if grain is at diverter gate 510 and the grain must be cleaned, it must be routed to conveyor belt 201. Subsequently, the grain is routed to a silo via conveyor belt 203.

The production rules described above can be extended in several ways. For example, in order to assert a premise, other relevant rules and facts may be used if they exist, or the user may be queried to acquire the missing fact.[22] Moreover, a conclusion, depending on an application domain, can be evaluated with a certainty factor using fuzzy sets.[28]

The inference program parses the production rules and builds a knowledge base. A syntax of the production rule can be defined using unary and binary operators with precedence. A parse tree of the production rule is given in Figure 8.

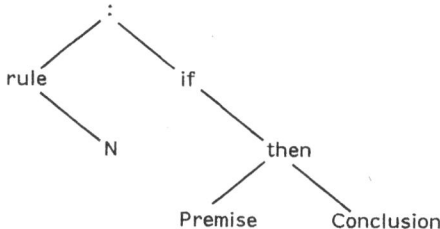

Fig. 8: A Parse Tree of an If-Then Rule

The Prolog built-in predicate "op" (operator) supports this parsing technique. The following operator declarations suffice to parse the above production rule.

```
?- op( 170, xfy, ':').
?- op( 150,  fx, rule).
?- op( 150,  fx, if).
```

```
?- op( 149, xfy, then).
?- op( 145, xfy, and).
?- op(  40, xfx, is).   (predefined in Prolog⁴)
```

The entire parse tree, together with operators, can now be stored in the Prolog database by the built-in predicates "assert". Thus, the following Prolog query determines the premise and conclusion of rule 2 in the database, where PREMISE and CONCLUSION are two variables to be matched with the content of the database.

```
?- rule 2: if PREMISE then CONCLUSION.
PREMISE = current_component is diverter_gate_510 and
         transport_service is silo
CONCLUSION = route is conveyor_belt_202
```

It is desired to find a route for grain which is to be cleaned when it is sent to a silo. Assuming the grain is currently at diverter gate 510 and needs to be cleaned, it may be reasoned (forward), based on the rule 1, that the grain is routed to conveyor belt 201. In this case an alternative route indicated by rule 2 will not be used. Conversely, assuming the grain needs to be routed to a cleaner, it may be reasoned (backward), based on rule 3, that it must be at conveyor belt 201.

A route-finding program can be formulated recursively as follows. We first construct a program which proceeds from a start component to a goal forward. The predicate route(X,Y,S,P) succeeds if X can be routed to Y through a path P with a service S. If the current position of grain is already at a destination, the route is found in P and we have in Prolog

```
route(X,X,S,P) :- write(P).
```

Otherwise, to route from one component X to other Y, find a rule which fulfills the routing condition at X, route the next component Z induced by the rule, make sure that the component is not yet contained in the path P, and route from Z to Y appending the current component Z to the path P. We have

```
route(X,Y,S,P) :- check(X,S,Z), not(member(Z,P)), route(Z,Y,S,[Z|P]).
check(X,S,Z) :- rule N: if current_component is X
                        and transport_service is S
                     then route is Z.
check(X,S,Z) :- rule N: if current_component is X
                     then route is Z.
```

If the predicate check reveals more than one rule satisfying the routing condition, the alternative routes can be searched by the backtracking mechanism of Prolog. Thus, a Prolog query

    ?- route(diverter_gate_510, silo, cleaning, []).

prints out a path

    [silo, conveyor_belt_203, cleaner_301, conveyor_belt_201, diverter_gate_510].

Now, we construct a program which works from a goal to a starting point backward. The second clause above becomes

    route(X,Y,S,P) :- check(Z,S,Y), not(member(Z,P)), route(X,Z,S,[Z|P]).

The predicate "check" inspects rules based on the destination component Y, and the predicate "route" works backwards towards the starting component X.

5.2 Frame-Based Systems

In a frame-based system, the knowledge base is made up of frames of the form

    frame_name: slot_name1 = slot_value1
                 & slot_name2 = slot_value2
                 & ...

A frame corresponds to a node of a semantic network. Attributes of the node as well as links to other nodes are represented in slots. A slot value can be an expression stating a structured value. Generally, frame systems additionally incorporate facets which elaborate the value of each slot, e.g., through constraints, defaults, demons. As an example of a very simple frame formalism, we represent the components of the routing network of Figure 7 in frames.

    dg510:   name = 'diverter gate 510'
           & kind = discriminator
           & routes = cb201 if [cleaning] or cb202 if [silo].

    cb201:   name = 'conveyor belt 201'
           & kind = carrier
           & routes = cl301.

```
cl301:     name = 'cleaner 301'
           & kind = server(cleaning)
           & routes = cb203.
```

The frame dg510 represents the combined routing knowledge of the if-then rules 1 and 2. The slot "kind" indicates that a plant component is either a simple carrier, or a server, or a discriminator where a routing decision must be made. As shown above, frames are suitable for structuring concepts and attributes of a domain knowledge. Note that the knowledge representation in frames is more general than the if-then rule. For example, the production rule can be represented in a frame with two slots, i.e., if_part = premise and then_part = conclusion.

A syntax of the frame can be defined similarly by a precedence grammar. Figure 9 illustrates the parse tree.

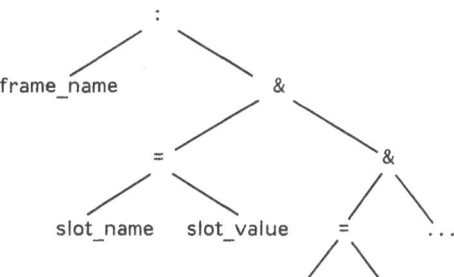

Fig. 9: A Parse Tree of a Frame

The following operator declarations suffice to parse the frames for the transport network.

```
?- op(160, xfy, :).
?- op(150, xfy, &).
?- op( 40, xfx, =).     (predefined in Prolog⁴)
?- op( 30, xfy, or).
?- op( 25, xfy, if).
```

Frames can now be parsed and translated into Prolog facts. The frame dg510, for example, is translated into the following four predicates:

```
dg510( name( 'diverter gate 510') ).
dg510( kind( discriminator) ).
```

```
dg510( routes( cb201 if [cleaning]) ).
dg510( routes( cb202 if [silo]) ).
```

Compared to the rule base, where all rules are looked upon for applicability, the translated frame base permits searching of the relevant information specific to a frame only. To find a route in the network, the same logic as described above can be used.

6. A KNOWLEDGE-BASED SYSTEM FOR DIAGNOSING ANALOG CIRCUITS

In process-control systems, analog and digital circuits are used for steering and controlling physical processes. In contrast to the diagnostic tests used for digital circuits, the complexity of analog components inhibits the application of the combinatorial test method. Furthermore, analog circuits contain components such as switches, jumpers, and trimmer condensors that have to be set manually or adjusted for correct operation. Thus, analog circuits have long resisted rationalization of fault diagnosis. Long-term repair is also a problem. Analog circuit boards used for process-control systems need to be maintained over ten to twenty years after design, which implies that the aid of an expert diagnostician or a designer with knowledge of the analog circuits may often be no longer available.

The knowledge-based approach has already been applied to the diagnosis of digital systems.[6,7] In this section we describe a knowledge-based diagnostic system for analog printed-circuit boards. Diagnostic knowledge is represented as a diagnostic graph based on the cause-and-effect relationship and expressed in frames. The diagnostic graph can incorporate causalities obtained from the structural and functional description of analog circuits as well as those obtained from empirical tests. Fault detection and deduction are performed by inference procedures using the diagnostic graph. A prototype system has been developed in Modula--Prolog[17] and runs on VAX/VMS and PC-AT/MS-DOS computers.

6.1 Model Approximation

The information that propagates through the components of an analog circuit is frequency (including phase), voltage, and current. The complexity of an analog circuit can be reduced by considering its functionality, just as a designer uses a functional block diagram for circuit design. Structural dependency can be simplified at the level of the functional block diagram. Figure 10 shows as an example a block diagram of an IF-HF frequency converter.

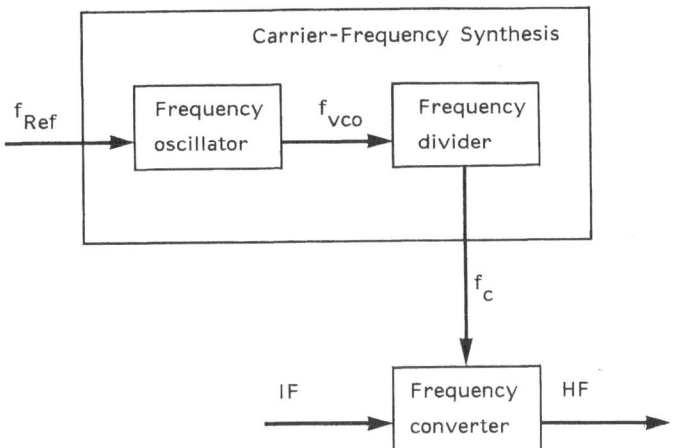

IF: Intermediate Frequency

HF: High Frequency

Fig. 10: IF-HF Frequency Converter

The voltage-controlled frequency oscillator generates a raw frequency f_{vco} using an input reference signal f_{Ref}. The f_{vco} is then divided into a desired carrier frequency f_c through a frequency divider. Finally, the input IF signal is converted to an HF signal based on the carrier frequency. The logic of the structure and functions of the diagram are

    If 'f_{Ref} signal is correct' and 'frequency oscillator is correct'
    then 'f_{vco} signal is correct'.   (1)

    If 'f_{vco} signal is correct' and 'frequency divider is correct'
    then 'f_c signal is correct'.   (2)

    If 'IF signal is correct' and 'f_c signal is correct' and
      'frequency converter is correct'
    then 'HF signal is correct'.   (3)

    If 'HF signal is correct' then 'IF-HF converter is correct'.   (4)

In other words, if a component and its input signals are correct, then its output is correct. Rule (4) is a goal stating that the IF-HF frequency converter is correct if the output HF signal is correct.

To test this circuit, we start by asserting that the circuit is faulty. The induction Rules (1) through (4) can now be used to deduce faults (i.e., causes). The corresponding causality rules can be obtained by negating the if-then rules. We first negate goal Rule (4) and obtain

If 'IF-HF converter is faulty'
then 'HF signal is faulty'.   (5)

Similarly,

If 'HF signal is faulty'
then 'IF signal is faulty' or 'f_c signal is faulty' or
    'frequency converter is faulty'.   (6)

If 'f_c signal is faulty'
then 'f_{vco} signal is faulty' or 'frequency divider is faulty'.   (7)

If 'f_{vco} signal is faulty'
then 'f_{Ref} signal is faulty' or 'frequency oscillator is faulty'.   (8)

The causality relationships of Rules (5) through (8) can be represented as a diagnostic graph (Figure 11). When the graph is used from atomic to derived faults, it induces effects (symptoms) of faults. Conversely, when the graph is used from derived to atomic faults, it deduces causes.

The deduction Rule (6) investigates three possible causes: the correctness condition for IF signal deals with the test environment of the printed circuit board; the other

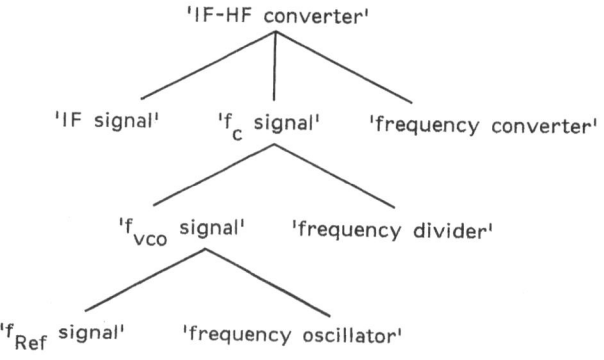

Fig. 11: Diagnostic Graph

two, for carrier synthesis and frequency conversion, deal with the board itself. In order to assert the correctness of the frequency converter, its input signals must first be tested against faults since the frequency converter alone cannot be tested separately. This constraint can be used to determine the ordering of diagnostic test steps. Thus, the diagnostic graph represents a model approximating the cause-and-effect relationships of an IF-HF frequency converter.

6.2 Knowledge Representation

At a node of the diagnostic graph, two kinds of knowledge are represented: the cause-and-effect relationship of a fault and partial ordering of faults to be investigated. Additionally, we give a short name to each fault, a required setup for test, a correctness test for a fault to see whether the fault exists or not, and the necessary actions if a fault is repairable. Since these attributes are specific to each fault, we use frames to represent them uniformly. Figure 12 shows a frame representation of the diagnostic graph of IF-HF converter (only the top fault and its causes).

The slot "type" indicates that a fault is either an atomic one or a derived one which

```
if_hf:           name = 'IF-HF frequency converter not working'
                 & type = derived
                 & causes = if_signal or fc_signal or freq_converter.

if_signal:       name = 'Input IF signal faulty'
                 & type = atomic
                 & test = '1.2'
                 & action = 'Check the input IF signal to the board'.

fc_signal:       name = 'Carrier frequency synthesizer not working'
                 & type = derived
                 & setups = 'Mount the test board in the test setup 1'
                 & causes = fvco_signal or freq_divider.

freq_converter:  name = 'Frequency converter fcv not working'
                 & type = atomic
                 & setups = 'Mount the test board in the test setup 2'
                 & test = '2.1'
                 & action = 'Frequency convertor is defect. Replace it'.
```

Fig. 12: Knowledge Representation in Frames

can be further refined to atomic faults. The slot "setups" assures that the test environment is correctly set. The slot "test" states the name of a test function to be
performed. The test function prepares test signals (voltage, current, frequency) for
the test board, measures the signals at output points, and compares them to see
whether they are in an acceptable range. Existence of a fault is confirmed if the test
fails. Similarly, an observable test can be expressed as two slots stating a question
to the operator and the expected answer to confirm the existence of the fault, e.g.,

      & question = 'Is the switch 701 open ?'

      & confirmed if no

The slot "causes" lists the possible causes of a fault.

6.3 Fault Diagnosis

To diagnose a board, it is assumed that the board is faulty. The state of all faults is
initialized to "may_exist". Then the causes of the top fault are investigated to see if
a fault exists. These deduction steps continue until either a fault is detected or the
absence of a fault can be asserted. If the fault node has a slot for the correctness
test, the existence or absence of a fault can be determined either by observing
components or by performing a function test. In this deduction phase there are three
fault states: "may exist", "exists", and "cannot exist". A fault may exist if it is
assumed (no correctness test), or unanswerable in the case of an observable test. A
fault exists if a test function fails or a question is answered affirmatively; otherwise,
a fault cannot exist.

The purpose of fault deduction is to investigate a board as faulty (i.e., a fault may
exist or exists) and to conclude either that faults cannot exist or that some unrepair-
able faults exist. For example, a fault F can be caused by causes $C_1, C_2, \ldots, C_n$.

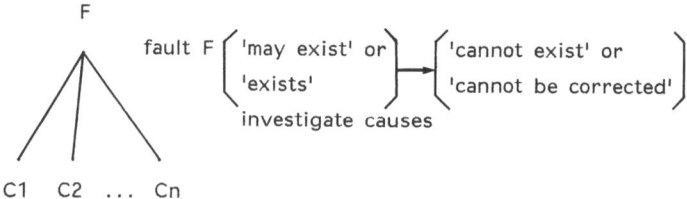

The deduced state of the fault F can be determined (recursively) from the deduced
states of its causes C_i, $1 \leq i \leq n$, in the following way:

Case 1:   There exists a cause C_j whose state is "cannot be corrected". Then the
          deduced state of F is "cannot be corrected".

Case 2:   The state of all causes C_i, $1 \leq i \leq n$, is either "corrected" or "cannot
          exit".
    a) If the current state of the fault F is "may exist" then the deduced state of
       F is "cannot exist".
    b) If the current state of the fault F is "exists" then the deduced state of F
       is not defined. In this case, the diagnostic model is incomplete and unable
       to deduce a fault although the function test implies its existence. The model
       is subject to correction.

6.4 Knowledge Refinement

Diagnostic knowledge represented in the diagnostic graph can be refined in the following situations. First, the diagnostic model is found to be incomplete. This is the
situation in which not all causes of a fault are covered by the range of its correctness test; the missing causes must be added.

Second, an atomic fault is found to be too "general" in its context and needs to be
refined. This is because the diagnostic model approximates the structure and functions of an analog circuit. For example, a frequency-divider fault, represented as a
fault frame

        freq_divider: name = 'Frequency divider is faulty'
                               & type = atomic
                               & test = '2.2'
                               & action = 'Frequency divider is defect. Replace it'.

is checked. The test 2.2 fails. The diagnostician finds out that the switch used to
progam a constant number (divider) is wrongly set. The frame freq_divider is now
refined to

        freq_divider: name = 'Frequency divider is faulty'
                               & type = derived
                               & test = '2.2'
                               & causes = freq_divider_switch.

    freq_divider_switch: name = 'Frequency divider switch is wrongly set'
                               & type = atomic
                               & question = 'Is the switch correctly set?'

```
& confirmed if no
& action = 'Set the switch correctly'.
```

The modified frame freq_divider might be incomplete: it covers only one specific component as a cause of the fault. The knowledge refinement in the domain of diagnosing analog circuits exploits the fact that the kinds of faults being discovered are few but recurrent in nature.

Third, the cause-and-effect relationship for diagnosis goes beyond the structural and functional description of an analog circuit. For example, when analog components are mounted in two or three layers, they may be subject to electromagnetic disturbances. Or, because of the peculiarity of the positional assignment of components to the circuit board, certain components tend to be mounted in an incorrect orientation. Such empirically found causes can be also added to the diagnostic graph.

The knowledge-based approach provides an economic and qualitative solution to the diagnosis of analog-circuit boards. The model in the form of a diagnostic graph approximates the diagnostic knowledge of an analog circuit. The constructed diagnostic graph is based on the structural and functional description of an analog circuit at the level of a functional block diagram and is then refined in the course of diagnosis.

7. CONCLUSION

Logic programming is based on precise and well-understood concepts of mathematical logic, particularly first-order predicate logic. Logic programming is characterized by a new, declarative style of programming. The programming activities are moved from implementing a procedure more toward problem solving. As pointed out by Kowalski,[12] an algorithm may be considered to consist of a logic component and a control component. At the first stage, the programmer can concentrate on the logic of the problem. The problem is specified by facts and rules that are known about the objects under consideration. This specification by means of a logic program can be executed as a fast prototype. Only at further stages of the development, is attention paid to optimization of the execution by rewriting the logic program and providing information on the control.

The programming language Prolog is a pragmatic realization of the concepts of logic programming. In order to be a practical language for a von Neumann computer, it was necessary to introduce various compromises to pure logic programming with Horn clauses. In spite of various deficiencies, Prolog has become a widely used language in the Artificial-Intelligence community and has been chosen as the core language for Fifth Generation Computer Systems. Current research is concerned with pure logic systems, parallelism in logic programs, and the design of inference machines for efficient execution of logic programs.

At the Brown Boveri Research Center, Prolog has been successfully applied to fast prototyping and to the realization of prototypes of knowledge-based systems. The system Modula--Prolog, which allows the combination of Prolog and Modula-2 programs, has been developed in cooperation with the Federal Institute of Technology in Zurich and is being extended to a flexible Prolog environment based on Modula-2. Modula--Prolog is being applied to the construction of a knowledge-based system for the diagnosis of analog printed circuits.

REFERENCES

1. Arnoldi, M., "Integration of the Grid File System in the Programming Language Prolog," Diploma Thesis, Federal Institute of Technology, Zurich, Switzerland, August 1985.

2. Biagioni, E.S., Hinrichs, K., Muller, C., and Nievergelt, J., "Interactive Deductive Data Management - the Smart Data Integration Package," Proc. GI-Kongress 1985, Wissensbasierte Systeme, Munich, October 1985, Informatik Fachberichte, Vol. 112, W. Brauer and B. Rudig, eds., Springer-Verlag, Berlin, 1985, pp. 208-220.

3. Boyer, R.S., and Moore, J.S., "The Sharing of Structure in Theorem-Proving Programs," in Machine Intelligence, Vol. 7, B. Meltzer and D. Michie, eds., Edinburgh Univ. Press, 1972, pp. 101-116.

4. Clocksin, W.F., and Mellish, C.S., Programming in Logic, 2nd ed., Springer-Verlag, Berlin, 1984, p. 250.

5. Colmerauer, A., Kanoui, H., Roussel, P., and Pasero, R., "Un Systeme de Communication Homme-Machine en Francais," Groupe de Recherche en Intelligence Artificielle, Université d'Aix-Marseille, 1973.

6. Davis, R., "Diagnostic Reasoning Based on Structure and Behavior," Artificial Intelligence, Vol. 24, December 1984, pp. 347-410.

7. Davis, R., Shrobe, H., Hamscher, W., Wieckert, K., Shirley, M., and Polit, S., "Diagnosis Based on Description of Structure and Function," Proc. National Conf. on Artificial Intelligence, AAAI-82, 1982, pp. 137-142.

8. Hammond, P., "Representation of DHSS Regulations as a Logic Program," Technical Report 82/26, Department of Computing, Imperial College, London, 1983.

9. Hammond, P., and Sergot, M., "A Prolog Shell for Logic Based Expert Systems," Proc. 3rd BCS Expert System Conf., 1983, pp. 95-104.

10. ICOT, "Outline of Research and Development Plans for Fifth Generation Computer Systems," Project Report, Institute for New Generation Computer Technology, Tokyo, May 1982.

11. Kowalski, R.A., "Predicate Logic as Programming Language," Proc. IFIP 74, North-Holland, 1974, pp. 569-574.

12. Kowalski, R.A., "Algorithm = Logic + Control," Commun. ACM, Vol. 22, No. 7, 1979, pp. 424-436.

13. Kowalski, R.A., "Logic Programming," Proc. IFIP 83, R.E.A. Mason, ed., North-Holland, Amsterdam, 1983, pp. 133-145.

14. Lloyd, J.W., Foundations of Logic Programming, Springer-Verlag, Berlin, 1984.

15. Merry, M., "APEX3: An Expert System Shell for Fault Diagnosis," The GEC Journal of Research, Vol. 1, No. 1, 1983, pp. 39-47.

16. Muller, C., "A Prolog Front End to the Grid File," Diploma Thesis, Institute for Informatics, Federal Institute of Technology, Zurich, 1984.

17. Muller, C., "Modula--Prolog User Manual," Report Brown Boveri Research Center, KLR 85-107 C. Also: Report 63, Institute for Informatics, Federal Institute of Technology, Zurich, 1985.

18. Nievergelt, J., Hinterberger, H., and Sevcik, K.C., "The Grid File: An Adaptable, Symmetric Multi-Key File Structure," Technical Report 46, Institute for Informatics, Federal Institute of Technology, Zurich, 1981. Also: ACM Trans. Database Syst., Vol. 9, No. 1, 1984, pp. 38-71.

19. Robinson, J.A., "A Machine-Oriented Logic Based on the Resolution Principle," J. ACM, Vol. 12, 1965, pp. 23-41.

20. Robinson, J.A., "Computational Logic: The Unification Computation," in Machine Intelligence, Vol. 6, B. Meltzer and D. Michie, eds., Edinburgh Univ. Press, 1971, pp. 63-72.

21. Roussel, P., "Prolog: Manuel de Reference et d'Utilization," Groupe de Recherche en Intelligence Artificielle, Université d'Aix-Marseille, 1975.

22. Sergot, M., "A Query-the-User Facility for Logic Programming," in Integrated Interactive Computing Systems, P. Degano and E. Sandewall, eds., North-Holland, Amsterdam, 1983, pp. 27-41.

23. Shapiro, E.Y., "A Subset of Concurrent Prolog and Its Interpreter," ICOT, Technical Report TR-003, Tokyo, 1983.

24. Sugaya, H., "Some Improvements to the Unification Algorithm of ETH-Prolog," Technical Memo, Brown Boveri Research Center, 1985.

25. Sugaya, H., "Prolog--Modula-2 Interface," Technical Memo, Brown Boveri Research Center, 1985.

26. Takei, K., Chikayama, T., and Takagi, S., "ESP - Extended Self-Contained Prolog - An Object Oriented Logic Programming Language," Technical Memo, TM-0075, Institute for New Generation Computing Technology, Tokyo, 1984.

27. Wirth, N., "Modula-2," Technical Report 36, Institute for Informatics, Federal Institute of Technology, Zurich, 1980.

28. Zadeh, L.A., "Fuzzy Sets," Information and Control, Vol. 8, 1965, pp. 338-353.

DISCUSSION

Chairman: O. Klammer (Brown Boveri, Copenhagen, Denmark)

J. Encarnaçao (Technical University of Darmstadt, Darmstadt, Fed. Rep. Germany)
You mentioned that you can have computer graphics in connection with Prolog. How do you handle picture generation at the Prolog interface?

J. Kriz
We use the facilities of our Modula--Prolog and program advanced input and output routines in the procedural language Modula-2. So graphics is done in Modula-2. The combination of declarative and procedural programming is a great advantage of our Modula--Prolog.

J. Goldberg (SRI International, Menlo Park, CA, USA)
Have you found any difficulty in applying your techniques to the recording and maintenance of real-time state information in process control, for example, in connection with dynamic databases?

J. Kriz
We are not currently using Prolog for real-time applications. Of course, there exist attractive applications in real-time process control, for example, knowledge-based consulting systems as a decision support for the plant operator. With respect to real-time applications, however, we still see severe problems due to the program execution speed.

E. Handschin (University of Dortmund, Dortmund, Fed. Rep. Germany)
In 1971 Brown Boveri organized a symposium entitled "Real-Time Control of Electric Power Systems" and, fourteen years later, of course, we now see that many powerful tools have been developed for real-time control. The gap, in my opinion, widens between the plant and the operator. Control systems become more and more complex. To make full use of the powerful control tools now available, the operator needs training. The user of the control system is the operator who is responsible for the plant. He does not want to know about the bits and boards of the control system. His main interest is in the process.

J. Kriz
I completely agree with your comment. Since the control systems and also the controlled processes become more complex and more difficult to handle by the operator, we need an advanced man-machine dialog, which supports the operator in an intelligent way without releasing him from his final responsibility. Such a system could be based on knowledge of the physical process, its current state and a user model. Hence, I think there is a great potential for the methods of Artificial Intelligence and cognitive science in this field.

AUTHOR INDEX

The underlined page numbers indicate full-length papers; all other page numbers refer to contributions to discussions.